LINEAR AND
COMPLEX ANALYSIS
FOR APPLICATIONS

Advances in Applied Mathematics

Series Editor: Daniel Zwillinger

Published Titles

Advances in Applied Mathematics

LINEAR AND COMPLEX ANALYSIS FOR APPLICATIONS

John P. D'Angelo

University of Illinois
Urbana-Champaign, USA

CRC Press
Taylor & Francis Group
Boca Raton   London   New York

CRC Press is an imprint of the
Taylor & Francis Group, an **informa** business

A CHAPMAN & HALL BOOK

CRC Press
Taylor & Francis Group
6000 Broken Sound Parkway NW, Suite 300
Boca Raton, FL 33487-2742

First issued in paperback 2022

© 2017 by Taylor & Francis Group, LLC
CRC Press is an imprint of Taylor & Francis Group, an Informa business

No claim to original U.S. Government works

Version Date: 20161123

ISBN 13: 978-1-03-247702-2 (pbk)
ISBN 13: 978-1-4987-5610-5 (hbk)

DOI: 10.1201/9781315118840

Publisher's Note
The publisher has gone to great lengths to ensure the quality of this reprint but points out that some imperfections in the original copies may be apparent.

Visit the Taylor & Francis Web site at
http://www.taylorandfrancis.com

and the CRC Press Web site at
http://www.crcpress.com

Contents

Preface

This book evolved from several of the author's teaching experiences, his research in complex analysis in several variables, and many conversations with friends and colleagues. It has been also influenced by a meeting he attended (Santa Barbara, 2014) on the revitalization of complex analysis in the curriculum. That meeting suggested that complex analysis was losing its luster as a gem in the curriculum. This book therefore aims to unify various parts of mathematical analysis, including complex variables, in an engaging manner and to provide a diverse and unusual collection of applications, both to other fields of mathematics and to physics and engineering.

The first draft of the book was based on a course taught in Spring 2014 at the University of Illinois. The course, Electrical and Computer Engineering (ECE) 493, is cross-listed as Math 487. Twenty-one students, including several graduate students, completed the class. Areas of specialization were electrical, mechanical, and civil engineering, engineering physics, and mathematics. Background and interests varied considerably; communicating with all these students at the same time and in the same notation was challenging.

The course began with three reviews. We discussed elementary linear algebra and differential equations, basic complex variables, and some multi-variable calculus. The students took two diagnostic exams on these topics. The first three chapters include these reviews but also introduce additional material. The author feels that spiral learning works well here; in the classroom he prefers (for example) to introduce differential forms when reviewing line and surface integrals rather than to start anew. The course used these reviews to gradually seduce the student into a more abstract point of view. This seduction is particularly prominent in the material on Hilbert spaces. This book aims to develop understanding of sophisticated tools by using them rather than by filling in all the details, and hence it is not completely rigorous. The review material helped students relearn things better while creating a desire to see powerful ideas in action.

After the reviews, the course began anew, with about 4 weeks on complex analysis, 1 week on the Laplace transform, and 3 weeks on Hilbert spaces and orthonormal expansion. We discussed Sturm–Liouville theory to give examples of orthonormal expansion, and we briefly discussed generating functions (and the so-called Z-transform) to provide a discrete analogue of the Laplace transform. The unifying theme of the class remained linearity and related spectral methods. The course sketched a proof of the Sturm–Liouville theorem, but filling in the details was not possible because student backgrounds in analysis were so varied.

After typing notes for the course, the author began adding material. The primary influences in selecting the added material have been honors linear algebra classes, applied complex variable classes, and the excitement coming from seeing the same mathematical principles in diverse settings. The author is happy with the many novel features, the wide variety of exercises, and the manner in which the text unifies topics.

As a whole, the book might be too mathematical for engineers, too informal for mathematicians, and too succinct for all but the top students. Below, however, we offer six specific course outlines that overcome these issues. The appendix, a late addition to the text, might help some readers overcome discomfort with the use of mathematical jargon. It also provides a diverse list of operations which can be regarded as functions; this list glimpses the vast scope of topics covered in the book.

Mathematical prerequisites for the book include three semesters of calculus, a beginning course in differential equations, and some elementary linear algebra. Complex variable theory is developed in the book, and hence is not a prerequisite. First year courses in Physics are useful but not strictly necessary.

Several possible courses can be based upon this book. We offer six possibilities:

(1) The first five chapters form a coherent course on engineering mathematics. Such a course should appeal to engineering faculty who want an integrated treatment of linear algebra and complex analysis, including applications, that also reviews vector analysis. The course could then conclude with Section 8 from Chapter 7 on linear time-invariant systems.

(2) The first five chapters (omitting parts of Chapter 3) together with the last three sections of Chapter 7 form a standard applied complex variables course, including transform methods.

(3) Each section in Chapter 7 begins by stating approximate prerequisites from the book. Some instructors might wish to decide a priori which applications in Chapter 7 they wish to discuss, and then cover Chapters 1 and 2 together with those topics needed for these applications.

(4) After covering the first four chapters, students in mathematics and physics can continue with the Hilbert space chapter and conclude with the sections on probability and quantum mechanics.

(5) Students with strong backgrounds can read the first three chapters on their own as review, and take a one-semester course based on Chapters 4 through 7. Such a course would be ideal for many graduate students in applied mathematics, physics, and engineering.

(6) Chapters 1, 6 and Sections 1, 5, 7 of Chapter 7 make a nice second course in linear algebra.

The book contains more than 450 exercises of considerably varying difficulty and feel. They are placed at the ends of sections, rather than at the end of chapters, to encourage the reader to solve them as she reads. Many of the exercises are routine and a few are difficult. Others introduce new settings where the techniques can be used. Most of the exercises come from mathematics; instructors may wish to provide additional exercises coming from engineering and physics. The text weaves abstract mathematics, computations, and applications into a coherent whole, whose unifying theme is linear systems.

The book itself has three primary goals. One goal is to develop enough linear analysis and complex variable theory to prepare students in engineering or applied mathematics for advanced work. The second goal is to unify many distinct and seemingly isolated topics. The third goal is to reveal mathematics as both interesting and useful, especially via the juxtaposition of examples and theorems. We give some examples of how we achieve these goals:

- Many aspects of the theory of linear equations are the same in finite and infinite dimensions; we glimpse both the similarities and the differences in Chapter 1 and develop the ideas in detail in Chapter 6.

- One recurring theme concerns functions of operators. We regard diagonalization of a matrix and taking Laplace or Fourier transforms as similar (pun intended) techniques. Early in the 493 class, and somewhat out of the blue, the author asked what it would mean to take "half of a derivative." One bright engineering student said "multiply by $\sqrt{s}$". A year and a half later the author asked his applied complex variables class the same question. A bright physics student answered "first take Laplace transforms and then $\sqrt{s}$". He did not actually say *multiply* but surely he understood.

- Choosing coordinates and notation in which computations are easy is a major theme in this book. Complex variables illustrate this theme throughout. For example, we use complex exponentials rather than trig functions nearly always.

- A Fourier series is a particular type of orthonormal expansion; we include many examples of orthonormal expansion as well as the general theory.

- We state the spectral theorem for compact operators on Hilbert space as part of our discussion of Sturm–Liouville equations, hoping that many readers will want to study more operator theory.

- We express Maxwell's equations in terms of both vector fields and differential forms, hoping that many readers will want to study differential geometry.
- We discuss root finding for polynomials in one variable and express the Jury stabilization algorithm in a simple manner.

Many of the examples will be accessible yet new to most readers. Here are some examples of unusual approaches and connections among different topics.

- To illustrate the usefulness of changing basis, we determine which polynomials with rational co-efficients map the integers to the integers. Later on, the same idea helps us to find the generating function of an arbitrary polynomial sequence.
- We mention the scalar field consisting of two elements, and include several exercises about lights-out puzzles to illuminate linear algebra involving this field. At the very end of the book, in the section on root finding, we give the simple exercise: show that the fundamental theorem of algebra fails in a finite field. (In Chapter 4 we prove this theorem for complex polynomials.)
- We compute line and surface integrals using both differential forms and the classical notation of vector analysis. We express Maxwell's equations using differential forms in a later section.
- To illustrate sine series, we prove the Wirtinger inequality relating the L^2 norms of a function and its derivative.
- We develop connections between finite- and infinite-dimensional linear algebra, and many of these ideas will be new to most readers. For example, we compute $\zeta(2)$ by finding the trace of the Green's operator for the second derivative.
- We introduce enough probability to say a bit about quantum mechanics; for example, we prove the Heisenberg uncertainty principle. We also include exercises on dice with strange properties.
- Our discussion of generating functions (Z-transforms) includes a derivation of a general formula for the sums $\sum_{n=1}^{N} n^p$ (using Bernoulli numbers) as well as (the much easier) Binet's formula for the Fibonacci numbers. The section on generating functions contains several unusual exercises, including a test for rationality and the notion of Abel summability. There we also discuss the precise meaning of the notorious but dubious formula $1 + 2 + 3 + \cdots = \frac{-1}{12}$, and we mention the concept of zeta function regularization. This section helps unify many of the ideas in the book, while at the same time glimpsing more sophisticated ideas.
- The section on Fourier series includes a careful discussion of summation by parts.

Several features distinguish this book from other *applied* math books. Perhaps most important is how the book **unifies** concepts regarded as distinct in many texts. Another is the extensive mathematical treatment of Hilbert spaces. The scope and diversity of our examples distinguish it from other *pure* math books.

Formal theoretical items such as theorems, definitions, lemmas, and propositions are numbered as a unit within each section. Examples are numbered separately, also within each section. Thus Lemma 3.2 must follow Theorem 3.1 or Definition 3.1 or whatever is the first theoretical item in the section, whereas Example 3.2 could precede Theorem 3.1. Remarks are not numbered.

This book is not an encyclopedia and it does not cover all significant topics in either analysis or engineering mathematics. Omissions include numerical methods and mathematical modeling. For example, although we write an inversion formula for the Laplace transform involving an integral, we never consider numerical evaluation of such integrals. We discuss eigenvalues throughout the book, well aware that techniques exist for finding them that are far better than using the characteristic polynomial; nonetheless we do not develop such techniques. We do show that the eigenvalues of a self-adjoint matrix can be found via optimization. Our discussion of the condition number of a matrix or linear map is quite brief; we do provide some exercises about the Hilbert matrix. Although we discuss how to solve a few linear differential equations, we say little about specific circumstances whose modeling leads to these equations. One omission concerns partial differential equations (PDE). We discuss the one-dimensional wave equation, the heat equation, and the Schrödinger equation, primarily to illustrate separation of variables, but we include little general theory. Section 6 of Chapter 7 discusses the Dirichlet problem and conformal mapping, but four pages cannot do justice to this material. Although we mention approximate identities at several places, we do not discuss distribution theory. The author recommends [T] for an excellent treatment of PDE that nicely weaves together the theory and important examples. Perhaps the most significant omission is wavelets. Although this topic nicely fits with much of what appears here, there is simply too much information. To quote rhythm and blues singer Smokey Robinson, "a taste of honey is worse than none at all." We therefore refer to [Dau], [M] and [N] for the subject of wavelets.

The author acknowledges the contributions of many people. People involved with the original ECE class include Jont Allen and Steve Levinson (both in ECE), Sarah Robinson (graduate student in ECE), Tej Chajed (then a senior in computer science), Pingyu Wang (a student in the class who took class notes on a tablet), and Jimmy Shan (a mathematics graduate student who typed these notes into a LaTeX file). A few of the topics resulted directly from questions asked by Jont Allen. Sam Homiller (then a student in engineering physics) made some of the pictures used in this book. Several of these students were partially supported by a small grant from IMSE, an initiative at the University of Illinois which is attempting to bring mathematicians and engineers closer together. Bruce Hajek and Erhan Kudeki (both of ECE) invited the author to several meetings of their curriculum committee where stimulating discussions ensued. Bruce Hajek gave the following wonderful advice: take the reader on a guided tour of some infinite forest, pointing out various highlights and landmarks, but it is not possible to give a complete view in one excursion. Physicist Simon Kos and the author had several productive discussions about Maxwell's equations and related topics. Physicist Mike Stone has provided numerous useful ideas to the author over the years. Ki Yeun (Eunice) Kim worked with the author in spring 2015 as a research assistant supported by the National Science Foundation. She provided many of the pictures as well as significantly many useful criticisms and comments. Upon reading a preliminary draft, Sergei Gelfand asked for a clarification of the goals of the book, thereby leading to significant improvements and revisions, especially in the last two chapters. Dan Zwillinger provided a useful balance of praise and criticism of an early draft and also helped me polish a late draft. Robert Ross offered both encouragement and many useful comments. Charlie Epstein suggested several significant improvements, while emphatically stating that the book is too mathematical for engineers. Students in other classes, especially Applied Complex Variables, have indirectly helped the author organize some of the material. Martin Ostoja-Starzewski (Mechanical Science and Engineering) provided valuable comments and encouragement while discussing the different perspectives of mathematics and engineering. Mathematics graduate student Zhenghui Huo influenced the treatment of several topics. Anil Hirani, Rick Laugesen, and Bill Helton each provided good ideas for bridging the gap between mathematics and engineering. Phil Baldwin patiently explained to the author what real world scientists actually do; his comments led to several improvements. The author's then thirteen-year-old son Henry suggested Exercise 5.1 of Chapter 3.

The author acknowledges support provided by NSF Grant DMS 13-61001.

CHAPTER 1

Linear algebra

Linear algebra arises throughout mathematics and the sciences. The most basic problem in linear algebra is to solve a system of linear equations. The unknown can be a vector in a finite-dimensional space, or an unknown function, regarded as a vector in an infinite-dimensional space. One of our goals is to present unified techniques which apply in both settings. The reader should be to some extent already familiar with most of the material in this chapter. Hence we will occasionally use a term before defining it precisely.

1. Introduction to linear equations

We begin by reviewing methods for solving both matrix equations and linear differential equations. Consider the linear equation $L\mathbf{x} = \mathbf{b}$, where $\mathbf{b}$ is a given vector, and $\mathbf{x}$ is an unknown vector. To be precise, $\mathbf{x}$ lives in a vector space V and $\mathbf{b}$ lives in a vector space W; in some but not all situations, $V = W$. The mapping $L : V \to W$ is linear. We give precise definitions of vector spaces and *linear* maps a bit later.

There are many books on linear algebra; we recommend [HK] for its abstract treatment and [St] for its applications and more concrete treatment. This chapter includes several topics not generally found in any linear algebra book. To get started we consider three situations, two just fun, where linear algebra makes an appearance. After that we will be a bit more formal.

When there are finitely many equations in finitely many unknowns, we solve using *row operations*, or *Gaussian elimination*. We illustrate by discussing a simple Kakuro puzzle. Kakuro is a puzzle where one is given a grid, blank squares, the row sums, the column sums, and blacked-out squares. One is supposed to fill in the blanks in the grid, using the digits 1 through 9. There are two added constraints; no repeats (within a fixed group) are allowed and the solution is known to be unique. In difficult puzzles, knowing that the solution is unique can sometimes help. We ignore these constraints in the following simple example.

Consider a three-by-three grid, as shown in Figure 1. From left to right, the column sums are given to be 10, 13, and 13. From top to bottom, the row sums are given to be 17, 7, and 12. Since the solution is known to use only the whole numbers from 1 through 9, one can easily solve it. In this example, there are two possible solutions. Starting from the top and going across, the possible solutions are $(9, 8, 1, 2, 4, 3, 9)$ and $(8, 9, 2, 1, 4, 3, 9)$.

FIGURE 1. An easy Kakuro puzzle

Label the unknown quantities $x_1, x_2, \ldots, x_7$. The given row and column sums lead to the following system of equations of six equations in seven unknowns:

$$x_1 + x_2 = 17$$
$$x_3 + x_4 + x_5 = 7$$
$$x_6 + x_7 = 12$$
$$x_1 + x_3 = 10$$
$$x_2 + x_4 + x_6 = 13$$
$$x_5 + x_7 = 13$$

To solve this system of equations, we first write it in matrix form

$$\begin{pmatrix} 1 & 1 & 0 & 0 & 0 & 0 & 0 & 17 \\ 0 & 0 & 1 & 1 & 1 & 0 & 0 & 7 \\ 0 & 0 & 0 & 0 & 0 & 1 & 1 & 12 \\ 1 & 0 & 1 & 0 & 0 & 0 & 0 & 10 \\ 0 & 1 & 0 & 1 & 0 & 1 & 0 & 13 \\ 0 & 0 & 0 & 0 & 1 & 0 & 1 & 13 \end{pmatrix}. \tag{1}$$

The last column is the right-hand side of the system of equations, the other columns are the coefficients of the equations, and each row corresponds to an equation. It is convenient to reorder the equations, obtaining:

$$\begin{pmatrix} 1 & 1 & 0 & 0 & 0 & 0 & 0 & 17 \\ 1 & 0 & 1 & 0 & 0 & 0 & 0 & 10 \\ 0 & 1 & 0 & 1 & 0 & 1 & 0 & 13 \\ 0 & 0 & 1 & 1 & 1 & 0 & 0 & 7 \\ 0 & 0 & 0 & 0 & 1 & 0 & 1 & 13 \\ 0 & 0 & 0 & 0 & 0 & 1 & 1 & 12 \end{pmatrix}.$$

Reordering the equations does not change the solutions. It is easy to see that we can achieve any order by several steps in which we simply interchange two rows. Such an interchange is an example of an *elementary row operation*. These operations enable us to simplify a system of linear equations expressed in matrix form. There are three types of elementary row operations. We can add a multiple of one row to another row, we can interchange two rows, and we can multiply a row by a non-zero constant.

This simple example does not require much effort. By row operations, we get the equivalent system

$$\begin{pmatrix} 1 & 1 & 0 & 0 & 0 & 0 & 0 & 17 \\ 0 & -1 & 1 & 0 & 0 & 0 & 0 & -7 \\ 0 & 0 & 1 & 1 & 0 & 1 & 0 & 6 \\ 0 & 0 & 1 & 1 & 1 & 0 & 0 & 7 \\ 0 & 0 & 0 & 0 & 1 & 0 & 1 & 13 \\ 0 & 0 & 0 & 0 & 0 & 1 & 1 & 12 \end{pmatrix}.$$

After a few more simple steps we obtain

$$\begin{pmatrix} 1 & 0 & 0 & -1 & 0 & 0 & 1 & 16 \\ 0 & 1 & 0 & 1 & 0 & 0 & -1 & 1 \\ 0 & 0 & 1 & 1 & 0 & 0 & -1 & -6 \\ 0 & 0 & 0 & 0 & 0 & 0 & 0 & 0 \\ 0 & 0 & 0 & 0 & 1 & 0 & 1 & 13 \\ 0 & 0 & 0 & 0 & 0 & 1 & 1 & 12 \end{pmatrix}. \tag{2}$$

The matrix equation in (2) has the same solutions as the matrix equation in (1), but it is expressed in a manner where one can see the solutions. We start with the last variable and work backwards. No row in (2) determines the last variable x_7, and hence this variable is arbitrary. The bottom row of (2) gives $x_6 + x_7 = 12$ and hence determines x_6 in terms of x_7. The second row from the bottom determines x_5 in terms of x_7. We proceed, working backwards, and note that x_4 is also arbitrary. The third row then determines x_3 in terms of both x_4 and x_7. Using the information from all the rows, we write the general solution to (1) in vector form as follows:

$$\mathbf{x} = \begin{pmatrix} x_1 \\ x_2 \\ x_3 \\ x_4 \\ x_5 \\ x_6 \\ x_7 \end{pmatrix} = \begin{pmatrix} 16 \\ 1 \\ -6 \\ 0 \\ 13 \\ 12 \\ 0 \end{pmatrix} + x_4 \begin{pmatrix} 1 \\ -1 \\ -1 \\ 1 \\ 0 \\ 0 \\ 0 \end{pmatrix} + x_7 \begin{pmatrix} -1 \\ 1 \\ 1 \\ 0 \\ -1 \\ -1 \\ 1 \end{pmatrix} \tag{3}$$

Notice that there are two arbitrary parameters in the solution (3): the variables x_4 and x_7 can be any real numbers. Setting them both equal to 0 yields one *particular solution*. Setting $x_7 = 9$ and $x_4 = 1$, for example, gives the solution $\mathbf{x} = (8, 9, 2, 1, 4, 3, 9)$, which solves the Kakuro puzzle. Setting $x_7 = 9$ and $x_4 = 2$ gives the other Kakuro solution.

When solving a system of linear equations, three things can happen. There can be no solutions, one solution, or infinitely many solutions. See Theorem 2.5 below. Before proving it, we will need to introduce some useful concepts. We conclude this section with an example.

EXAMPLE 1.1. Consider the system of three equations in three unknowns given by

$$x_1 + 5x_2 - 3x_3 = 2$$
$$2x_1 + 11x_2 - 4x_3 = 12$$
$$x_2 + cx_3 = b.$$

Here c and b are constants, and the form of the answer depends on them. In matrix form the system is

$$\begin{pmatrix} 1 & 5 & -3 & 2 \\ 2 & 11 & -4 & 12 \\ 0 & 1 & c & b \end{pmatrix}.$$

Two row operations yield the equivalent system

$$\begin{pmatrix} 1 & 5 & -3 & 2 \\ 0 & 1 & 2 & 8 \\ 0 & 0 & c-2 & b-8 \end{pmatrix}.$$

If $c = 2$, the last row says $0 = 0x_1 + 0x_2 + 0x_3 = b - 8$. If $b \neq 8$, then we get a contradiction and the system has no solution. If $c = 2$ and $b = 8$, then we can choose x_3 arbitrarily and satisfy all three equations. If $c \neq 2$, then we can solve the system whether or not $b = 8$, and the solution is unique.

EXERCISE 1.1. Put $c = 1$ in Example 1.1 and solve the resulting system.

EXERCISE 1.2. Solve the Kakuro puzzle in Figure 2. (Do not use linear equations!)

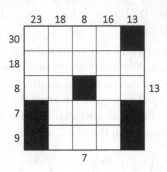

FIGURE 2. A kakuro puzzle

EXERCISE 1.3. Consider the linear system

$$x + y + z = 11$$

$$2x - y + z = 13$$

$$y + 3z = 19.$$

Convert it to matrix form, solve by row operations, and check your answer.

EXERCISE 1.4. Consider the (augmented) linear system, expressed in matrix form, of four equations in six unknowns.

$$\begin{pmatrix} 1 & 2 & 3 & 0 & 0 & 1 & 30 \\ 0 & 0 & 1 & 0 & 1 & 2 & 8 \\ 0 & 0 & 0 & 1 & 0 & 0 & 0 \\ 0 & 0 & 0 & 0 & 1 & 5 & 0 \end{pmatrix}$$

Find the most general solution, and write it in the form

$$\mathbf{x} = \mathbf{x_0} + \sum c_j \mathbf{v_j}.$$

2. Vectors and linear equations

We begin an informal discussion of linear equations in general circumstances. This discussion needs to be somewhat abstract in order to allow us to regard linear differential equations and linear systems (matrix equations) on the same footing.

Definition 2.1 (informal). A real vector space is a collection of objects called *vectors*. These objects can be added together, or multiplied by real scalars. These operations satisfy the axioms from Definition 5.2.

Definition 2.2. Let V and W be real vector spaces, and suppose $L : V \to W$ is a function. Then L is called a **linear transformation** (or a linear map or linear) if for all $u, v \in V$, and for all $c \in \mathbb{R}$, we have

$$L(u + v) = Lu + Lv \qquad (L.1)$$

$$L(cv) = c\,L(v). \qquad (L.2)$$

Note that the addition on the left-hand side of (L.1) takes place in V, but the addition on the right-hand side in (L.1) takes place in W. The analogous comment applies to the scalar multiplications in (L.2).

The reader is surely familiar with the vector space $\mathbb{R}^n$, consisting of n-tuples of real numbers. Both addition and scalar multiplication are formed component-wise. See (7.1) and (7.2) below. Each linear map from $\mathbb{R}^n$ to $\mathbb{R}^m$ is given by matrix multiplication. The particular matrix depends on the basis chosen, as we discuss in Section 8. Let A be a matrix of real numbers with n columns and m rows. If $\mathbf{x}$ is an element of $\mathbb{R}^n$, regarded as a column vector, then the matrix product $A\mathbf{x}$ is an element of $\mathbb{R}^m$. The mapping $x \mapsto A\mathbf{x}$ is linear. The notation I can mean either the identity matrix or the identity linear transformation; in both cases $I(\mathbf{x}) = \mathbf{x}$ for each $\mathbf{x} \in \mathbb{R}^n$.

While $\mathbb{R}^n$ is fundamental, it will be insufficiently general for the mathematics in this book. We will also need both $\mathbb{C}^n$ and various infinite-dimensional spaces of functions.

In Example 1.1, V is the vector space $\mathbb{R}^3$. The linear map $L : \mathbb{R}^3 \to \mathbb{R}^3$ is defined by

$$L(x, y, z) = (x + 5y - 3z, 2x + 11y - 4z, y + cz).$$

We are solving the linear equation $L(x, y, z) = (2, 12, b)$.

When V is a space of functions, and $L : V \to V$ is linear, we often say L is a *linear operator* on V. The simplest linear operator is the identity mapping, sending each v to itself; it is written I. Other well-known examples include differentiation and integration. For example, when V is the space of all polynomials in one variable x, the map $D : V \to V$ given by $D(p) = p'$ is linear. The map $J : V \to V$ defined by $(Jp)(x) = \int_0^x p(t)dt$ is also linear.

EXAMPLE 2.1. We consider the vector space of polynomials of degree at most three, and we define L by $L(p) = p - p''$. We show how to regard L as a matrix, although doing so is a bit clumsy. Put

$$p(x) = a_0 + a_1 x + a_2 x^2 + a_3 x^3.$$

Then we have

$$(Lp)(x) = p(x) - p''(x) = (a_0 - 2a_2) + (a_1 - 6a_3)x + a_2 x^2 + a_3 x^3.$$

There is a linear operator on $\mathbb{R}^4$, which we still write as L, defined by

$$L(a_0, a_1, a_2, a_3) = (a_0 - 2a_2, a_1 - 6a_3, a_2, a_3).$$

In matrix form, we have

$$L = \begin{pmatrix} 1 & 0 & -2 & 0 \\ 0 & 1 & 0 & -6 \\ 0 & 0 & 1 & 0 \\ 0 & 0 & 0 & 1 \end{pmatrix}. \qquad (*)$$

Example 2.1 illustrates several basic points. The formula $L(p) = p - p''$ is much simpler than the matrix representation in formula (*). In this book we will often use simple but abstract notation. One benefit is that different applied situations become mathematically identical. In this example we can think of the space V_3 of polynomials of degree at most three as essentially the same as $\mathbb{R}^4$. Thus Example 2.1 suggests the notions of **dimension** of a vector space and isomorphism.

Definition 2.3. Let $L : V \to V$ be a linear transformation. Then L is called **invertible** if there are linear transformations A and B such that $I = LB$ and $I = AL$. A square matrix M is **invertible** if there are square matrices A and B such that $I = MB$ and $I = AM$.

REMARK. A linear map L or a matrix M is invertible if such A and B exist. If they both exist, then $A = B$:

$$A = AI = A(LB) = (AL)B = IB = B.$$

Note that we have used associativity of composition in this reasoning. When both A and B exist, we therefore write (without ambiguity) the inverse mapping as L^{-1}.

Examples where only one of A or B exists arise only in infinite-dimensional settings. See Example 5.3, which is well known to calculus students when expressed in different words. It clarifies why we write $+C$ when we do indefinite integrals.

Definition 2.4. Let $L : V \to W$ be a linear transformation. The **null space** of L, written $\mathcal{N}(L)$, is the set of v in V such that $Lv = 0$. The **range** of L, written $\mathcal{R}(L)$, is the set of $w \in W$ such that there is some $v \in V$ with $Lv = w$.

REMARK. Let $L : \mathbb{R}^n \to \mathbb{R}^n$ be linear. Then $\mathcal{N}(L) = \{0\}$ if and only if L is invertible. The conclusion does not hold if the domain and target spaces have different dimensions.

Theorem 2.5. *Let V, W be real vector spaces, and suppose $L : V \to W$ is linear. For the linear system $L\mathbf{x} = \mathbf{b}$, exactly one of three possibilities holds:*

(1) *For each $\mathbf{b} \in V$, the system has a unique solution. When $W = V$, this possibility occurs if and only if L is invertible, and the unique solution is $\mathbf{x} = L^{-1}\mathbf{b}$. In general, this possibility occurs if and only if $\mathcal{N}(L) = \{0\}$ (uniqueness) and $\mathcal{R}(L) = W$ (existence).*

(2) *$\mathbf{b} \notin \mathcal{R}(L)$. Then there is no solution.*

(3) *$\mathbf{b} \in \mathcal{R}(L)$, and $\mathcal{N}(L) \neq \{0\}$. Then there are infinitely many solutions. Each solution satisfies*

$$\mathbf{x} = \mathbf{x}_{\text{part}} + \mathbf{v}, \tag{$*$}$$

where $\mathbf{x}_{\text{part}}$ is a particular solution and $\mathbf{v}$ is an arbitrary element of $\mathcal{N}(L)$.

PROOF. Consider $\mathbf{b} \in W$. If $\mathbf{b}$ is not in the range of L, then (2) holds. Thus we assume $\mathbf{b} \in W$ is an element of $\mathcal{R}(L)$. Then there is an $\mathbf{x}_{\text{part}}$ with $L(\mathbf{x}_{\text{part}}) = \mathbf{b}$. Let $\mathbf{x}$ be an arbitrary solution to $L(\mathbf{x}) = \mathbf{b}$. Then $\mathbf{b} = L(\mathbf{x}) = L(\mathbf{x}_{\text{part}})$. By linearity, we obtain

$$0 = L(\mathbf{x}) - L(\mathbf{x}_{\text{part}}) = L(\mathbf{x} - \mathbf{x}_{\text{part}}).$$

Hence $\mathbf{x} - \mathbf{x}_{\text{part}} \in \mathcal{N}(L)$. If $\mathcal{N}(L) = \{0\}$, then $\mathbf{x} = \mathbf{x}_{\text{part}}$ and (1) holds. If $\mathcal{N}(L) \neq \{0\}$, then we claim that (3) holds. We have verified (*); it remains to show that there are infinitely many solutions. Since $\mathcal{N}(L)$ is not 0 alone, and it is closed under scalar multiplication, it is an infinite set, and hence the number of solutions to $L(\mathbf{x}) = \mathbf{b}$ is infinite as well. $\square$

The main point of Theorem 2.5 is that a system of linear equations with real coefficients (in finitely or infinitely many variables) can have no solutions, one solution, or infinitely many solutions. There are no other possibilities. Analogous results hold for systems with complex coefficients and even more generally. See Exercises 2.5 through 2.8 for easy examples illustrating Theorem 2.5.

This idea of particular solution plus an element of the null space should be familiar from differential equations. We illustrate by solving an inhomogeneous, constant-coefficient, linear, ordinary differential equation (ODE).

EXAMPLE 2.2. We solve the differential equation $y'' + 5y' + 6y = 6x - 6$. Here we regard y as a vector in an infinite-dimensional vector space of functions. Let $A(y) = y'' + 5y' + 6y$ and $\mathbf{b} = 6x - 6$. Then A is linear and we wish to solve $Ay = \mathbf{b}$.

We first find the null space of A by considering the homogeneous equation $A(y) = 0$. The standard approach to solve it is to assume a solution of the form $e^{\lambda x}$. We obtain the equation

$$(\lambda^2 + 5\lambda + 6)e^{\lambda x} = 0.$$

Since $e^{\lambda x} \neq 0$, we can divide by it and obtain $\lambda^2 + 5\lambda + 6 = 0$. This equation is known as the **characteristic equation** for the differential equation. Here its solutions are $\lambda = -2, -3$. Hence the null space of A is spanned by the two functions e^{-2x} and e^{-3x}. Thus there are constants c_1 and c_2 such that $Ay = 0$ if and only if $y(x) = c_1 e^{-2x} + c_2 e^{-3x}$.

Next we need to find a particular solution. To do so, we seek a particular solution of the form $y = cx + d$. Then $y' = c$ and $y'' = 0$. Plugging in the equation and doing simple algebra yields $c = 1$ and $d = -\frac{11}{6}$.

The general solution to the differential equation is therefore

$$y(x) = c_1 e^{-2x} + c_2 e^{-3x} + x - \frac{11}{6}.$$

In a real world problem, we would have additional information, such as the values of y and y' at some initial point. We would then use these values to find the constants c_1 and c_2.

REMARK. In engineering, the current is often written $i(t)$ or $I(t)$. For us i will denote the imaginary unit ($i^2 = -1$) and I will denote the identity operator or identity matrix. Hence we write $\mathcal{I}$ for the current, both in the next example and in later chapters.

EXAMPLE 2.3. The current in an RLC circuit (Figure 3) is governed by a second-order, constant-coefficient, linear ordinary differential equation.

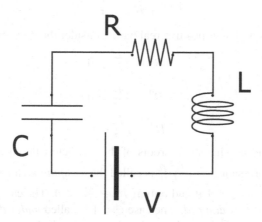

FIGURE 3. An RLC circuit

Three circuit elements, or impedances, arise in the coefficients of this ODE. Let $E(t)$ be the voltage drop at time t, and let $\mathcal{I}(t)$ be the current at time t. For a resistor, Ohm's law yields

$$E_{\text{resistor}}(t) = R\mathcal{I}(t).$$

The constant R is called the **resistance**. Hence $E'_{\text{resistor}} = R\mathcal{I}'$.

For an inductor, the voltage drop is proportional to the time derivative of the current:

$$E_{\text{inductor}}(t) = LI'(t).$$

The constant of proportionality L is called the **inductance**. Hence $E'_{\text{inductor}} = LI''$.

For a capacitor, the voltage drop is proportional to the charge $Q(t)$ on the capacitor, and the current is the time derivative of the charge:

$$E_{\text{capacitor}}(t) = \frac{1}{C}Q(t).$$

$$E'_{\text{capacitor}}(t) = \frac{1}{C}I(t).$$

The constant C is called the **capacitance**.

Using Kirchhoff's laws, which follow from conservation of charge and conservation of energy, and assuming linearity, we add the three contributions to obtain the time derivative of the voltage drop:

$$LI''(t) + RI'(t) + \frac{1}{C}I(t) = E'(t). \tag{4}$$

The differential equation (4) expresses the unknown current in terms of the voltage drop, the inductance, the resistance, and the capacitance. The equation is linear and of the form $A(I) = \mathbf{b}$. Here A is the operator given by

$$A = L\frac{d^2}{dt^2} + R\frac{d}{dt} + \frac{1}{C}I.$$

The domain of A consists of a space of functions. The current $t \mapsto I(t)$ lives in some infinite-dimensional vector space of functions. We find the current by solving the linear system $A(I) = \mathbf{b}$.

REMARK. The characteristic equation of the differential equation for an RLC circuit is

$$L\lambda^2 + R\lambda + \frac{1}{C} = 0.$$

Since $L, R, C \geq 0$, the roots have non-positive real part. Consider the three cases

$$R^2 - 4\frac{L}{C} > 0$$

$$R^2 - 4\frac{L}{C} = 0$$

$$R^2 - 4\frac{L}{C} < 0.$$

These three cases correspond to whether the roots of the characteristic equation are real and distinct, repeated, or complex. The expression $\zeta = \sqrt{\frac{C}{L}}\frac{R}{2}$ is called the *damping factor* of this circuit. Note that $\zeta > 1$ if and only if $R^2 - \frac{4L}{C} > 0$ and $\zeta < 1$ if and only if $R^2 - \frac{4L}{C} < 0$. The case $\zeta > 1$ is called *over-damped*. The case $\zeta = 1$ is called *critically damped*. The case $\zeta < 1$ is called *under-damped* and we get oscillation. See Figure 4. The discriminant $R^2 - 4\frac{L}{C}$ has units ohms squared; the damping factor is a dimensionless quantity.

EXERCISE 2.1. What is the null space of the operator $D^3 - 3D^2 + 3D - I$? Here D denotes differentiation, defined on the space of infinitely differentiable functions.

EXERCISE 2.2. Find the general solution to the ODE $y'' - y = x$.

EXERCISE 2.3. Find the general solution to the ODE $y' - ay = e^x$. What happens when $a = 1$?

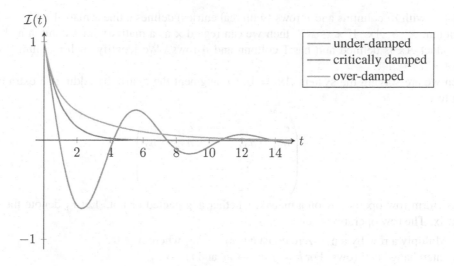

FIGURE 4. The damping factor for a circuit

EXERCISE 2.4. For what value of C is the circuit in eqn. (4) critically damped if $L = 1$ and $R = 4$?

EXERCISE 2.5. Consider the system of two equations in two unknowns given by $x + 2y = 7$ and $2x + 4y = 14$. Find all solutions and express them in the form (3) from Section 1. For what values of a, b does the system $x + 2y = a$ and $2x + 4y = b$ have a solution?

EXERCISE 2.6. Consider the system $x + 2y = 7$ and $2x + 6y = 16$. Find all solutions. For what values of a, b does the system $x + 2y = a$ and $2x + 6y = b$ have a solution?

EXERCISE 2.7. In this exercise i denotes $\sqrt{-1}$. Consider the system $z + (1+i)w = -1$ and $iz - w = 0$, for z, w unknown complex numbers. Find all solutions.

EXERCISE 2.8. Again i denotes $\sqrt{-1}$. Consider the system $z + (1+i)w = -1$ and $-iz + (1-i)w = i$. Find all solutions and express in the form (3) from Section 1.

3. Matrices and row operations

We wish to extend our discussion of row operations. We begin by recalling the definitions of matrix addition and matrix multiplication. We add two matrices of the same size by adding the corresponding elements. The i, j entry in $A + B$ is therefore given by

$$(A + B)_{ij} = A_{ij} + B_{ij}.$$

We can multiply matrices A and B to form BA only when the number of columns of B equals the number of rows of A. In that case, the i, j entry in BA is defined by

$$(BA)_{ij} = \sum_{k=1}^{n} B_{ik} A_{kj}.$$

Here A has n columns and B has n rows. It follows easily from this definition that matrix multiplication is *associative*: $A(BC) = (AB)C$ whenever all the products are defined. Matrix multiplication is not generally commutative; even if both AB and BA are both defined, they need not be equal. We return to these issues a bit later.

A matrix with m columns and n rows (with real entries) defines a linear map $A : \mathbb{R}^m \to \mathbb{R}^n$ by matrix multiplication: $\mathbf{x} \mapsto A\mathbf{x}$. If $\mathbf{x} \in \mathbb{R}^m$, then we can regard $\mathbf{x}$ as a matrix with 1 column and m rows. The matrix product $A\mathbf{x}$ is defined and has 1 column and n rows. We identify such a column vector with an element of $\mathbb{R}^n$.

When we are solving the system $A\mathbf{x} = \mathbf{b}$, we augment the matrix by adding an extra column corresponding to $\mathbf{b}$:

$$
\begin{pmatrix}
a_{11} & a_{12} & \cdots & a_{1m} & b_1 \\
a_{21} & a_{22} & \cdots & a_{2m} & b_2 \\
\vdots & \vdots & \cdots & \vdots & \vdots \\
a_{n1} & a_{n2} & \cdots & a_{nm} & b_n
\end{pmatrix}.
$$

We perform row operations on a matrix, whether augmented or not. Let r_j denote the j-th row of a given matrix. The row operations are

- Multiply a row by a non-zero constant: $r_j \to \lambda r_j$ where $\lambda \neq 0$.
- Interchange two rows. For $k \neq j$, $r_j \to r_k$ and $r_k \to r_j$.
- Add a multiple of a row to another row. For $\lambda \neq 0$ and $k \neq j$, $r_j \to r_j + \lambda r_k$.

Doing a row operation on a matrix A corresponds to multiplying on the left by a particular matrix, called an *elementary row matrix*. We illustrate each of the operations in the 2-dimensional case:

$$
\begin{pmatrix} 1 & 0 \\ 0 & \lambda \end{pmatrix} \tag{ER.1}
$$

$$
\begin{pmatrix} 0 & 1 \\ 1 & 0 \end{pmatrix} \tag{ER.2}
$$

$$
\begin{pmatrix} 1 & \lambda \\ 0 & 1 \end{pmatrix} \tag{ER.3}.
$$

Multiplying the second row of a (two-by-two) matrix corresponds to multiplying on the left by the matrix in (ER.1). Switching two rows corresponds to multiplying on the left by the matrix in (ER.2). Adding λ times the second row to the first row corresponds to multiplying on the left by the matrix in (ER.3).

The language used to describe linear equations varies considerably. When mathematicians say that $Lu = b$ is a *linear equation*, L can be an arbitrary linear map between arbitrary vector spaces. We regard b as known, and we seek all solutions u. Here u is the variable, or the unknown. It could be a column vector of n unknown numbers, or an unknown function, regarded as a vector in an infinite-dimensional space of functions. When u is a column vector of n unknown numbers and b consists of m known numbers, we can regard the linear equation $Lu = b$ as m linear equations in n unknowns. The term **system of linear equations** is often used in this case. The author prefers thinking of a system as a single equation for an unknown vector.

Row operations enable us to replace a linear equation with another linear equation whose solutions are the same, but which is easier to understand.

Definition 3.1. Let $L : V \to W_1$ and $M : V \to W_2$ be linear transformations. The linear equations $Lu = b$ and $Mv = c$ are **equivalent** if they have the same solution sets.

In Definition 3.1, the linear maps L and M must have the same domain, but not necessarily the same target space. The reason is a bit subtle; the definition focuses on the solution set. Exercise 3.3 shows for example that a system of two linear equations in three unknowns can be equivalent to a system of three linear equations in three unknowns. In such a situation one of the three equations is redundant. These ideas will be clarified when we formalize topics such as linear dependence and dimension.

Row operations on a matrix enable us to replace a given linear equation with an equivalent equation, which we hope is easier to understand. When L is invertible, the equation $Lu = b$ is equivalent to the equation $u = L^{-1}b$, which exhibits the unique solution. We repeat that a row operation amounts to multiplying on the left by an invertible matrix, and doing so does not change the set of solutions to the equation.

When solving a system, one performs enough row operations to be able to read off the solutions. One need not reach either of the forms in the following definition, but the terminology is nonetheless useful.

Definition 3.2. A matrix is **row-reduced** if the first entry (from the left) in each non-zero row equals 1, and all the other entries in the column corresponding to such an entry equal 0. For each non-zero row, the 1 in the first entry is called a **leading** 1. A matrix is in **row echelon form** if it is row-reduced and, in addition, the following hold:

- Each zero row is below all of the non-zero rows.
- The non-zero rows are ordered as follows: if the leading 1 in row j occurs in column c_j, then $c_1 < c_2 < \cdots < c_k$.

EXAMPLE 3.1. The following matrices are in row echelon form:

$$\begin{pmatrix} 0 & 1 & 2 & 0 & 5 \\ 0 & 0 & 0 & 1 & 4 \\ 0 & 0 & 0 & 0 & 0 \end{pmatrix}$$

$$\begin{pmatrix} 1 & 1 & 2 & 0 & 0 \\ 0 & 0 & 0 & 1 & 0 \\ 0 & 0 & 0 & 0 & 1 \end{pmatrix}$$

The following matrix is not in row-echelon form:

$$\begin{pmatrix} 0 & 1 & 2 & 0 & 5 \\ 0 & 1 & 0 & 1 & 4 \end{pmatrix}.$$

The matrix (1) from the Kakuro problem is not row-reduced; the matrix (2) is row-reduced but not in row echelon form.

It is important to understand that AB need not equal BA for matrices A and B. When $AB = BA$ we say that A and B **commute**. The difference $AB - BA$ is called the **commutator** of A and B; it plays a significant role in quantum mechanics. See Chapter 7. An amusing way to regard the failure of commutivity is to think of B as putting on your socks and A as putting on your shoes. Then the order of operations matters. Furthermore, we also see (when A and B are invertible) that $(AB)^{-1} = B^{-1}A^{-1}$. Put on your socks, then your shoes. To undo, first take off your shoes, then your socks.

We have observed that matrix multiplication is associative. The identity

$$(AB)C = A(BC)$$

holds whenever the indicated matrix products are defined; it follows easily from the definition of matrix multiplication. If we regard matrices as linear maps, then associativity holds because composition of functions is always associative.

If some sequence of elementary row operations takes A to B, then we say that A and B are **row-equivalent**. Since each elementary row operation has an inverse, it follows that row-equivalence is an equivalence relation. We discuss equivalence relations in detail in the next section.

EXERCISE 3.1. Prove that matrix multiplication is associative.

EXERCISE 3.2. Write the following matrix as a product of elementary row matrices and find its inverse:

$$\begin{pmatrix} 1 & 2 & 3 \\ 2 & 4 & 7 \\ 0 & 1 & 0 \end{pmatrix}$$

EXERCISE 3.3. Determine whether the linear equations are equivalent:

$$\begin{pmatrix} 1 & 2 & 3 \\ 1 & 2 & 4 \\ 0 & 0 & 1 \end{pmatrix} \begin{pmatrix} x \\ y \\ z \end{pmatrix} = \begin{pmatrix} 6 \\ 7 \\ 1 \end{pmatrix}$$

$$\begin{pmatrix} 1 & 2 & 3 \\ 0 & 0 & 1 \end{pmatrix} \begin{pmatrix} x \\ y \\ z \end{pmatrix} = \begin{pmatrix} 6 \\ 1 \end{pmatrix}.$$

EXERCISE 3.4. In each case, factor A into a product of elementary matrices. Check your answers.

$$A = \begin{pmatrix} 6 & 1 \\ -3 & -1 \end{pmatrix}$$

$$A = \begin{pmatrix} 0 & 1 & 0 & 0 \\ 0 & 0 & 1 & 0 \\ 0 & 0 & 0 & 1 \\ 1 & 0 & 0 & 0 \end{pmatrix}$$

EXERCISE 3.5. Find two-by-two matrices A, B such that $AB = 0$ but $BA \neq 0$.

EXERCISE 3.6. When two impedances Z_1 and Z_2 are connected in series, the resulting impedance is $Z = Z_1 + Z_2$. When they are connected in parallel, the resulting impedance is

$$Z = \frac{1}{\frac{1}{Z_1} + \frac{1}{Z_2}} = \frac{Z_1 Z_2}{Z_1 + Z_2}.$$

A traveler averages v_1 miles per hour going one direction, and returns (exactly the same distance) averaging v_2 miles per hour. Find the average rate for the complete trip. Why is the answer analogous to the answer for parallel impedances?

4. Equivalence relations

In order to fully understand linear systems, we need a way to say that two linear systems, perhaps expressed very differently, provide the same information. Mathematicians formalize this idea using **equivalence relations**. This concept appears in both pure and applied mathematics; we discuss it in detail and provide a diverse collection of examples.

Sometimes we are given a set, and we wish to regard different members of the set as the same. Perhaps the most familiar example is parity. Often we care only whether a whole number is even or odd. More generally, consider the integers $\mathbb{Z}$ and a *modulus* p larger than 1. Given integers m, n we regard them as the same if $m - n$ is divisible by p. For example, p could be 12 and we are thinking of *clock arithmetic*. To make this sort of situation precise, mathematicians introduce equivalence relations and equivalence classes.

First we need to define **relation**. Let A and B be arbitrary sets. Their **Cartesian product** $A \times B$ is the set of ordered pairs (a, b) where $a \in A$ and $b \in B$. A **relation** R is an arbitrary subset of $A \times B$. A function $f : A \to B$ can be regarded as a special kind of relation; it is the subset consisting of pairs of the form $(a, f(a))$. Another example of a relation is inequality. Suppose $A = B = \mathbb{R}$; we say that $(a, b) \in R$

if, for example, $a < b$. This relation is not a function, because for each a there are many b related to a. A function is a relation where, for each a, there is a unique b related to a.

We use the term **relation on** S for a subset of $S \times S$. We often write $x \sim y$ instead of $(x, y) \in R$.

Definition 4.1. An **equivalence relation** on a set S is a relation $\sim$ such that

- $x \sim x$ for all $x \in S$. (reflexive property)
- $x \sim y$ implies $y \sim x$. (symmetric property)
- $x \sim y$ and $y \sim z$ implies $x \sim z$. (transitive property)

An equivalence relation partitions a set into **equivalence classes**. The equivalence class containing x is the collection of all objects equivalent to x. Thus an equivalence class is a set.

Equality of numbers (or other objects) provides the simplest example of an equivalence relation, but it is too simple to indicate why the concept is useful. We give several additional examples next, emphasizing those from linear algebra. The exercises provide additional insight.

EXAMPLE 4.1. Congruence provides a good example of an equivalence relation. Let $\mathbb{Z}$ be the set of integers, and let p be a modulus. We write $m \sim n$ if $m - n$ is divisible by p. For example, if $p = 2$, then there are two equivalence classes, the odd numbers and the even numbers. For example, if $p = 3$, then there are three equivalence classes: the set of whole numbers divisible by three, the set of whole numbers one more than a multiple of three, and the set of whole numbers one less than a multiple of three.

EXAMPLE 4.2. Let A and B be matrices of the same size. We say A and B are **row-equivalent** if there is a sequence of row operations taking A to B.

EXAMPLE 4.3. Square matrices A and B are called **similar** if there is an invertible matrix P such that $A = PBP^{-1}$. Similarity is an equivalence relation.

EXAMPLE 4.4. In Chapter 6 we say that two functions are equivalent if they agree except on a small set (a set of measure zero). An element of a Hilbert space will be an equivalence class of functions, rather than a function itself.

In this book we presume the real numbers are known. One can however regard the natural numbers $\mathbb{N}$ as the starting point. Then one constructs the integers $\mathbb{Z}$ from $\mathbb{N}$, the rational numbers $\mathbb{Q}$ from $\mathbb{Z}$, and finally the real number system $\mathbb{R}$ from $\mathbb{Q}$. Each of these constructions involves equivalence classes. We illustrate for the rational number system.

EXAMPLE 4.5 (Fractions). To construct $\mathbb{Q}$ from $\mathbb{Z}$, we consider ordered pairs of integers (a, b) with $b \neq 0$. The equivalence relation here is that $(a, b) \sim (c, d)$ if $ad = bc$. Then $(1, 2) \sim (-1, -2) \sim (2, 4)$ and so on. Let $\frac{a}{b}$ denote the equivalence class of all pairs equivalent to (a, b). Thus a rational number is an equivalence class of pairs of integers! We define addition and multiplication of these equivalence classes. For example, $\frac{a}{b} + \frac{m}{n}$ is the class containing $(an + bm, bn)$. See Exercise 4.5.

An equivalence class is a set, but thinking that way is sometimes too abstract. We therefore often wish to choose a particular element, or *representative*, of a given equivalence class. For example, if we think of a rational number as an equivalence class (and hence infinitely many pairs of integers), then we might select as a representative the fraction that is in lowest terms.

In linear algebra, row equivalence provides a compelling example of an equivalence relation. Exercise 4.1 asks for the easy proof. Since an invertible matrix is row equivalent to the identity, we could represent the equivalence class of all invertible matrices by the identity matrix. Later in this chapter we show that a matrix is diagonalizable if and only if it is similar to a diagonal matrix. In this case, the diagonal matrix is often the most useful representative of the equivalence class.

Most linear algebra books state the following result, although the language used may differ. The row-equivalence class containing the identity matrix consists of the invertible matrices.

Theorem 4.2. *Let A be a square matrix of real (or complex numbers). Then the following statements all hold or all fail simultaneously.*

- *A is invertible.*
- *A is row equivalent to the identity matrix.*
- *A is a product of elementary matrices.*

We illustrate this theorem by factoring and finding the inverse of the following matrix:

$$A = \begin{pmatrix} 2 & 1 \\ 1 & 1 \end{pmatrix}.$$

We perform row operations on A until we reach the identity:

$$\begin{pmatrix} 2 & 1 \\ 1 & 1 \end{pmatrix} \rightarrow \begin{pmatrix} 1 & 1 \\ 2 & 1 \end{pmatrix} \rightarrow \begin{pmatrix} 1 & 1 \\ 0 & -1 \end{pmatrix} \rightarrow \begin{pmatrix} 1 & 0 \\ 0 & -1 \end{pmatrix} \rightarrow \begin{pmatrix} 1 & 0 \\ 0 & 1 \end{pmatrix}$$

At each step we multiply on the left by the corresponding elementary matrix and express the process by a product of matrices:

$$\begin{pmatrix} 1 & 0 \\ 0 & -1 \end{pmatrix} \begin{pmatrix} 1 & 1 \\ 0 & 1 \end{pmatrix} \begin{pmatrix} 1 & 0 \\ -2 & 1 \end{pmatrix} \begin{pmatrix} 0 & 1 \\ 1 & 0 \end{pmatrix} \begin{pmatrix} 2 & 1 \\ 1 & 1 \end{pmatrix} = \begin{pmatrix} 1 & 0 \\ 0 & 1 \end{pmatrix}. \tag{5}$$

We think of (5) as $E_4 E_3 E_2 E_1 A = I$, where the E_j are elementary row matrices. Thus $A^{-1} = E_4 E_3 E_2 E_1$. Hence

$$A^{-1} = \begin{pmatrix} 1 & -1 \\ -1 & 2 \end{pmatrix}.$$

Taking the inverse matrices and writing the product in the reverse order yields $A = E_1^{-1} E_2^{-1} E_3^{-1} E_4^{-1}$ as the factorization of A into elementary matrices:

$$\begin{pmatrix} 2 & 1 \\ 1 & 1 \end{pmatrix} = \begin{pmatrix} 0 & 1 \\ 1 & 0 \end{pmatrix} \begin{pmatrix} 1 & 0 \\ 2 & 1 \end{pmatrix} \begin{pmatrix} 1 & -1 \\ 0 & 1 \end{pmatrix} \begin{pmatrix} 1 & 0 \\ 0 & -1 \end{pmatrix} \begin{pmatrix} 1 & 0 \\ 0 & 1 \end{pmatrix}.$$

When finding inverses alone, one can write the same steps in a more efficient fashion. Augment A by including the identity matrix to the right, and apply the row operations to this augmented matrix. When the left half becomes the identity matrix, the right half is A^{-1}. In this notation we get

$$\begin{pmatrix} 2 & 1 & 1 & 0 \\ 1 & 1 & 0 & 1 \end{pmatrix} \rightarrow \begin{pmatrix} 1 & 1 & 0 & 1 \\ 2 & 1 & 1 & 0 \end{pmatrix} \rightarrow \begin{pmatrix} 1 & 1 & 0 & 1 \\ 0 & -1 & 1 & -2 \end{pmatrix} \rightarrow$$

$$\begin{pmatrix} 1 & 0 & 1 & -1 \\ 0 & -1 & 1 & -2 \end{pmatrix} \rightarrow \begin{pmatrix} 1 & 0 & 1 & -1 \\ 0 & 1 & -1 & 2 \end{pmatrix}. \tag{6}$$

Thus row operations provide a method both for finding A^{-1} and for factoring A into elementary matrices. This reasoning provides the proof of Theorem 4.2, which we leave to the reader.

We conclude this section with an amusing game. The solution presented here illustrates the power of geometric thinking; orthogonality and linear independence provide the key idea.

EXAMPLE 4.6. There are two players A and B. Player A names the date January 1. Player B must then name a date, later in the year, changing only one of the month or the date. Then player A continues in the same way. The winner is the player who names December 31. If player B says January 20, then player B can force a win. If player B says any other date, then player A can force a win.

We work in two-dimensional space, writing (x, y) for the month and the day. The winning date is $(12, 31)$. Consider the line given by $y = x + 19$. Suppose a player names a point (x, y) on this line. According to the rules, the other player can name either $(x + a, y)$ or $(x, y + b)$, for a and b positive. Neither of these points lies on the line. The player who named (x, y) can then always return to the line. Thus a player who names a point on this line can always win. Hence the only way for player B to guarantee a win on the first turn is to name January 20, the point $(1, 20)$.

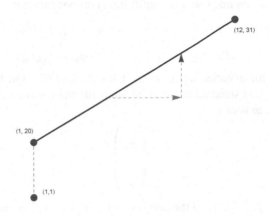

FIGURE 5. Solution of the game

EXERCISE 4.1. Prove that row equivalence is an equivalence relation.

EXERCISE 4.2. Prove Theorem 4.2.

EXERCISE 4.3. Describe all row equivalence classes of two-by-two matrices.

EXERCISE 4.4. Verify that the three properties of an equivalence relation hold in Example 4.5. Why is this definition used? What is the definition of multiplication of rational numbers?

EXERCISE 4.5. Let S be the set of students at a college. For $a, b \in S$ write $a \sim b$ if a and b have taken a class together. Is this relation an equivalence relation?

EXERCISE 4.6. (Dollars and sense). For real numbers, define $(x, y) \sim (a, b)$ if $x + 100y = a + 100b$. Is this relation an equivalence relation? Give a simple interpretation!

EXERCISE 4.7. Determine all two-by-two matrices that are row equivalent to $\begin{pmatrix} 1 & 0 \\ 0 & 0 \end{pmatrix}$.

EXERCISE 4.8. There is a pile of 1000 pennies for a two-person game. The two players alternate turns. On your turn, you may remove 1, 2, or 3 pennies. The person who removes the last penny wins the game. Which player wins? What if the original pile had 1001 pennies?

EXERCISE 4.9. Which of the following subsets of $\mathbb{R}^2$, thought of as relations on $\mathbb{R}$, define functions?
- The set of (x, y) with $x + y = 1$.
- The set of (x, y) with $x^2 + y^2 = 1$.
- The set of (x, y) with $x^3 + y^3 = 1$.

5. Vector spaces

We recall that $\mathbb{R}^n$ consists of n-tuples of real numbers. Thus we put

$$\mathbb{R}^n = \{(x_1, x_2, \cdots, x_n) : x_i \in \mathbb{R}\}$$
$$\mathbb{C}^n = \{(z_1, z_2, \cdots, z_n) : z_i \in \mathbb{C}\}$$

In both cases, we define addition and scalar multiplication componentwise:

$$(u_1, u_2, \cdots, u_n) + (v_1, v_2, \cdots, v_n) = (u_1 + v_1, u_2 + v_2, \cdots, u_n + v_n) \tag{7.1}$$

$$c(u_1, u_2, \cdots, u_n) = (cu_1, cu_2, \cdots, cu_n) \tag{7.2}$$

In both cases, the axioms for a vector space hold. Thus $\mathbb{R}^n$ and $\mathbb{C}^n$ will be two of our most important examples of vector spaces. It is standard notation within mathematics to regard the n-tuple $(x_1, \ldots, x_n)$ as the same object as the *column vector*

$$\begin{pmatrix} x_1 \\ x_2 \\ \vdots \\ x_n \end{pmatrix}.$$

The (column) vector $(x_1, x_2, \ldots, x_n)$ and the *row vector* $(x_1 \ x_2 \ \ldots \ x_n)$ are **not** considered the same object. We return to this point in Chapter 3 when we introduce dual spaces.

For $\mathbb{R}^n$, the scalars are real numbers. For $\mathbb{C}^n$, the scalars will be complex numbers. The collection of scalars forms what is known in mathematics as the *ground field* or *scalar field*.

We give the formal definition of vector space. First we need to define **field**. The word *field* has two distinct uses in mathematics. Its use in terms such as vector field is completely different from its use in the following definition. In nearly all our examples, the ground field will be $\mathbb{R}$ or $\mathbb{C}$, although other fields also arise in various applications.

Definition 5.1. A **field** $\mathbf{F}$ is a mathematical system consisting of a set of objects (often called *scalars*) together with two operations, addition and multiplication, satisfying the following axioms. As usual, we write xy for multiplication, instead of a notation such as $x * y$.

- There are distinct elements of $\mathbf{F}$ written 0 and 1.
- For all x, y we have $x + y = y + x$.
- For all x, y, z we have $(x + y) + z = x + (y + z)$
- For all x we have $x + 0 = x$
- For all x there is a $-x$ such that $x + (-x) = 0$.
- For all x, y we have $xy = yx$.
- For all x, y, z we have $(xy)z = x(yz)$.
- For all x we have $x1 = x$.
- For all $x \neq 0$ there is an x^{-1} such that $xx^{-1} = 1$.
- For all x, y, z we have $x(y + z) = xy + xz$.

Although these axioms might seem tedious, they are easy to remember. A field is a mathematical system in which one can add, subtract, multiply, and divide (except that one cannot divide by zero), and the usual laws of arithmetic hold. The basic examples for us are the real numbers $\mathbb{R}$ and the complex numbers $\mathbb{C}$. The rational numbers $\mathbb{Q}$ also form a field, as does the set $\{0, 1, \ldots, p-1\}$ under modular arithmetic, when p is a prime number. When p is not prime, factors of p (other than 1) do not have a multiplicative inverse.

REMARK. Two elementary facts about fields often arise. The first fact states that $x0 = 0$ for all x. This result follows by writing

$$x0 = x(0 + 0) = x0 + x0.$$

Adding the additive inverse $-(x0)$ to both sides yields $0 = x0$. The second fact is that $xy = 0$ implies that at least one of x and y is itself 0. The proof is simple. If $x = 0$, the conclusion holds. If not, multiplying both sides of $0 = xy$ by x^{-1} and then using the first fact yield $0 = y$.

REMARK. Both the rational numbers and the real numbers form *ordered* fields. In an ordered field, it makes sense to say $x > y$ and one can work with inequalities as usual. The complex numbers are a field but not an ordered field. It does not make sense to write $z > w$ if z and w are non-real complex numbers.

Definition 5.2. Let $\mathbf{F}$ be a field. A **vector space** V over $\mathbf{F}$ is a set of vectors together with two operations, addition and scalar multiplication, such that

(1) There is an object $\mathbf{0}$ such that, for all $v \in V$, we have $v + \mathbf{0} = v$.
(2) For all $u, v \in V$ we have $u + v = v + u$.
(3) For all $u, v, w \in V$ we have $u + (v + w) = (u + v) + w$.
(4) For all $v \in V$ there is a $-v \in V$ such that $v + (-v) = \mathbf{0}$.
(5) For all $v \in V$, we have $1v = v$.
(6) For all $c \in \mathbf{F}$ and all $v, w \in V$ we have $c(v + w) = cv + cw$.
(7) For all $c_1, c_2 \in \mathbf{F}$ and all $v \in V$, we have $(c_1 + c_2)v = c_1 v + c_2 v$.
(8) For all $c_1, c_2 \in \mathbf{F}$ and for all $v \in V$, we have $c_1(c_2 v) = (c_1 c_2)v$.

The list of axioms is again boring but easy to remember. If we regard vectors as arrows in the plane, or as forces, then these axioms state obvious compatibility relationships. The reader should check that, in the plane, the definitions

$$(x, y) + (u, v) = (x + u, y + v)$$

$$c(x, y) = (cx, cy)$$

correspond to the usual geometric meanings of addition and scalar multiplication.

REMARK. The axioms in Definition 5.2 imply additional basic facts, most of which we use automatically. For example, for all $v \in V$, we have $0v = \mathbf{0}$. To illustrate proof techniques, we pause to verify this conclusion. Some readers might find the proof pedantic and dull; others might enjoy the tight logic used.

Since $0 + 0 = 0$ in a field, we have $0v = (0 + 0)v$. By axiom (7), we get $0v = (0 + 0)v = 0v + 0v$. By axiom (4), there is a vector $-(0v)$ with $\mathbf{0} = -(0v) + 0v$. We add this vector to both sides and use axioms (3) and (1):

$$\mathbf{0} = -(0v) + 0v = -(0v) + (0v + 0v) = (-(0v) + 0v)) + 0v = \mathbf{0} + 0v = 0v.$$

We have replaced the standard but vague phrase "a vector is an object with both magnitude and direction" with " a vector is an element of a vector space." Later on we will consider situations in which we can measure the length (or *norm*, or *magnitude*) of a vector.

The diversity of examples below illustrates the power of abstraction. The ideas of linearity apply in all these cases.

EXAMPLE 5.1 (vector spaces). The following objects are vector spaces:

(1) $\mathbb{R}^n$ with addition and scalar multiplication defined as in (7.1) and (7.2).
(2) $\mathbb{C}^n$ with addition and scalar multiplication defined as in (7.1) and (7.2).

(3) Let S be any set and let $V_S = \{$functions $f : S \to \mathbb{R}\}$. Then V_S is a vector space. The zero vector is the function that is identically 0 on S. Here we *define* addition and scalar multiplication by

$$(f + g)(s) = f(s) + g(s) \text{ and } (cf)(s) = cf(s). \tag{8}$$

(4) The set of real- or complex-valued continuous functions on a subset S of $\mathbb{R}^n$. Addition and scalar multiplication are defined as in (8).

(5) The set of differentiable functions on an open subset S of $\mathbb{R}^n$.

(6) The solutions to a linear homogeneous differential equation.

(7) The collection of *states* in quantum mechanics.

(8) For any vector space V, the collection of linear maps from V to the scalar field is a vector space, called the **dual space** of V.

We mention another example with a rather different feel. Let $\mathbf{F}$ be the field consisting only of the two elements 0 and 1 with $1 + 1 = 0$. Exercises 5.6-5.10 give interesting uses of this field. Consider n-tuples consisting of zeroes and ones. We add componentwise and put $1 + 1 = 0$. This vector space $\mathbf{F}^n$ and its analogues for prime numbers larger than 2 arise in computer science and cryptography.

EXAMPLE 5.2. Consider the lights-out puzzle; it consists of a five-by-five array of lights. Each light can be on or off. Pressing a light toggles the light itself and its neighbors, where its neighbors are those locations immediately above, to the right, below, and to the left. For example, pressing the entry P_{11} toggles the entries P_{11}, P_{12}, P_{21}. (There is no neighbor above and no neighbor to the left.) Pressing the entry P_{33} toggles $P_{33}, P_{23}, P_{34}, P_{43}, P_{32}$. We can regard a configuration of lights as a point in a 25-dimensional vector space over the field of two elements. Toggling amounts to adding 1, where addition is taken modulo 2. Associated with each point P_{ij}, there is an operation A_{ij} of adding a certain vector of ones and zeroes to the array of lights. For example, A_{33} adds one to the entries labeled $P_{23}, P_{33}, P_{43}, P_{32}$, and P_{34} while leaving the other entries alone.

The following example from calculus shows one way in which linear algebra in infinite dimensions differs from linear algebra in finite dimensions.

EXAMPLE 5.3. Let V denote the space of polynomials in one real variable. Thus $p \in V$ means we can write $p(x) = \sum_{j=0}^{N} a_j x^j$. Define $D : V \to V$ by $D(p) = p'$ and $J : V \to V$ by $J(p)(x) = \int_0^x p(t)dt$. Both D and J are linear. Moreover, $DJ = I$ since

$$(DJ)(p)(x) = \frac{d}{dx} \int_0^x p(t)dt = p(x).$$

Therefore $DJ(p) = p$ for all p and hence $DJ = I$. By contrast,

$$(JD)(p)(x) = \int_0^x p'(t)dt = p(x) - p(0).$$

Hence, if $p(0) \neq 0$, then $(JD)p \neq p$. Therefore $JD \neq I$.

In general, having a one-sided inverse does not imply having an inverse. In finite dimensions, however, a one-sided inverse is also a two-sided inverse.

In the previous section we showed how to use row operations to find the inverse of (an invertible) square matrix. There is a formula, in terms of the matrix entries, for the inverse. Except for two-by-two matrices, the formula is too complicated to be of much use in computations.

EXAMPLE 5.4. Suppose that $L : \mathbb{R}^2 \to \mathbb{R}^2$ is defined by

$$L \begin{pmatrix} x \\ y \end{pmatrix} = \begin{pmatrix} 2 & 1 \\ 1 & 1 \end{pmatrix} \begin{pmatrix} x \\ y \end{pmatrix} = \begin{pmatrix} 2x + y \\ x + y \end{pmatrix}.$$

Then L is invertible, and the inverse mapping L^{-1} is matrix multiplication by

$$\begin{pmatrix} 1 & -1 \\ -1 & 2 \end{pmatrix}.$$

A general two-by-two matrix

$$L = \begin{pmatrix} a & b \\ c & d \end{pmatrix}$$

has an inverse if and only if $ad - bc \neq 0$. The number $ad - bc$ is called the determinant of L. See Section 9. When $ad - bc \neq 0$, we have

$$L^{-1} = \frac{1}{ad - bc} \begin{pmatrix} d & -b \\ -c & d \end{pmatrix}. \tag{9}$$

Suppose $U \subseteq V$ and V is a vector space. If U is also a vector space under the same operations, then U is called a **subspace** of V. Many of the vector spaces arising in this book are naturally given as subspaces of known vector spaces. To check whether U is a subspace of V, one simply must check that U is closed under the operations. In particular, U must contain the zero vector.

EXAMPLE 5.5. Suppose $L : V \to W$ is linear. Then

$$\mathcal{N}(L) = \{v : Lv = 0\},$$

the null space of L, is a vector subspace of V. Also the range of L,

$$\mathcal{R}(L) = \{u : u = Lv, \text{for some } v \in V\},$$

is a vector subspace of W. The proof is simple: if $w_1 = Lv_1$ and $w_2 = Lv_2$, then (by the linearity of L)

$$w_1 + w_2 = Lv_1 + Lv_2 = L(v_1 + v_2).$$

If $w = Lv$, then (by the other aspect of linearity), $cw = c(Lv) = L(cv)$. The range is therefore closed under both operations and hence it is a subspace of W.

EXAMPLE 5.6. Eigenspaces (defined formally in Definition 7.1) are important examples of subspaces. Assume $L : V \to V$ is linear. Let

$$E_\lambda(L) = \{v : Lv = \lambda v\}$$

denote the eigenspace corresponding to λ. It is a subspace of V: if $L(v_1) = \lambda v_1$ and $L(v_2) = \lambda v_2$, then

$$L(v_1 + v_2) = Lv_1 + Lv_2 = \lambda v_1 + \lambda v_2 = \lambda(v_1 + v_2).$$

$$L(cv) = cL(v) = c\lambda v = \lambda(cv).$$

Thus E_λ is closed under both sum and scalar multiplication.

EXAMPLE 5.7. Let V be the space of continuous functions on the interval $[0, 1]$. The set of functions that vanish at a given point p is a subspace of V. The set of functions that are differentiable on the open interval $(0, 1)$ is also a subspace of V.

EXAMPLE 5.8. A line in the plane $\mathbb{R}^2$ is a subspace if and only if it contains $(0, 0)$. If the line does not go through $(0, 0)$, then the set of points on the line is not closed under scalar multiplication by 0. If the line does go through $(0, 0)$, then we parametrize the line by $(x, y) = t(a, b)$ for $t \in \mathbb{R}$. Here (a, b) is a non-zero vector giving the direction of the line. Closure under addition and scalar multiplication is then evident. The map $t \to (at, bt)$ is linear from $\mathbb{R}$ to $\mathbb{R}^2$.

We close this section by solving an exercise in abstract linear algebra. Our purpose is to help the reader develop skill in abstract reasoning.

EXAMPLE 5.9. Suppose $L : V \to V$ is linear. Consider the following two properties:

(1) $L(Lv) = 0$ implies $Lv = 0$.

(2) $\mathcal{N}(L) \cap \mathcal{R}(L) = \{0\}$.

Prove that (1) holds if and only if (2) holds.

PROOF. We must do **two** things. Assuming (1), we must verify (2). Assuming (2), we must verify (1).

First we assume (1). To establish (2) we pick an arbitrary element w in both the null space and the range. We must somehow use (1) to show that $w = 0$. We are given $Lw = 0$ and that $w = Lv$ for some v. Therefore $0 = Lw = L(Lv)$. By the hypothesis in (1), we conclude that $Lv = 0$. But $w = Lv$ and hence $w = 0$. We have established (2).

Second we assume (2). To establish (1) we must assume $L(Lv) = 0$ and somehow show that $Lv = 0$. But Lv is in the range of L and, if $L(Lv) = 0$, then Lv is also in the null space of L. Therefore Lv is in $\mathcal{N}(L) \cap \mathcal{R}(L)$. By (2), we have $Lv = 0$. We have established (1). $\square$

EXERCISE 5.1. For $d > 0$, the set of polynomials of degree d is not a vector space. Why not?

EXERCISE 5.2. Show that the set of m-by-n matrices with real entries is a vector space.

EXERCISE 5.3. Show that the set of functions $f : \mathbb{R} \to \mathbb{R}$ with $f(1) = 0$ is a vector space. Show that the set of functions $f : \mathbb{R} \to \mathbb{R}$ with $f(0) = 1$ is not a vector space.

EXERCISE 5.4. Is the set $\mathbb{R}$ of real numbers a vector space over the field $\mathbb{Q}$ of rational numbers?

EXERCISE 5.5. The set of square matrices A with real entries is a vector space. (See Exercise 5.2.) Show that the collection of A for which $\sum a_{jj} = 0$ is a subspace.

The next several exercises illustrate to some extent how linear algebra can change if we work over a field other than the real or complex numbers.

EXERCISE 5.6. Consider the matrix

$$A = \begin{pmatrix} 1 & -1 \\ 1 & 1 \end{pmatrix}.$$

Find the null space of A assuming the scalars are real numbers. Find a basis for the null space of A assuming the scalars come from the field $\mathbf{F}$ of two elements. ($1 + 1 = 0$ in this field.)

EXERCISE 5.7. Again suppose that $\mathbf{F}$ is the field of two elements. Consider the matrix

$$A = \begin{pmatrix} 1 & 1 \\ 0 & 1 \end{pmatrix}.$$

Show that $A^{-1} = A$. Find all the two-by-two matrices A with real entries and $A^{-1} = A$.

EXERCISE 5.8. This problem is useful in analyzing the lights-out puzzle from Example 5.2. Consider the five-by-five matrix

$$L = \begin{pmatrix} 0 & 1 & 1 & 0 & 1 \\ 1 & 1 & 1 & 0 & 0 \\ 1 & 1 & 0 & 1 & 1 \\ 0 & 0 & 1 & 1 & 1 \\ 1 & 0 & 1 & 1 & 0 \end{pmatrix}.$$

Determine the null space of L, assuming first that the scalar field is the real numbers. Next determine the null space of L, assuming that the scalar field is the field of two elements.

REMARK. The next exercise has an interesting consequence. In computer science the notion of *exclusive or* is more natural than the notion of union. *Intersection* corresponds to multiplication in the field of two elements; *exclusive or* corresponds to addition in this field.

EXERCISE 5.9. Let 0 denote the concept of *false* and 1 denote the concept of *true*. Given logical statements P and Q, we assign 0 or 1 to each of them in this way. Verify the following:

- The statement P and Q both hold, written $P \cap Q$, becomes $PQ \mod (2)$.
- The statement that P or Q holds but not both, written $P \oplus Q$, becomes $P + Q \mod (2)$.

EXERCISE 5.10. Let $\mathbf{F}$ be the field of two elements. Show that the function f given by $f(x) = 1 + x + x^2$ is a constant.

6. Dimension

We have discussed dimension in an informal way. In this section we formalize the notion of dimension of a vector space, using the fundamental concepts of linear independence and span. We begin with an important pedagogical remark.

REMARK. Beginning students sometimes wonder why mathematicians consider dimensions higher than three. One answer is that higher dimensions arise in many applied situations. Consider for example the position $r(t)$ at time t of a rectangular eraser thrown into the air by the professor. It takes *six* numbers to specify $r(t)$. It takes another six to describe its velocity vector. For another example, consider modeling baseball using probability. In a given situation, there are eight possible ways for baserunners to be located and three possibilities for the number of outs. Ignoring the score of the game, *twenty-four* pieces of information are required to specify the situation. One could also take balls and strikes into account and multiply this number by twelve. In general, dimension tells us how many independent pieces of information are needed to describe a given situation. The concept is closely related to the notion of *degrees of freedom* in statistics or mechanics.

The *dimension* of a vector space tells us how many independent parameters are needed to specify each vector. To define dimension, we need to understand the concept of a *basis*, which is defined formally in Definition 6.4. If a vector space V has a finite basis, then we define the **dimension** of V to be the number of elements in this basis. For this definition to be valid, we must check that this number does not depend upon the basis chosen. See Corollary 6.6. By convention, the vector space consisting of the zero vector alone has dimension 0. When no finite basis exists and the vector space does not consist of the zero vector alone, we say that the vector space has *infinite dimension*. We write $\dim(V)$ for the dimension of a vector space V. Soon it will become clear that $\dim(\mathbb{R}^n) = n$.

In the Kakuro example from Section 1, the null space had dimension 2, and the range had dimension 5. The sum is 7, which is the dimension of the domain, or the number of variables. The following general result is easily proved by row operations. We sketch its proof to illustrate the ideas that follow. We also illustrate the result with examples.

Theorem 6.1. *Assume $L : \mathbb{R}^n \to \mathbb{R}^k$ is linear. Then*

$$\dim \mathcal{N}(L) + \dim \mathcal{R}(L) = n.$$

PROOF. (Sketch) After bases are chosen, we may assume that L is defined via matrix multiplication. Note that n is the number of columns in this matrix. Each row operation preserves $\mathcal{N}(L)$ and $\mathcal{R}(L)$. When we reach row echelon form, the number of leading ones is the number of independent columns, which is the dimension of the range. The dimension of the null space is the number of arbitrary variables in the general solution, and hence is n minus the number of independent columns. In other words, the result is obvious when a matrix is in row-echelon form. Row operations preserve $\mathcal{N}(L)$ and $\mathcal{R}(L)$ and hence their dimensions. The result follows. $\square$

EXAMPLE 6.1. Define $L : \mathbb{R}^3 \to \mathbb{R}^3$, where L is matrix multiplication by

$$\begin{pmatrix} 1 & 2 & 3 \\ 0 & 1 & -1 \\ 1 & 0 & 1 \end{pmatrix}.$$

Put $\mathbf{b} = (31, 6, 7)$. To solve $L\mathbf{x} = \mathbf{b}$, we use row operations as usual. The system is the matrix equation:

$$\begin{pmatrix} 1 & 2 & 3 \\ 0 & 1 & -1 \\ 1 & 0 & 1 \end{pmatrix} \begin{pmatrix} x_1 \\ x_2 \\ x_3 \end{pmatrix} = \begin{pmatrix} 31 \\ 6 \\ 7 \end{pmatrix}.$$

We write the system as an augmented matrix and perform two row operations:

$$\begin{pmatrix} 1 & 2 & 3 & 31 \\ 0 & 1 & -1 & 6 \\ 1 & 0 & 1 & 7 \end{pmatrix} \to \begin{pmatrix} 1 & 2 & 3 & 31 \\ 0 & 1 & -1 & 6 \\ 0 & -2 & -2 & -24 \end{pmatrix} \to \begin{pmatrix} 1 & 2 & 3 & 31 \\ 0 & 1 & -1 & 6 \\ 0 & 0 & -4 & -12 \end{pmatrix}$$

We obtain $\mathbf{x} = (4, 9, 3)$. In this example, L is invertible and hence $\mathcal{N}(L) = \{0\}$. Also $\mathcal{R}(L) = \mathbb{R}^3$.

EXAMPLE 6.2. In this example the linear map is not invertible. Take

$$M = \begin{pmatrix} 1 & 2 & 3 \\ 0 & 1 & -1 \\ 1 & 3 & -2 \end{pmatrix}.$$

We use row operations to find the null space and range of the linear mapping corresponding to the matrix M. Assume (a, b, c) is in the range. We try to solve:

$$\begin{pmatrix} 1 & 2 & 3 & a \\ 0 & 1 & -1 & b \\ 1 & 3 & 2 & c \end{pmatrix} \to \begin{pmatrix} 1 & 2 & 3 & a \\ 0 & 1 & -1 & b \\ 0 & 0 & 0 & c-b-a \end{pmatrix} \to \begin{pmatrix} 1 & 0 & 5 & a-2b \\ 0 & 1 & -1 & b \\ 0 & 0 & 0 & c-b-a \end{pmatrix}. \tag{10}$$

For a solution to exist, we require $c - a - b = 0$. Conversely, when this condition is satisfied we solve the system by letting x_3 be arbitrary, putting $x_2 = b + x_3$, and putting $x_1 = a - 2b - 5x_3$. Therefore,

$$\mathcal{R}(M) = \left\{ \begin{pmatrix} a \\ b \\ a+b \end{pmatrix} \right\} = \left\{ a \begin{pmatrix} 1 \\ 0 \\ 1 \end{pmatrix} + b \begin{pmatrix} 0 \\ 1 \\ 1 \end{pmatrix} \right\}.$$

The range of M thus has dimension 2. The null space of M has dimension 1; it consists of all multiples of $(-5, 1, 1)$. Notice that $\dim(\mathcal{N}(M)) + \dim(\mathcal{R}(M)) = 1 + 2 = 3$ here, illustrating Theorem 6.1.

We formally introduce the basic concepts of linear independence and span.

Definition 6.2. A subset $\{v_1, \cdots, v_l\}$ of a vector space is **linearly independent** if

$$\sum_{j=0}^{l} c_j v_j = 0 \Rightarrow c_j = 0 \text{ for all } j.$$

The reader should be familiar with this concept from ODE.

EXAMPLE 6.3. Consider the ODE $L(y) = y'' - (\alpha + \beta)y' + \alpha\beta y = 0$. We regard L as a linear mapping defined on the space of twice differentiable functions on $\mathbb{R}$ or $\mathbb{C}$. Solving $Ly = 0$ is the same as finding the null space of L. We have

$$\mathcal{N}(L) = \begin{cases} c_1 e^{\alpha x} + c_2 e^{\beta x} & \text{if } \alpha \neq \beta \\ c_1 e^{\alpha x} + c_2 x e^{\alpha x} & \text{if } \alpha = \beta \end{cases}$$

To show that $e^{\alpha x}$ and $e^{\beta x}$ are independent for $\alpha \neq \beta$, consider the equations

$$\begin{cases} c_1 e^{\alpha x} + c_2 e^{\beta x} = 0 \\ c_1 \alpha e^{\alpha x} + c_2 \beta e^{\beta x} = 0. \end{cases}$$

The second equation is obtained by differentiating the first. Rewrite these equations as a matrix equation:

$$\begin{pmatrix} e^{\alpha x} & e^{\beta x} \\ \alpha e^{\alpha x} & \beta e^{\beta x} \end{pmatrix} \begin{pmatrix} c_1 \\ c_2 \end{pmatrix} = \begin{pmatrix} 0 \\ 0 \end{pmatrix}.$$

We claim that the null space of

$$\begin{pmatrix} e^{\alpha x} & e^{\beta x} \\ \alpha e^{\alpha x} & \beta e^{\beta x} \end{pmatrix}$$

is the zero vector $\{\mathbf{0}\}$. Since $e^{\alpha x} \neq 0$ and $e^{\beta x} \neq 0$, this statement holds if and only if

$$\begin{pmatrix} 1 & 1 \\ \alpha & \beta \end{pmatrix}$$

is invertible, which holds if and only if $\beta - \alpha \neq 0$. When $\alpha \neq \beta$, the functions $e^{\alpha x}$ and $e^{\beta x}$ are therefore linearly independent (also see Exercise 6.1). In case $\alpha = \beta$, one checks that the null space is spanned by $e^{\alpha x}$ and $x\,e^{\alpha x}$. These functions are linearly independent because

$$0 = c_1 e^{\alpha x} + c_2 x e^{\alpha x}$$

implies $0 = c_1 + c_2 x$ for all x, and hence $c_1 = c_2 = 0$.

Definition 6.3. A collection of vectors $v_1, \ldots, v_N$ **spans** a vector space V if, for each vector $v \in V$ there are scalars $c_1, \ldots, c_N$ such that

$$v - \sum_{j=1}^{N} c_j v_j.$$

For example, in Example 6.2, we found two independent vectors that span $\mathcal{R}(M)$. We do not define span in the infinite-dimensional case at this time.

Definition 6.4. A collection of vectors $v_1, \ldots, v_N$ is a **basis** for a vector space V if this collection both spans V and is linearly independent.

We observe a simple fact relating linear independence and span.

Proposition 6.5. *Let U be a subspace of V. If $u_1, u_2, \ldots, u_k$ span U, and $v_1, \ldots, v_n$ are linearly independent elements of U, then $k \geq n$. In particular,*

$$n \leq \dim(U) \leq k.$$

PROOF. We sketch the idea. Given a spanning set $u_1, \ldots, u_k$ for U and a linearly independent set $v_1, \ldots, v_n$ in U, we inductively *exchange* n of the u_j for the v_j. After renumbering the u_j's, we get a spanning set $v_1, \ldots, v_n, u_{n+1}, \ldots, u_k$. □

Corollary 6.6. *Each basis of (a finite-dimensional) vector space has the same number of elements.*

PROOF. Suppose both $w_1, \ldots, w_k$ and $u_1, \ldots, u_n$ are bases. Since the u_j are linearly independent and the w_j span, $n \leq k$. Since the w_j are linearly independent and the u_j span, $k \leq n$. Hence $k = n$.   □

We summarize. The dimension of a vector space V equals the positive integer k if V has a basis with k elements. By Corollary 6.6, the dimension is well-defined. The dimension of the vector space consisting of the origin alone has dimension 0; there are no linearly independent elements and hence no basis. The dimension is infinite otherwise.

Exercise 6.6 asks for a proof of a general dimension formula. This result precisely expresses ideas about degrees of freedom, interdependence of variables, and related ideas. To help prepare for that exercise, and to illustrate the basic ideas, we sketch the solution of a simple exercise in the next example.

EXAMPLE 6.4. The set W_1 of vectors of the form $(x, -x, y, z)$ is a subspace of $\mathbb{R}^4$, and the set W_2 of vectors of the form $(a, b, -a, c)$ is also a subspace. (The reader who does not see why should re-read the definition of subspace!) We verify the following general formula (See Exercise 6.6) in this case:

$$\dim(W_1) + \dim(W_2) = \dim(W_1 \cap W_2) + \dim(W_1 + W_2).$$

Recall that the dimension of a space is the number of elements in a basis. A basis for W_1 is given by the three vectors

$$(1, -1, 0, 0) \quad (0, 0, 1, 0) \quad (0, 0, 0, 1).$$

A basis for W_2 is given by the three elements

$$(1, 0, -1, 0) \quad (0, 1, 0, 0) \quad (0, 0, 0, 1).$$

The intersection $W_1 \cap W_2$ is the set of vectors of the form $(x, -x, -x, c)$. A basis for $W_1 \cap W_2$ is therefore given by the two vectors

$$(1, -1, -1, 0) \quad (0, 0, 0, 1).$$

Every vector in $\mathbb{R}^4$ is the sum of a vector in W_1 and W_2. Therefore $\dim(W_1 + W_2) = 4$. In this case the formula therefore reads $3 + 3 = 2 + 4$.

EXERCISE 6.1. Let V be the vector space of functions on $\mathbb{R}$. In each case, show that the functions are linearly independent:

- $1, x, x^2, \ldots, x^n$.
- $p_1(x), \ldots, p_n(x)$ if each p_j is a polynomial of different degree.
- $e^{a_1 x}, e^{a_2 x}, \ldots, e^{a_n x}$. Assume that the a_j are distinct.
- x and $|x|$.

EXERCISE 6.2. Let $a_1, \ldots, a_n$ be distinct real numbers. Define polynomials p_j of degree $n - 1$ by

$$p_j(x) = \prod_{j \neq i}(x - a_i).$$

Show that these polynomials are linearly independent.

EXERCISE 6.3. Let V be the vector space of all functions $f : \mathbb{R} \to \mathbb{R}$. Given $f, g \in V$, show that $\min(f, g)$ is in the span of $f + g$ and $|f - g|$. Determine an analogous result for the maximum.

EXERCISE 6.4. Let V be the vector space of all functions $f : \mathbb{R} \to \mathbb{R}$. True or false? $|x|$ is in the span of x and $-x$.

EXERCISE 6.5. Determine whether $(6, 4, 2)$ is in the span of $(1, 2, 0)$ and $(2, 0, 1)$. Determine for which values of c the vector $(4, 7, c)$ is in the span of $(1, 2, 3)$ and $(2, 3, 4)$.

EXERCISE 6.6. Let U and V be subspaces of a finite-dimensional vector space W. Write $U + V$ for the span of U and V. Follow the following outline to prove that

$$\dim(U + V) + \dim(U \cap V) = \dim(U) + \dim(V).$$

First find a basis $\{\alpha_j\}$ for $U \cap V$. Extend it to a basis $\{\alpha_j, \beta_j\}$ for U and to a basis $\{\alpha_j, \gamma_j\}$ for V. Then show that the vectors $\{\alpha_j, \beta_j, \gamma_j\}$ are linearly independent and span $U + V$.

EXERCISE 6.7. Fill in the details from Example 6.4.

EXERCISE 6.8. Use the language of linear algebra to explain the following situation. A circuit has k unknown currents and many known voltages and impedances. Kirchhoff's laws yield more than k equations. Yet there is a unique solution for the list of currents.

EXERCISE 6.9. Give an example of a linear map L for which both $\mathcal{N}(L)$ and $\mathcal{R}(L)$ are positive dimensional and also $\mathcal{N}(L) \cap \mathcal{R}(L) = \{0\}$.

EXERCISE 6.10. Find an analogue of the formula from Example 6.4 for three subspaces W_1, W_2, and W_3. Prove your formula. Suggestion: Consider the inclusion-exclusion principle.

7. Eigenvalues and eigenvectors

Definition 7.1. Let V be a real or complex vector space. Suppose $L : V \to V$ is linear. A scalar λ is called an **eigenvalue** of L if there is a non-zero vector v such that $Lv = \lambda v$. Such a vector v is called an **eigenvector** of L and λ is called the eigenvalue corresponding to the eigenvector v. When λ is an eigenvalue of L, the set

$$E_\lambda = \{v : Lv = \lambda v\}$$

is a subspace of V, called the **eigenspace** corresponding to λ.

Eigenvalues are also known as *spectral values* and arise throughout science; the Greek letter λ suggests wave length. We will provide many examples and explanations as we continue.

EXAMPLE 7.1. Why is the exponential function so important? One of several related reasons is that the function $x \mapsto e^{\lambda x} = f(x)$ has the following property: $Df = \lambda f$. Thus f is an eigenvector, with eigenvalue λ, for the linear operator known as differentiation. We will return to this idea in our section on Fourier series in Chapter 5.

In most of this section we work in finite-dimensional spaces. Our linear maps will be given by square matrices. It is nonetheless important to understand that many problems in applied mathematics involve infinite-dimensional versions of the same ideas.

Definition 7.2. A linear transformation $L : V \to V$ is called **diagonalizable** if V has a basis consisting of eigenvectors of L.

When L is diagonalizable, with $v_1, \ldots, v_N$ a basis of eigenvectors, the matrix of L with respect to this basis (in both the domain and target copies of V) is diagonal. The diagonal elements are the corresponding eigenvalues. We will see an infinite-dimensional analogue when we study the Sturm–Liouville differential equation in Chapter 6.

In Example 7.2 we compute eigenspaces in a simple situation to illustrate the useful notion of diagonalization. A matrix D_{ij} is *diagonal* when $D_{ij} = 0$ for $i \neq j$. Computing with diagonal matrices is easy. A matrix A is called **diagonalizable** if there is an invertible matrix P and a diagonal D such that $A = PDP^{-1}$. We will see that this condition means that A behaves like a diagonal matrix, when we choose the correct basis. When A is regarded as a linear transformation, the column vectors of P are the eigenvectors of A.

The two uses of the term diagonalizable (for a matrix and for a linear map) are essentially the same. Consider a diagonal matrix D, with elements $\lambda_1, \ldots, \lambda_n$ on the diagonal. Let $\mathbf{e}_j$ denote the vector which is 1 in the j-th slot and 0 in the other slots. For each j, we have $D(\mathbf{e}_j) = \lambda_j \mathbf{e}_j$. Hence $\mathbf{e}_j$ is an eigenvector of D with corresponding eigenvalue λ_j. Consider the matrix $A = PDP^{-1}$. Then

$$A(P(\mathbf{e}_j)) = (PDP^{-1})(P\mathbf{e}_j) = PD(\mathbf{e}_j) = P(\lambda_j \mathbf{e}_j) = \lambda_j P(\mathbf{e}_j).$$

Therefore $P(\mathbf{e}_j)$ is an eigenvector for A with corresponding eigenvalue λ_j. A linear map is diagonalizable if and only if its matrix, with respect to a basis of eigenvalues, is diagonal.

EXAMPLE 7.2. Let $L : \mathbb{R}^2 \to \mathbb{R}^2$ be defined by the matrix

$$L = \begin{pmatrix} -1 & 10 \\ -5 & 14 \end{pmatrix}. \tag{11}$$

We find the eigenvalues λ and the corresponding eigenvectors. Setting $v = (x, y)$ and $Lv = \lambda v$ yields

$$\begin{pmatrix} -1 & 10 \\ -5 & 14 \end{pmatrix} \begin{pmatrix} x \\ y \end{pmatrix} = \lambda \begin{pmatrix} x \\ y \end{pmatrix}.$$

Hence

$$\begin{pmatrix} -1-\lambda & 10 \\ -5 & 14-\lambda \end{pmatrix} \begin{pmatrix} x \\ y \end{pmatrix} = \begin{pmatrix} 0 \\ 0 \end{pmatrix}.$$

Since $\begin{pmatrix} a & b \\ c & d \end{pmatrix}$ has a non-trivial null space if and only if $ad - bc = 0$, we obtain the characteristic equation

$$0 = (\lambda+1)(\lambda-14) + 50 = \lambda^2 - 13\lambda + 36 = (\lambda-4)(\lambda-9).$$

Thus $\lambda_1 = 4$ and $\lambda_2 = 9$ are the eigenvalues of $\begin{pmatrix} -1 & 10 \\ -5 & 14 \end{pmatrix}$. To find the corresponding eigenvectors, we must find the null spaces of $L - 4I$ and $L - 9I$. If $Lv = 4v$, we find that v is a multiple of $\begin{pmatrix} 2 \\ 1 \end{pmatrix}$. Also $Lv = 9v$ implies that v is a multiple of $\begin{pmatrix} 1 \\ 1 \end{pmatrix}$. We can check the result of our calculations:

$$\begin{pmatrix} -1 & 10 \\ -5 & 14 \end{pmatrix} \begin{pmatrix} 2 \\ 1 \end{pmatrix} = \begin{pmatrix} 8 \\ 4 \end{pmatrix} = 4\begin{pmatrix} 2 \\ 1 \end{pmatrix}.$$

$$\begin{pmatrix} -1 & 10 \\ -5 & 14 \end{pmatrix} \begin{pmatrix} 1 \\ 1 \end{pmatrix} = \begin{pmatrix} 9 \\ 9 \end{pmatrix} = 9\begin{pmatrix} 1 \\ 1 \end{pmatrix}.$$

When the eigenvectors of a linear transformation $L : V \to V$ form a basis of V, we can **diagonalize** L. In this case we write $L = PDP^{-1}$, where the columns of P are the eigenvectors of L, and D is diagonal. See Figures 6 and 7 for the geometric interpretation of diagonalization of the matrix L from (11).

$$L = \begin{pmatrix} -1 & 10 \\ -5 & 14 \end{pmatrix} = \begin{pmatrix} 2 & 1 \\ 1 & 1 \end{pmatrix} \begin{pmatrix} 4 & 0 \\ 0 & 9 \end{pmatrix} \begin{pmatrix} 2 & 1 \\ 1 & 1 \end{pmatrix}^{-1} = PDP^{-1}. \tag{12}$$

REMARK. For a two-by-two-matrix L, there is a simple way to check whether L is diagonalizable and if so, to diagonalize it. A two-by-two matrix $\begin{pmatrix} a & b \\ c & d \end{pmatrix}$ has two invariants, its **determinant** $ad - bc$ and its **trace** $a + d$. The eigenvalues λ_1 and λ_2 must satisfy the system of equations

$$\lambda_1 + \lambda_2 = \text{trace}(L) = a + d$$
$$\lambda_1 \lambda_2 = \det L = ad - bc.$$

If this (quadratic) system has two distinct solutions, then the matrix is diagonalizable. If the roots are repeated, then the matrix is not diagonalizable unless it is already diagonal. See also Exercises 7.4 and 7.5.

EXAMPLE 7.3. We use the remark to verify (12). The trace of L is 13 and the determinant of L is 36. We have $4 + 9 = 13$ and $4 \cdot 9 = 36$. Hence 4 and 9 are the eigenvalues. Since $L \begin{pmatrix} 2 \\ 1 \end{pmatrix} = 4 \begin{pmatrix} 2 \\ 1 \end{pmatrix}$, the vector $\begin{pmatrix} 2 \\ 1 \end{pmatrix}$ is an eigenvector corresponding to the eigenvalue 4. Similarly $\begin{pmatrix} 1 \\ 1 \end{pmatrix}$ is an eigenvector corresponding to the eigenvalue 9. Therefore (12) holds.

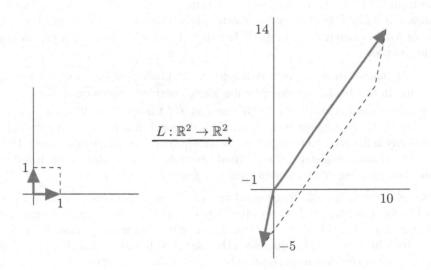

FIGURE 6. A linear map using one basis

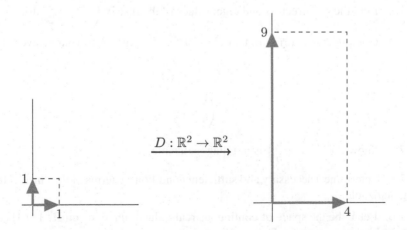

FIGURE 7. The same linear map using eigenvectors as a basis

REMARK. Not every square matrix with real entries has real eigenvalues. The following example both explains why and helps anticipate our work on complex numbers. Later in this book we will consider unitary and Hermitian matrices. Unitary matrices will have their eigenvalues on the unit circle and Hermitian matrices will have only real eigenvalues. Both circumstances have significant applications in physics and engineering.

EXAMPLE 7.4. Consider the two-by-two matrix

$$J_\theta = \begin{pmatrix} \cos(\theta) & -\sin(\theta) \\ \sin(\theta) & \cos(\theta) \end{pmatrix}$$

as a mapping from $\mathbb{R}^2$ to itself. The geometric interpretation of J_θ is a counterclockwise rotation by the angle θ. Unless θ is a multiple of π, such a transformation cannot map a non-zero vector to a multiple of itself. Hence J_θ will have no real eigenvalues. The special case when $\theta = \frac{\pi}{2}$ corresponds to multiplication by the complex number i.

Formula (12) enables us to compute functions of L. Finding functions of a linear operator will be a recurring theme in this book. To anticipate the ideas, consider a polynomial $f(x) = \sum_{j=0}^d c_j x^j$. We can substitute L for x and consider $f(L)$. We can find $f(L)$ using (12). When $L = PDP^{-1}$, we have $f(L) = P\,f(D)\,P^{-1}$. See Lemma 10.1. When D is diagonal, it is easy to compute $f(D)$. See also the section on similarity at the end of this chapter for an analogue of (12) and an application. The reader who has heard of Laplace or Fourier transforms should think about the sense in which these transforms *diagonalize* differentiation. This idea enables one to make sense of *fractional* derivatives.

REMARK. Assume V is finite-dimensional and $L : V \to V$ is linear. The collection of eigenvalues of L is known as the **spectrum** of L and is often denoted by $\sigma(L)$. In the finite-dimensional case, λ is an eigenvalue if and only if $L - \lambda I$ is not invertible. In the infinite-dimensional case, the spectrum consists of those scalars λ for which $(L - \lambda I)^{-1}$ does not exist. See [RS] for discussion about this distinction. In this book, we will encounter *continuous spectrum* only briefly, in the last chapter.

EXERCISE 7.1. With L as in (11), compute $L^2 - 3L$ in two ways. First find it directly. Then find it using (12). Use (12) to find (all possibilities for) $\sqrt{L}$.

EXERCISE 7.2. Find the eigenvectors and eigenvalues of the matrix $\begin{pmatrix} 20 & -48 \\ 8 & -20 \end{pmatrix}$.

EXERCISE 7.3. Compute the matrix product $P^{-1}QP$, and easily check your answer, if

$$Q = \begin{pmatrix} -1 & 8 \\ -4 & 11 \end{pmatrix}$$

$$P = \begin{pmatrix} 2 & 1 \\ 1 & 1 \end{pmatrix}.$$

EXERCISE 7.4. Show that $\begin{pmatrix} 1 & 4 \\ -1 & 5 \end{pmatrix}$ is not diagonalizable.

EXERCISE 7.5. Determine a necessary and sufficient condition (in terms of the entries) for a two-by-two matrix to be diagonalizable.

EXERCISE 7.6. Let V be the space of continuous real-valued functions on $\mathbb{R}$. Let W be the subspace for which the integral below converges. Define a linear map $L : W \to V$ by

$$(Lf)(x) = \int_{-\infty}^{x} f(t)dt.$$

Determine all eigenvalues and eigenvectors of L. Do the same problem if

$$(Lf)(x) = \int_{-\infty}^{x} tf(t)dt.$$

In both cases, what happens if we replace the lower limit of integration with 0?

EXERCISE 7.7. Let V be the space of infinitely differentiable complex-valued functions on $\mathbb{R}$. Let $D : V \to V$ denote differentiation. What are the eigenvalues of D?

EXERCISE 7.8. Find the complex eigenvalues of the matrix J_θ from Example 7.4.

EXERCISE 7.9. True or false? A two-by-two matrix of complex numbers always has a square root.

8. Bases and the matrix of a linear map

We often express linear transformations in finite dimensions as matrices. We emphasize that the matrix depends on a choice of basis. The ideas here are subtle. We begin by discussing different vector spaces of the same dimension and different bases of the same vector space.

Given vector spaces of the same dimension, to what extent can we regard them as the same? Bases allow us to do so, as follows. Vector spaces V, W (over the same field) are **isomorphic** if there is a linear map $L : V \to W$ such that L is both injective (one-to-one) and surjective (onto). The map L is called an **isomorphism**. It is routine to check that isomorphism is an equivalence relation.

REMARK. Assume A and B are sets and $f : A \to B$ is a function. Then f is called **injective** or **one-to-one** if $f(x) = f(y)$ implies $x = y$. Also f is called **surjective** or **onto** if, for each $b \in B$, there is an $a \in A$ such that $f(a) = b$. We can think of these words in the context of solving an equation. To say that f is surjective means for each b that we can solve $f(a) = b$. To say that f is injective means that the solution, if it exists, is unique. When both properties hold, we get both existence and uniqueness. Thus f is an isomorphism between A and B: to each $b \in B$ there is a unique $a \in A$ for which $f(a) = b$.

REMARK. In the context of linear algebra, in order for a map $L : V \to W$ to be an isomorphism, we also insist that L is linear.

Let us give a simple example. Consider the set V_3 of polynomials in one variable of degree at most 3. Then V_3 is a four-dimensional real vector space, spanned by the monomials $1, x, x^2, x^3$. The space V_3 is not the same as $\mathbb{R}^4$, but there is a natural correspondence: put $\phi(a + bx + cx^2 + dx^3) = (a, b, c, d)$. Then $\phi : V_3 \to \mathbb{R}^4$ is linear, injective, and surjective.

In the same manner, an ordered basis $v_1, \dots, v_n$ of V provides an isomorphism $\phi : V \to \mathbb{R}^n$. When $v = \sum c_j v_j$ we put $\phi(v) = (c_1, \dots, c_n)$. Since the inverse of an isomorphism is an isomorphism, and the composition of isomorphisms is itself an isomorphism, two vector spaces (with the same scalar field) of the same finite dimension are isomorphic.

Most readers should be familiar with the *standard* basis of $\mathbb{R}^3$. In physics one writes $\mathbf{i}, \mathbf{j}$, and $\mathbf{k}$ for the standard basis of $\mathbb{R}^3$. Other notations are also common:

$$\mathbf{i} = (1, 0, 0) = \vec{e_1} = \mathbf{e_1}$$
$$\mathbf{j} = (0, 1, 0) = \vec{e_2} = \mathbf{e_2}$$
$$\mathbf{k} = (0, 0, 1) = \vec{e_3} = \mathbf{e_3}.$$

We can therefore write

$$(a, b, c) = a(1, 0, 0) + b(0, 1, 0) + c(0, 0, 1) = a\mathbf{i} + b\mathbf{j} + c\mathbf{k}.$$

In n dimensions, we use the notation $\mathbf{e_1}, \ldots, \mathbf{e_n}$ to denote the standard basis. Here $\mathbf{e_1} = (1, 0, \ldots, 0)$ and so on. Given $v_1, v_2, \ldots, v_n \in \mathbb{R}^n$ (or in $\mathbb{C}^n$), there are scalars v_{jk} such that

$$v_1 = \sum_{j=1}^{n} v_{1j}\mathbf{e_j}$$

$$v_2 = \sum_{j=1}^{n} v_{2j}\mathbf{e_j}$$

$$v_n = \sum_{j=1}^{n} v_{nj}\mathbf{e_j}.$$

The vectors v_j themselves form a basis if and only if the matrix

$$\begin{pmatrix} v_{11} & v_{12} & \cdots & v_{1n} \\ \vdots & \vdots & \vdots & \vdots \\ v_{n1} & v_{n2} & \cdots & v_{nn} \end{pmatrix}$$

is invertible. An ordered basis of an n-dimensional real vector space V thus provides an isomorphism from V to $\mathbb{R}^n$. Different ordered bases provide different isomorphisms.

We are finally ready to discuss the matrix of a linear map. Let V and W be finite-dimensional vector spaces. Assume $v_1, v_2, \ldots, v_n$ is a basis of V and $w_1, w_2, \ldots, w_k$ is a basis of W. Suppose $L : V \to W$ is linear. What is the matrix of L with respect to these bases?

For each index l, the vector $L(v_l)$ is an element of W. Since the w_j are a basis for W, there are scalars such that we can write

$$L(v_1) = \sum_{j=1}^{k} c_{j1}w_j$$

$$L(v_2) = \sum_{j=1}^{k} c_{j2}w_j$$

$$\vdots =$$

$$L(v_n) = \sum_{j=1}^{k} c_{jn}w_j.$$

Put these scalars into a matrix:

$$C = \begin{pmatrix} c_{11} & c_{12} & \cdots & c_{1n} \\ c_{21} & c_{22} & \cdots & c_{2n} \\ \vdots & \vdots & & \vdots \\ c_{k1} & c_{k2} & \cdots & c_{kn} \end{pmatrix}.$$

Then C is the matrix of L with respect to these bases. The first column of C consists of the scalars needed to write the image of the first basis vector of V as a linear combination of the basis vectors of W. The same holds for each column. Thus the entries of C depend on the bases and on the order the basis elements are selected.

Two points are worth making here. First, if we choose bases and find the matrix of a linear map L with respect to them, then we need to know what properties of L are independent of the bases chosen. For example, whether L is invertible is independent of these choices. Second, when we choose bases, we should choose them in a way that facilitates the computations. For example, when $L : V \to V$ has a basis of eigenvectors, most computations are easiest when we choose this basis. See also Theorem 8.1 for an interesting example.

EXAMPLE 8.1. Suppose $L : \mathbb{R}^2 \to \mathbb{R}^2$ is defined by $L(x, y) = (x + y, x - y)$. The matrix of L with respect to the standard bases in both copies of $\mathbb{R}^2$ is

$$\begin{pmatrix} 1 & 1 \\ 1 & -1 \end{pmatrix}.$$

Consider the basis $(1, 1)$ and $(0, 1)$ of the domain copy of $\mathbb{R}^2$ and the standard basis in the image copy of $\mathbb{R}^2$. The matrix of L with respect to these bases is now

$$\begin{pmatrix} 2 & 1 \\ 0 & -1 \end{pmatrix}.$$

EXAMPLE 8.2. Let V be the vector space of polynomials of degree at most N. We define a linear map $L : V \to V$ by

$$Lp(x) = x \int_0^x p''(t)dt = x(p'(x) - p'(0)).$$

Computing $L(x^j)$ for $0 \le j \le N$ gives $L(1) = L(x) = 0$ and $L(x^j) = jx^j$ otherwise. Each monomial is an eigenvector, and the matrix of L (with respect to the monomial basis) is diagonal.

The next example is more elaborate:

EXAMPLE 8.3. Let V denote the space of polynomials of degree at most 3. Then $\dim(V) = 4$ and the polynomials $1, x, x^2, x^3$ form a basis. Let D denote differentiation. Then $D : V \to V$. The matrix of D with respect to the basis $1, x, x^2, x^3$ in both the domain and target spaces is

$$\begin{pmatrix} 0 & 1 & 0 & 0 \\ 0 & 0 & 2 & 0 \\ 0 & 0 & 0 & 3 \\ 0 & 0 & 0 & 0 \end{pmatrix}$$

If instead we use the basis $1, x + 1, \frac{(x+1)(x+2)}{2}, \frac{(x+1)(x+2)(x+3)}{6}$, then the matrix of D (with respect to this basis) is

$$\begin{pmatrix} 0 & 1 & \frac{1}{2} & \frac{1}{3} \\ 0 & 0 & 1 & \frac{1}{2} \\ 0 & 0 & 0 & 1 \\ 0 & 0 & 0 & 0 \end{pmatrix}.$$

The reader can check that

$$D(1) = 0$$

$$D(x + 1) = 1 = 1 + 0(x + 1) + 0\frac{(x + 1)(x + 2)}{2} + 0\frac{(x + 1)(x + 2)(x + 3)}{6}$$

$$D\left(\frac{(x + 1)(x + 2)}{2} \right) = x + \frac{3}{2} = \frac{1}{2}(1) + 1(x + 1)$$

$$D\left(\frac{(x + 1)(x + 2)(x + 3)}{6} \right) = \frac{1}{2}x^2 + 2x + \frac{11}{6} = \frac{1}{3} + \frac{1}{2}(x + 1) + \frac{(x + 1)(x + 2)}{2}.$$

We now discuss an interesting situation where choosing the correct basis simplifies things. Consider polynomials with rational coefficients. Thus

$$p(x) = a_0 + a_1 x + a_2 x^2 + \cdots + a_k x^k,$$

where each $a_j \in \mathbb{Q}$. What is the condition on these coefficients such that $p(n)$ is an integer for every integer n? For example, $p(x) = \frac{1}{3}x^3 + \frac{1}{2}x^2 + \frac{1}{6}x$ has this property, even though its coefficients are not whole numbers.

The key idea is to consider the combinatorial polynomials $\binom{x+j}{j}$. For $j = 0$ this notation means the constant polynomial 1. For $j \geq 1$, the notation $\binom{x+j}{j}$ means the polynomial

$$\frac{(x+1)(x+2)\ldots(x+j)}{j!}.$$

These polynomials are of different degree and hence linearly independent over the rational numbers $\mathbb{Q}$. By linear algebra, using $\mathbb{Q}$ as the scalar field, we can write p as a linear combination with rational coefficients of these polynomials. For example, we have

$$\frac{1}{3}x^3 + \frac{1}{2}x^2 + \frac{1}{6}x = 2\frac{(x+3)(x+2)(x+1)}{6} - 3\frac{(x+2)(x+1)}{2} + \frac{x+1}{1}.$$

These combinatorial polynomials map the integers to the integers. Using them as a basis, we obtain the decisive answer.

Theorem 8.1. *A polynomial with rational coefficients maps $\mathbb{Z}$ to itself if and only if, when writing*

$$p(x) = \sum_{j=0}^{k} m_j \binom{x+j}{j},$$

each m_j is an integer.

PROOF. By Exercise 8.1, each combinatorial polynomial $\binom{x+j}{j}$ maps the integers to the integers. Hence an integer combination of them also does, and the *if* direction follows. To prove the *only if* direction, consider a polynomial p of degree k with rational coefficients. There are rational numbers c_j such that

$$p(x) = \sum_{j=0}^{k} c_j \binom{x+j}{j},$$

because the polynomials $\binom{x+j}{j}$ for $0 \leq j \leq k$ are linearly independent, and hence form a basis for the vector space of polynomials of degree at most k with rational coefficients. We must show that each c_j is an integer. To do so, first evaluate at $x = -1$. We get $p(-1) = c_0$. Since $p(-1)$ is an integer, c_0 also is an integer. Then evaluate at -2. Since $p(-2)$ is an integer, $c_1 = p(-2) - c_0$ is also an integer. Proceeding in this way, we show inductively that all the coefficients are integers. See Exercise 8.3. $\square$

The next theorem provides another basis for the space V_d of polynomials of degree at most d in one real (or complex) variable. Note that all the polynomials in the basis are of the same degree, and each appears in factored form.

Theorem 8.2. *Let $\lambda_1, \cdots, \lambda_n$ be distinct real numbers. For $1 \leq j \leq n$ define polynomials $p_j(x)$ by*

$$p_j(x) = a_j \prod_{k \neq j}(x - \lambda_k),$$

where

$$a_j = \frac{1}{\prod_{k \neq j}(\lambda_j - \lambda_k)}.$$

These polynomials are linearly independent and thus form a basis for V_{n-1}. Also, the map $\phi : V_{n-1} \to \mathbb{R}^n$ defined by

$$\phi(p) = (p(\lambda_1), \cdots, p(\lambda_n))$$

is an isomorphism.

PROOF. First note that $p_j(\lambda_k) = 0$ if $j \neq k$ and $p_j(\lambda_j) = 1$ for each j. To verify linear independence, assume that

$$s(x) = \sum_{j=1}^{n} c_j p_j(x) = 0.$$

Then $0 = s(\lambda_j) = c_j$ for each j. Hence the p_j are independent. Since there are n of them in the n-dimensional space V_{n-1}, they form a basis for this space. To show that ϕ is an isomorphism it suffices to show that ϕ is injective. If $\phi(p) = 0$, then each λ_j is a root of p. Since p is of degree at most $n-1$, but has n distinct roots, p must be the zero polynomial. Thus $\mathcal{N}(\phi) = \{0\}$, and ϕ is injective. $\square$

EXERCISE 8.1. Show that each $\binom{x+j}{j}$ maps the integers to the integers.

EXERCISE 8.2. Given α, β, γ, find A, B, C such that

$$\alpha x^2 + \beta x + \gamma = A\frac{(x+1)(x+2)}{2} + B(x+1) + C.$$

EXERCISE 8.3. Write out a formal inductive proof of the *only if* part of Theorem 8.1.

EXERCISE 8.4. Consider the matrix $A = \begin{pmatrix} 2 & 9 \\ -1 & 8 \end{pmatrix}$. Find a basis of $\mathbb{R}^2$ (in both the domain and range) such that the matrix of A in this basis is $\begin{pmatrix} 5 & 1 \\ 0 & 5 \end{pmatrix}$. Suggestion: Use an eigenvector as the first basis element.

EXERCISE 8.5. Assume $L : \mathbb{R}^4 \to \mathbb{R}^4$ and L is represented by the matrix

$$\begin{pmatrix} 3 & 1 & 0 & 0 \\ 0 & 3 & 1 & 0 \\ 0 & 0 & 3 & 0 \\ 0 & 0 & 0 & 3 \end{pmatrix}$$

with respect to the standard basis. Verify that 3 is the only eigenvalue of L. Find the corresponding eigenvectors. Find the null space of $(L - 3I)^2$ and of $(L - 3I)^3$.

9. Determinants

Determinants arise throughout mathematics, physics, and engineering. Many methods exist for computing them. Row operations provide the most computationally useful method for finding determinants, but numerical computation of determinants should usually be avoided. Example 5.3 of Chapter 5 indicates one problematic issue. Theorem 9.5 at the end of this section summarizes much of our discussion on linear algebra; the determinant provides one of many tests for the invertibility of a matrix. We state this important result as Theorem 9.1. The rest of this section reveals other uses for determinants.

Theorem 9.1. *A linear map on a finite-dimensional space is invertible if and only if its determinant is non-zero.*

As usual, we will work with determinants of square matrices. Let A be an n-by-n matrix, whose elements are in $\mathbb{R}$ or $\mathbb{C}$. How do we define its determinant, $\det(A)$? We want $\det(A)$ to be a scalar that equals the oriented n-dimensional volume of the box spanned by the rows (or columns) of A. In particular, if A is a diagonal matrix, then $\det(A)$ will be the product of the diagonal elements.

The key idea is to regard $\det$ as a function of the rows of a matrix. For this function to agree with our sense of oriented volume, it must satisfy certain properties. It should be linear in each row when the other rows are fixed. When the matrix fails to be invertible, the box degenerates, and the volume should be 0. To maintain the sense of orientation, the determinant should get multiplied by -1 when we interchange two rows. Finally, we assume that the identity matrix corresponds to the unit box, which has volume 1. These ideas uniquely determine the determinant function and tell us how to compute it.

We apply row operations to a given matrix. We find the determinant using the following properties. In (13.1) we interchange two rows:

$$\det \begin{pmatrix} r_1 \\ \vdots \\ r_j \\ \vdots \\ r_l \\ \vdots \end{pmatrix} = -\det \begin{pmatrix} r_1 \\ \vdots \\ r_l \\ \vdots \\ r_j \\ \vdots \end{pmatrix} \tag{13.1}$$

In (13.2) we multiply a row by a constant:

$$\det \begin{pmatrix} r_1 \\ \vdots \\ c\,r_j \\ \vdots \\ r_l \\ \vdots \end{pmatrix} = c \, \det \begin{pmatrix} r_1 \\ \vdots \\ r_j \\ \vdots \\ r_l \\ \vdots \end{pmatrix} \tag{13.2}$$

In (13.3) we assume that $k \neq j$ and we add a multiple of the k-th row to the j-th row:

$$\det \begin{pmatrix} r_1 \\ \vdots \\ r_j + \lambda r_k \\ \vdots \\ r_l \\ \vdots \end{pmatrix} = \det \begin{pmatrix} r_1 \\ \vdots \\ r_j \\ \vdots \\ r_l \\ \vdots \end{pmatrix} \tag{13.3}$$

Finally we know the determinant of the identity matrix is 1:

$$\det(I) = 1. \tag{13.4}$$

REMARK. When finding determinants, one usually row reduces until one reaches upper triangular form. The determinant of a matrix in upper triangular form is the product of the diagonal elements.

EXAMPLE 9.1.

$$\det \begin{pmatrix} 1 & 2 & 3 \\ 4 & 5 & 6 \\ 7 & 7 & 9 \end{pmatrix} = \det \begin{pmatrix} 1 & 2 & 3 \\ 0 & -3 & -6 \\ 0 & -7 & -12 \end{pmatrix} = (-3)(-7)\det \begin{pmatrix} 1 & 2 & 3 \\ 0 & 1 & 2 \\ 0 & 1 & \frac{12}{7} \end{pmatrix}$$

$$= 21\det \begin{pmatrix} 1 & 2 & 3 \\ 0 & 1 & 2 \\ 0 & 0 & -\frac{2}{7} \end{pmatrix} = (-\frac{2}{7})(21)\det \begin{pmatrix} 1 & 2 & 3 \\ 0 & 1 & 2 \\ 0 & 0 & 1 \end{pmatrix} = -6.$$

The determinant has several standard uses:

- finding the volume of an n-dimensional box,
- finding cross products,
- testing for invertibility,
- finding eigenvalues.

REMARK. Eigenvalues are fundamental to science and hence many methods exist for finding them numerically. Using the determinant to find eigenvalues is generally impractical. See [St] for information comparing methods and illustrating their advantages and disadvantages. At this stage of the book we are more concerned with theoretical issues than with computational methods. Row operations generally provide the best way to compute determinants, but determinants do not provide the best way to compute eigenvalues.

The following theorem summarizes the basic method of computing determinants. The multi-linearity property includes both (13.2) and (13.3). The alternating property implies (13.1).

Theorem 9.2. *Let $M(n)$ denote the space of n-by-n matrices with elements in a field $\mathbf{F}$. There is a unique function $D : M(n) \to \mathbf{F}$, called the determinant, with the following properties:*

- *D is linear in each row. (multi-linearity)*
- *$D(I) = 1$.*
- *$D(A) = 0$ if two rows of A are the same. (alternating property)*

See [HK] (or any other good book on linear algebra) for a proof of this theorem. The idea of the proof is simple; the three properties force the determinant to satisfy the Laplace expansion from Theorem 9.3.

Over the real or complex numbers, for example, the third property from Theorem 9.2 can be restated as $D(A') = -D(A)$ if A' is obtained from A by interchanging two rows. For general fields, however, the two properties are not identical!

EXAMPLE 9.2. Suppose the field of scalars consists only of the two elements 0 and 1. Hence $1 + 1 = 0$. The determinant of the two-by-two identity matrix is 1. If we interchange the two rows, then the determinant is still 1, since $-1 = 1$ in this field. This example suggests why we state the alternating property as above.

The next result (Theorem 9.3) is of theoretical importance, but it should not be used for computing determinants. Recall that a **permutation** on n-letters is a bijection of the set $\{1, \ldots, n\}$. The **signum** of a permutation σ, written $\mathrm{sgn}(\sigma)$, is a measure of orientation: $\mathrm{sgn}(\sigma) = 1$ if it takes an even number of transpositions to obtain the identity permutation, and $\mathrm{sgn}(\sigma) = -1$ if it takes an odd number to do so. Most books on linear or abstract algebra discuss why this number is well-defined. In other words, given a permutation, no matter what sequence of transpositions we use to convert it to the identity, the parity will be the same. Perhaps the easiest approach to this conclusion is to *define* sgn as in (14) below. Formula (14) implies that $\mathrm{sgn}(\sigma \circ \eta) = \mathrm{sgn}(\sigma)\,\mathrm{sgn}(\eta)$, from which the well-defined property follows. See Exercise 9.10.

$$\mathrm{sgn}(\sigma) = \frac{\prod_{1 \le i < j \le n} \sigma(i) - \sigma(j)}{\prod_{1 \le i < j \le n}(i - j)}. \tag{14}$$

Theorem 9.3 (Laplace expansion). *Let A be an n-by-n matrix. Then*

$$\det(a_{ij}) = \sum_{\sigma} \operatorname{sgn}(\sigma) \prod_{j=1}^{n} a_{j\sigma(j)}.$$

The sum is taken over all permutations σ of $\{1, \ldots, n\}$.

The next theorem is of major theoretical importance. It is also important in various applications, such as the $ABCD$ matrices used in electrical engineering for transmission lines and the ray transfer matrices used in optical design. See Section 7 of Chapter 2.

Theorem 9.4. *Suppose A, B are square matrices of the same size. Then $\det(AB) = \det(A) \det(B)$.*

One particular proof of this result is quite interesting. We indicate this proof, without filling in details. First one proves that $\det(B) = 0$ if and only if B is not invertible. If $\det(B) = 0$, then B is not invertible, and hence (for every A) AB is not invertible. Therefore $\det(AB) = 0$ for all A, and the result holds. If B is invertible, then $\det(B) \neq 0$, and we can consider the function taking the rows of A to

$$\frac{\det(AB)}{\det(B)}.$$

One then shows that this function satisfies the properties of the determinant function, and hence (by the previous theorem) must be $\det(A)$.

Theorem 9.4 is of fundamental importance, for several reasons. A matrix is invertible if and only if it is a product or elementary row matrices. This statement allows us to break down the effect of an arbitrary invertible linear map into simple steps. The effect of these steps on volumes is easy to understand. Using Theorem 9.4 we can then prove that the determinant of an n-by-n matrix is the oriented n-dimensional volume of the box spanned by the row or column vectors. In the next section we will discuss *similarity*, and we will use Theorem 9.4 to see that similar matrices have the same determinant.

Given a linear map on $\mathbb{R}^n$ or $\mathbb{C}^n$, we define its determinant to be the determinant of its matrix representation with respect to the standard basis. For a general finite-dimensional vector space V and linear map $L : V \to V$, the determinant depends on the choice of bases. Whether the determinant is 0, however, is independent of the basis chosen. We can now summarize much of linear algebra in the following result. We have already established most of the equivalences, and hence we relegate the proof to the exercises.

Theorem 9.5. *Let V be a finite-dimensional vector space. Assume $L : V \to V$ is linear. Then the following statements hold or fail simultaneously:*

(1) *L is invertible.*
(2) *L is row equivalent to the identity.*
(3) *There are elementary matrices $E_1, E_2, \cdots, E_k$ such that $E_k E_{k-1} \cdots E_1 L = I$.*
(4) *There are elementary matrices $F_1, F_2, \cdots, F_k$ such that $L = F_1 F_2 \cdots F_k$.*
(5) *$\mathcal{N}(L) = \{0\}$.*
(6) *L is injective.*
(7) *$\mathcal{R}(L) = V$.*
(8) *For any basis, the columns of the matrix of L are linearly independent.*
(9) *For any basis, the rows of the matrix of L are linearly independent.*
(10) *For any basis, the columns of the matrix of L span V.*
(11) *For any basis, the rows of the matrix of L span V.*
(12) *For any basis, $\det(L) \neq 0$.*

EXERCISE 9.1. Show that the determinant of an elementary matrix is non-zero.

EXERCISE 9.2. Prove that items (5) and (6) in Theorem 9.5 hold simultaneously for any linear map L, even in infinite dimensions.

EXERCISE 9.3. Derive the equivalence of (5) and (7) from Theorem 6.1.

EXERCISE 9.4. Give an example (it must be infinite-dimensional) where item (5) holds in Theorem 9.5, but items (1) and (7) fail.

EXERCISE 9.5. Use row operations and Exercise 9.1 to prove that items (1) through (7) from Theorem 9.5 are equivalent.

EXERCISE 9.6. It is obvious that (8) and (10) from Theorem 9.5 are equivalent, and that (9) and (11) are equivalent. Why are all four statements equivalent?

EXERCISE 9.7. Graph the parallelogram with vertices at $(0,0)$, $(-1,10)$, $(-5,14)$, and $(-6,24)$. Find its area by finding the determinant of $\begin{pmatrix} -1 & 10 \\ -5 & 14 \end{pmatrix}$.

EXERCISE 9.8. Use row operations to find the determinant of A:

$$A = \begin{pmatrix} 1 & 2 & 3 & 4 \\ 4 & 3 & 2 & 1 \\ 2 & 4 & 7 & 8 \\ 3 & 6 & 9 & 9 \end{pmatrix}.$$

EXERCISE 9.9. Given n distinct numbers x_j, form the matrix with entries x_j^k for $0 \le j, k \le n-1$. Find its determinant. This matrix is called a *Vandermonde* matrix. Suggestion: Regard the matrix as a polynomial in the entries and determine where this polynomial must vanish.

EXERCISE 9.10. Use (14) to show that $\mathrm{sgn}(\sigma \circ \eta) = \mathrm{sgn}(\sigma)\,\mathrm{sgn}(\eta)$.

EXERCISE 9.11. Consider the permutation σ that reverses order. Thus $\sigma(j) = n+1-j$ for $1 \le j \le n$. Find $\mathrm{sgn}(\sigma)$.

10. Diagonalization and generalized eigenspaces

A linear transformation L is called *diagonalizable* if there is basis consisting of eigenvectors. In this case, the matrix of L with respect to this basis is diagonal, with the eigenvalues on the diagonal. The matrix of L with respect to an arbitrary basis will not generally be diagonal. If a matrix A is diagonalizable, however, then there is a diagonal matrix D and an invertible matrix P such that

$$A = PDP^{-1}. \tag{15}$$

The columns of P are the eigenvectors. See Example 5.1 for an illustration. More generally, matrices A and B are called **similar** if there is an invertible P for which $A = PBP^{-1}$. Similarity is an equivalence relation. Note that A is diagonalizable if and only if A is similar to a diagonal matrix.

Formula (15) has many consequences. One application is to define and easily compute functions of A. For example, if $p(x) = \sum_j c_j x^j$ is a polynomial in one variable x, we naturally define $p(A)$ by

$$p(A) = \sum_j c_j A^j.$$

The following simple result enables us to compute polynomial functions of matrices.

Lemma 10.1. *Let $f(x)$ be a polynomial in one real (or complex) variable x. Suppose A and B are similar matrices with $A = PBP^{-1}$. Then $f(A)$ and $f(B)$ are similar, with*

$$f(A) = Pf(B)P^{-1}. \tag{16}$$

PROOF. The key step in the proof is that $(PBP^{-1})^j = PB^jP^{-1}$. This step is easily verified by induction; the case when $j = 1$ is trivial, and passing from j to $j + 1$ is simple:

$$(PBP^{-1})^{j+1} = (PBP^{-1})^j PBP^{-1} = PB^jP^{-1}PBP^{-1} = PB^{j+1}P^{-1}. \tag{17}$$

The second step uses the induction hypothesis, and we have used associativity of matrix multiplication. Formula (16) easily follows:

$$f(A) = \sum c_j A^j = \sum c_j PB^jP^{-1} = P(\sum c_j B^j)P^{-1} = Pf(B)P^{-1}.$$

$\square$

REMARK. Once we know (16) for polynomial functions, we can conclude an analogous result when f is a convergent power series. A proof of the following Corollary requires complex variable theory.

Corollary 10.2. *Suppose the power series $f(z) = \sum a_n z^n$ converges for $|z| < R$. Let $A : \mathbb{C}^n \to \mathbb{C}^n$ be a linear map whose eigenvalues λ all satisfy $|\lambda| < R$. Then the series $f(A) = \sum c_n A^n$ converges and defines a linear map. Furthermore, if $A = PBP^{-1}$, then $f(A) = Pf(B)P^{-1}$.*

Two special cases are particularly useful. The series $\sum_0^\infty \frac{z^n}{n!}$ converges for all complex numbers z and defines the complex exponential function. Hence, if A is a square matrix, we define e^A by

$$e^A = \sum_0^\infty \frac{A^n}{n!}. \tag{18}$$

The second example comes from the geometric series. If all the eigenvalues of A have magnitude smaller than 1, then $I - A$ is invertible, and

$$(I - A)^{-1} = \sum_{n=0}^\infty A^n.$$

See Definition 5.8 of Chapter 4 for an alternative approach, using the Cauchy integral formula, to define functions of operators (linear transformations).

The next remark gives one of the fundamental applications of finding functions of operators. All constant coefficient homogeneous ODEs reduce to exponentiating matrices.

REMARK. Consider an n-th-order constant coefficient ODE for a function y:

$$y^{(n)} = c_{n-1}y^{(n-1)} + \cdots + c_1 y' + c_0 y.$$

We can rewrite this equation as a first-order system, by giving new names to the derivatives, to obtain

$$\mathbf{Y}' = \begin{pmatrix} y' \\ y'' \\ \cdots \\ y^{(n-1)} \\ y^{(n)} \end{pmatrix} = \begin{pmatrix} 0 & 1 & 0 & \cdots & 0 \\ 0 & 0 & 1 & \cdots & 0 \\ \cdots & \cdots & \cdots & \cdots & \cdots \\ 0 & 0 & 0 & \cdots & 1 \\ c_0 & c_1 & c_2 & \cdots & c_{n-1} \end{pmatrix} \begin{pmatrix} y \\ y' \\ \cdots \\ y^{(n-2)} \\ y^{(n-1)} \end{pmatrix} = A\mathbf{Y}.$$

The solution to the system is given by $\mathbf{Y}(t) = e^{At}\mathbf{Y}(0)$. See Exercises 10.4 through 10.7 for examples.

How can we exponentiate a matrix? If A is diagonal, with diagonal elements λ_k, then e^A is diagonal with diagonal elements e^{λ_k}. If A is *diagonalizable*, that is, similar to a diagonal matrix, then finding e^A is also easy. If $A = PDP^{-1}$, then $e^A = Pe^D P^{-1}$. Not all matrices are diagonalizable. The simplest example is the matrix A, given by

$$A = \begin{pmatrix} \lambda & 1 \\ 0 & \lambda \end{pmatrix}. \tag{19}$$

Exercise 10.1 shows how to exponentiate this particular A.

The matrix in (19) satisfies $A(e_1) = \lambda e_1$ and $A(e_2) = \lambda e_2 + e_1$. Thus $(A - \lambda I)(e_2) = e_1$ and hence

$$(A - \lambda I)^2(e_2) = (A - \lambda I)(e_1) = 0.$$

The vector e_2 is an example of a *generalized eigenvector*. We have the following definition.

Definition 10.3. Let $L : V \to V$ be a linear map. The *eigenspace* E_λ is the subspace defined by

$$E_\lambda = \{v : Lv = \lambda v\} = \mathcal{N}(L - \lambda I).$$

The *generalized eigenspace* K_λ is the subspace defined by

$$K_\lambda = \{v : (L - \lambda I)^k v = 0 \text{ for some } k\}.$$

REMARK. Both E_λ and K_λ are subspaces of V. Furthermore, $L(E_\lambda) \subseteq L(E_\lambda)$ and $L(K_\lambda) \subseteq K_\lambda$.

The following result clarifies both the notion of generalized eigenspace and what happens when the characteristic polynomial of a constant coefficient linear ODE has multiple roots.

Theorem 10.4. *Let $L : V \to V$ be a linear map. Assume $(L - \lambda I)^{k-1}(v) \neq 0$, but $(L - \lambda I)^k v = 0$. Then, for $0 \leq j \leq k - 1$, the vectors $(L - \lambda I)^j(v)$ are linearly independent.*

PROOF. As usual, to prove that vectors are linearly independent, we assume that (the zero vector) 0 is a linear combination of them, and we show that all the coefficient scalars must be 0. Therefore assume

$$0 = \sum_{j=0}^{k-1} c_j (L - \lambda)^j(v). \tag{20}$$

Apply $(L - \lambda I)^{k-1}$ to both sides of (20). The only term not automatically 0 is the term $c_0(L - \lambda I)^{k-1}v$. Hence this term is 0 as well; since $(L - \lambda I)^{k-1}(v) \neq 0$, we must have $c_0 = 0$. Now we can apply $(L - \lambda I)^{k-2}$ to both sides of (20) and obtain $c_1 = 0$. Continue in this fashion to obtain $c_j = 0$ for all j. $\square$

Consider the subspace spanned by the vectors $(L - \lambda I)^j v$ for $0 \leq j \leq k - 1$. For simplicity we first consider the case $k = 2$. Suppose $A(e_1) = \lambda e_1$ and $A(e_2) = \lambda e_2 + e_1$. Then $(A - \lambda I)e_1 = 0$ whereas $(A - \lambda I)(e_2) \neq 0$ but $(A - \lambda I)^2(e_2) = 0$. Thus e_1 is an eigenvector and e_2 is a generalized eigenvector. Using this basis, we obtain the matrix

$$A = \begin{pmatrix} \lambda & 1 \\ 0 & \lambda \end{pmatrix}.$$

To make the pattern evident, we also write down the analogous matrix when $k = 5$.

Using the generalized eigenvectors as a basis, again in the reverse order, we can write L as the matrix

$$L = \begin{pmatrix} \lambda & 1 & 0 & 0 & 0 \\ 0 & \lambda & 1 & 0 & 0 \\ 0 & 0 & \lambda & 1 & 0 \\ 0 & 0 & 0 & \lambda & 1 \\ 0 & 0 & 0 & 0 & \lambda \end{pmatrix}. \tag{21}$$

Such a matrix is called a *Jordan block*; it cannot be diagonalized. Jordan blocks arise whenever a matrix has (over the complex numbers) too few eigenvectors to span the space. In particular, they arise in the basic theory of linear differential equations. We illustrate with the differential equation

$$(D - \lambda I)^2 y = y'' - 2\lambda y' + \lambda^2 y = 0.$$

We know that $y(x) = e^{\lambda x}$ is one solution. We obtain the linearly independent solution $xe^{\lambda x}$ by solving the equation $(D - \lambda I)(y) = e^{\lambda x}$. Thus $e^{\lambda x}$ is an eigenvector of D, and $xe^{\lambda x}$ is a generalized eigenvector. More generally, consider the operator $(D - \lambda I)^k$. The solution space to $(D - \lambda I)^k y = 0$ is k-dimensional and is spanned by $x^j e^{\lambda x}$ for $0 \le j \le k - 1$. This well-known fact from differential equations illustrates the notion of generalized eigenvector. The independence of the functions $x^j e^{\lambda x}$ follows from Theorem 8.4, but can also be checked directly. See Exercise 10.3.

To further develop the connection between linear algebra and constant coefficient differential equations, we will need to discuss more linear algebra.

EXERCISE 10.1. Find e^A where A is as in (19), by explicitly summing the series. Suggestion: Things work out easily if one writes $A = \lambda I + N$ and notes that $N^2 = 0$.

EXERCISE 10.2. Verify that E_λ and K_λ, as defined in Definition 8.3, are invariant subspaces under L. A subspace U is invariant under L if $L(U) \subseteq U$.

EXERCISE 10.3. Prove that the functions $x^j e^{\lambda x}$ for $0 \le j \le n$ are linearly independent.

EXERCISE 10.4. Consider the ODE $y''(t) + 4y(t) = 0$. By letting $x(t) = y'(t)$, rewrite the ODE as a coupled system of first-order ODE in the form

$$\begin{pmatrix} x' \\ y' \end{pmatrix} = A \begin{pmatrix} x \\ y \end{pmatrix}.$$

EXERCISE 10.5. Use linear algebra to find all solutions of the coupled system of ODE given by

$$\begin{pmatrix} x' \\ y' \end{pmatrix} = \begin{pmatrix} 0 & 4 \\ -1 & 4 \end{pmatrix} \begin{pmatrix} x \\ y \end{pmatrix}.$$

Also solve it by eliminating one of the variables and obtaining a second-order ODE for the other.

EXERCISE 10.6. Consider the coupled system of ODE given by $x' = -y$ and $y' = x$. The general solution is given by

$$x(t) = a \cos(t) + b \sin(t)$$
$$y(t) = a \sin(t) - b \cos(t).$$

First, derive this fact by eliminating x and y. In other words, write the first-order system as a second-order equation and solve it. Next, derive this fact by writing

$$\begin{pmatrix} x' \\ y' \end{pmatrix} = \begin{pmatrix} 0 & -1 \\ 1 & 0 \end{pmatrix} \begin{pmatrix} x \\ y \end{pmatrix}$$

and then exponentiating the matrix $\begin{pmatrix} 0 & -1 \\ 1 & 0 \end{pmatrix}$.

EXERCISE 10.7. Solve the previous exercise using complex variables, by putting $z = x + iy$ and noting that the equation becomes $z' = iz$. Comment: This exercise suggests the connection between the exponential function and trig functions elaborated in Chapter 2.

11. Characteristic and minimal polynomials

This section assumes some familiarity with complex numbers. We noted in Example 7.4 that a real matrix need not have real eigenvalues. We will therefore work with n-by-n matrices with complex entries; such matrices always have complex eigenvalues. Some readers may wish to read Section 1 from Chapter 2 before reading this section.

Definition 11.1. Suppose $L : \mathbb{C}^n \to \mathbb{C}^n$ is linear. Its **characteristic polynomial** $p_L(z)$ is defined by $p_L(z) = \det(L - zI)$.

The characteristic polynomial is intrinsically associated with a linear map. Let A and B be square matrices. If B is similar to A, then A and B have the same characteristic polynomial. See Lemma 9.2.

We have seen that the roots of p_A are the eigenvalues of A. Hence the set of eigenvalues of B, including multiplicities, is the same as that of A when B is similar to A. More generally, any symmetric function of the eigenvalues is unchanged under a similarity transformation $A \mapsto PAP^{-1}$.

We benefit from working over $\mathbb{C}$ here because each polynomial $p(z)$ with complex coefficients splits into linear factors. The proof of this statement, called the **fundamental theorem of algebra** for historical reasons, requires complex analysis, and we postpone it until Chapter 4. If we work over the real numbers, then even simple examples such as $J = \begin{pmatrix} 0 & -1 \\ 1 & 0 \end{pmatrix}$ have no real eigenvalues. The matrix J rotates a vector by ninety degrees, and hence has no eigenvalues. See the discussion in Example 7.4 and at the end of the first section of Chapter 2.

Since each polynomial factors over $\mathbb{C}$, we can write the characteristic polynomial of A as

$$p_A(z) = \pm \prod_j (z - \lambda_j)^{m_j}.$$

The exponent m_j is called the *multiplicity* of the eigenvalue λ_j.

The famous Cayley–Hamilton theorem states for each L that $p_L(L) = 0$; in other words, p_L annihilates L. The characteristic polynomial need not be the polynomial of smallest degree that annihilates L. The **minimal polynomial** of L is the unique monic polynomial m_L of smallest degree for which $m_L(L) = 0$. It follows that m_L divides p_L. In abstract linear algebra, the definition of m_L is sometimes stated a bit differently, but the meaning is the same. We give several simple examples.

EXAMPLE 11.1. Put $L = \begin{pmatrix} 3 & 1 \\ 0 & 3 \end{pmatrix}$. Then $p_L(z) = (z-3)^2$ and $m_L(z) = (z-3)^2$.

Put $L = \begin{pmatrix} 3 & 0 \\ 0 & 3 \end{pmatrix}$. Then $p_L(z) = (z-3)^2$ but $m_L(z) = (z-3)$.

EXAMPLE 11.2. Put

$$L = \begin{pmatrix} 4 & 1 & 0 & 0 & 0 & 0 \\ 0 & 4 & 1 & 0 & 0 & 0 \\ 0 & 0 & 4 & 0 & 0 & 0 \\ 0 & 0 & 0 & 9 & 0 & 0 \\ 0 & 0 & 0 & 0 & 3 & 0 \\ 0 & 0 & 0 & 0 & 0 & 3 \end{pmatrix}.$$

Then $p_L(z) = (z-4)^3(z-9)(z-3)^2$ whereas $m_L(z) = (z-4)^2(z-9)(z-3)$.

The following simple result about similarity shows that the characteristic and minimal polynomials are properties of the equivalence class under similarity.

Lemma 11.2. *Assume A and B are square matrices with $B = PAP^{-1}$. Then $p_A = p_B$ and $m_A = m_B$.*

PROOF. Assume $B = PAP^{-1}$. Then:

$$\det(B - \lambda I) = \det\left(P(A - \lambda I)P^{-1}\right) = \det(P)\det(A - \lambda I)\det(P^{-1}) = \det(A - \lambda I).$$

Hence $p_B = p_A$. Suppose next that q annihilates A. Then

$$q(B) = \sum c_k B^k = \sum c_k PA^k P^{-1} = P(\sum c_k A^k)P^{-1} = Pq(A)P^{-1} = 0.$$

Therefore q annihilates B also. We could have also applied Lemma 8.1. By the same argument, if q annihilates B, then q annihilates A. Hence the set of polynomials annihilating B is the same as the set of polynomials annihilating A, and therefore $m_A = m_B$. $\square$

Definition 11.3. The **trace** of an n-by-n matrix is the sum of its diagonal elements.

The following important result is quite simple to prove; see Exercise 11.4.

Theorem 11.4. *For square matrices A, B of the same size, $\operatorname{trace}(AB) = \operatorname{trace}(BA)$. Furthermore, if A and C are similar, then $\operatorname{trace}(A) = \operatorname{trace}(C)$.*

Corollary 11.5. *If A is an n-by-n matrix of complex numbers, then $\operatorname{trace}(A)$ is the sum of its eigenvalues.*

The following theorem is a bit harder to prove. See Exercise 11.5.

Theorem 11.6. *Assume that $L : \mathbb{C}^n \to \mathbb{C}^n$ is linear. Suppose that its distinct eigenvalues are $\lambda_1, \ldots, \lambda_k$. Then L is diagonalizable if and only if*

$$m_L(z) = (z - \lambda_1)(z - \lambda_2)\ldots(z - \lambda_k).$$

For L to be diagonalizable, the multiplicities of the eigenvalues in its characteristic polynomial can exceed 1, but the multiplicities in the minimal polynomial must each equal 1.

EXERCISE 11.1. Give an example of a matrix whose minimal polynomial is $(z - 4)^3(z - 9)^2(z - 3)$ and whose characteristic polynomial is $(z - 4)^3(z - 9)^2(z - 3)^3$.

EXERCISE 11.2. Find the characteristic polynomial of the three-by-three matrix

$$\begin{pmatrix} 1 & 2 & 3 \\ 4 & 5 & 6 \\ 7 & 8 & 9 \end{pmatrix}.$$

(Harder) Find the characteristic polynomial of the n-by-n matrix

$$\begin{pmatrix} 1 & 2 & \ldots & n \\ n+1 & n+2 & \ldots & 2n \\ \vdots & \vdots & \vdots & \vdots \\ n^2 - n + 1 & n^2 - n + 2 & \ldots & n^2 \end{pmatrix}.$$

Suggestion: First note that 0 is an eigenvalue of multiplicity $n - 2$. Then find the eigenvectors corresponding to the non-zero eigenvectors. You might also need to evaluate $\sum_{j=1}^{n} j^2$.

EXERCISE 11.3. Find the determinant of an n-by-n matrix whose diagonal entries all equal x and whose other entries all equal 1. Suggestion: consider a matrix all of whose entries are 1.

EXERCISE 11.4. Prove Theorem 11.4. The first statement is a simple computation using the definition of matrix multiplication. The second statement follows from the first and the definition of similarity.

EXERCISE 11.5. Prove Theorem 11.6.

EXERCISE 11.6. Let the characteristic polynomial for some L be

$$p_L(z) = \sum_{j=0}^{n} c_j z^j = (-1)^n \prod_{k=1}^{n} (z - \lambda_k).$$

What are c_0 and c_{n-1}? Can you explain what c_j is for other values of j?

12. Similarity

Consider the equation $A = PDP^{-1}$ defining similarity. This equation enables us to express polynomial functions (or convergent power series) of A by $f(A) = Pf(D)P^{-1}$. When D is easier to understand than A, computations using D become easier than those using A. The author calls this idea *Einstein's principle*; when doing any problem, choose coordinates in which computation is easiest. We illustrate this idea with a problem from high-school math.

Suppose $f(x) = mx + b$ is an affine function on the real line. What happens if we iterate f many times? We write $f^{(n)}(x)$ for the n-th iterate of f at x. The reader can easily check that

$$f^{(2)}(x) = f(f(x)) = m(f(x) + b) + b = m(mx + b) + b = m^2 x + (m + 1)b$$

$$f^{(3)}(x) = f(f^2(x)) = m(m^2 x + (m + 1)b) + b = m^3 + (m^2 + m + 1)b.$$

From these formulas one can guess the general situation and prove it by induction.

There is a much better approach to this problem. We try to write $f(x) = m(x - c) + c$. Doing so is possible if and only if $m \neq 1$, and then c satisfies $b = c(1 - m)$. If $m = 1$, then $f(x) = x + b$; thus f is a translation by b. Iterating n times results in translating by nb. Hence $f^{(n)}(x) = x + nb$ in this case.

If $m \neq 1$, then we can solve $c - mc = b$ for c and write $f(x) = m(x - c) + c$. Then f translates by $-c$, multiplies by m, and then translates back by c. Thus $f = TMT^{-1}$, where T translates and M multiplies. We immediately obtain

$$f^{(n)}(x) = TM^n T^{-1}(x) = m^n(x - c) + c = m^n x + b \frac{1 - m^n}{1 - m}. \tag{22}$$

Notice that we can recover the case $m = 1$ by letting m tend to 1 in (22).

Figure 8 illustrates the idea. Iterating f amounts to going across the top line. We can also go down, multiply, go up, go back down, multiply, etc. Going up and back down cancel out. Hence $f^{(n)} = TM^n T^{-1}$. Iterating M amounts to going across the bottom line.

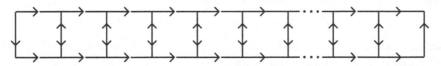

FIGURE 8. Iteration

We summarize the idea. Whenever an operation can be expressed as TMT^{-1}, iterating the operation is easy. We need only iterate M. More generally we can consider functions of the operation, by applying those functions to the operator M. This idea explains the usefulness of many mathematical concepts in engineering, including diagonalizing matrices, the Fourier and Laplace transforms, and the use of linear fractional transformations in analyzing networks.

In the context of linear algebra, certain objects (characteristic polynomial, eigenvalues, trace, determinant, etc.) are invariant under similarity. These objects are *intrinsically* associated with the given linear map, and hence are likely to have physical meaning. Other expressions (the sum of all the entries of a matrix, how many entries are 0, etc.) are artifacts of the coordinates chosen. They distract the observer from the truth.

REMARK. Consider a linear transformation between spaces of different dimensions. It does not then make sense to consider eigenvectors and eigenvalues. Nonetheless, these notions are so crucial that we need to find their analogues in this more general situation. Doing so leads to the notions of singular values and the **singular value decomposition**. Although we could discuss this situation now, we postpone this material until Section 5 of Chapter 6, after we have defined adjoints.

EXERCISE 12.1. Suppose $f(x) = Ax + b$ for $x \in \mathbb{C}^n$, where $A : \mathbb{C}^n \to \mathbb{C}^n$ is linear. Under what condition does the analogue of (22) hold?

13. Additional remarks on ODE

We saw in Section 8 how to solve a constant coefficient linear ordinary differential equation, by exponentiating a matrix. In this section we revisit such equations and provide an alternative manner for solving them, in order to apply the ideas on similarity.

Consider the general such equation

$$y^{(n)} + c_{n-1}y^{(n-1)} + \cdots + c_1 y' + c_0 y = f. \tag{23}$$

Here we allow the coefficients c_j to be real or complex numbers. We rewrite (23) in the form

$$p(D)y = f,$$

where p is the polynomial

$$p(z) = z^n + c_{n-1}z^{n-1} + \cdots + c_0,$$

and $p(D)$ means that we substitute the operator D of differentiation for z. In Chapter 4 we prove the fundamental theorem of algebra, which implies that we can factor a polynomial into linear factors:

$$p(z) = \prod_{j=1}^n (z - \lambda_j). \tag{24}$$

Here the λ_j are complex numbers, which can be repeated. Hence $p(D) = \prod_{j=1}^n (D - \lambda_j)$. Hence to solve $p(D)y = f$, it suffices to solve $(D - \lambda)g = h$ and then iterate. In other words

$$p(D)^{-1} = \prod (D - \lambda_j)^{-1}. \tag{25}$$

In (25), the order of the factors does not matter, because

$$(D - \lambda_1)(D - \lambda_2) = (D - \lambda_2)(D - \lambda_1).$$

Proposition 13.1. *Let $D - \lambda$ be a first-order ordinary differential operator. Let M_λ denote the operation of multiplication by $e^{\lambda t}$. Let J denote integration. Then, $(D - \lambda)^{-1} = M_\lambda J M_\lambda^{-1}$. In more concrete language, we solve $g' - \lambda g = h$ by writing*

$$g(t) = e^{\lambda t} \int_{t_o}^t e^{-\lambda u} h(u)\,du. \tag{26}$$

PROOF. One discovers formula (26) by variation of parameters. To prove the proposition, it suffices to check that $g' - \lambda g = h$, which is a simple calculus exercise. $\square$

Corollary 13.2. *We solve $p(D)y = f$ by factoring p and iterating Proposition 13.1.*

EXERCISE 13.1. Verify that (26) solves $g' - \lambda g = h$.

EXERCISE 13.2. Discover (26) as follows. Assume that $g(t) = c(t)e^{\lambda t}$, plug in the equation, and determine c.

EXERCISE 13.3. Solve $(D - \lambda_1)(D - \lambda_2)y = f$ using Corollary 13.2.

EXERCISE 13.4. Use Proposition 13.1 to easily solve $(D - \lambda)^k y = f$. In particular, revisit Example 4.3 using this approach.

CHAPTER 2

Complex numbers

The extent to which complex numbers arise in physics and engineering is almost beyond belief. We start by defining the complex number field $\mathbb{C}$. Since $\mathbb{C}$ is a field, in the rest of the book we can and will consider complex vector spaces. In this chapter, however, we focus on the geometric and algebraic properties of the complex numbers themselves. One of the main points is the relationship between the complex exponential function and the trig functions.

1. Basic definitions

A complex number is a pair (x, y) of real numbers, or simply the point (x, y) in $\mathbb{R}^2$. By regarding this pair as a single object and introducing appropriate definitions of addition and multiplication, remarkable simplifications arise. As sets, we have $\mathbb{C} = \mathbb{R}^2$. We add componentwise as usual, but we introduce a definition of multiplication which has profound consequences. Figures 1 and 2 provide geometric intuition.

$$(x, y) + (a, b) = (x + a, y + b) \tag{1}$$

$$(x, y) * (a, b) = (xa - yb, xb + ya). \tag{2}$$

The idea behind formula (2) comes from seeking a square root of -1. Write $(x, y) = x + iy$ and $(a, b) = a + ib$. If we assume the distributive law is valid, then we get

$$(x + iy)(a + ib) = xa + i^2 yb + i(xb + ya). \tag{3}$$

If we further assume in (3) that $i^2 = -1$, then we get

$$(x + iy)(a + ib) = xa - yb + i(xb + ya). \tag{4}$$

This result agrees with (2). Why do we start with (2) rather than with (4)? The answer is somewhat philosophical. If we start with (2), then we have not presumed the existence of an object whose square is -1. Nonetheless, such an object exists, because $(0, 1) * (0, 1) = (-1, 0) = -(1, 0)$. Therefore no unwarranted assumptions have been made. If we started with (4), then one could question whether the number i exists.

Henceforth we will use the complex notation, writing $(x, y) = x + iy$ and $(a, b) = a + ib$. Then

$$(x + iy) + (a + ib) = x + a + i(y + b)$$

$$(x + iy)(a + ib) = xa - yb + i(xb + ya).$$

Put $z = x + iy$. We call x the **real part** of z and y the **imaginary part** of z. We write $x = \text{Re}(z)$ and $y = \text{Im}(z)$. Note that the imaginary part of z is a real number.

One sees immediately that the additive identity 0 is the complex number $0 + i0$ and that the multiplicative identity 1 is the complex number $1 + i0$. The additive inverse of $x + iy$ is $-x + i(-y)$. The multiplicative inverse is noted in (6.1) and (6.2). Verifying the field axioms is routine but tedious, and we leave it to the interested reader.

We regard the imaginary unit i as the point $(0,1)$ in the plane. The theory would develop in the same way had we started with $(0,-1)$ as the imaginary unit. This symmetry naturally leads to the notion of complex conjugation.

Definition 1.1. Put $z = x + iy$. Its **complex conjugate** $\overline{z}$ is the complex number given by $\overline{z} = x - iy$.

We note the following simple but important formulas:

$$z\overline{z} = x^2 + y^2$$
$$z + \overline{z} = 2x$$
$$z - \overline{z} = 2iy$$
$$\overline{z + w} = \overline{z} + \overline{w}.$$
$$\overline{zw} = \overline{z}\,\overline{w}.$$

We put $|z|^2 = x^2 + y^2 = z\overline{z}$, the squared Euclidean distance from the point (x, y) to the origin. We call $|z|$ the *magnitude* or *absolute value* of z. See Exercise 1.8 for an elegant proof that

$$|zw| = |z|\,|w|.$$

Furthermore, $|z - w|$ equals the Euclidean distance between z and w. This distance function satisfies the *triangle inequality*:

$$|z - w| \leq |z - \zeta| + |\zeta - w|. \tag{5}$$

Exercise 1.4 suggests an algebraic proof of (5). The reader should draw a picture to see why it is called the triangle inequality. When $\zeta = 0$, inequality (5) gives $|z - w| \leq |z| + |w|$, which is also called the triangle inequality.

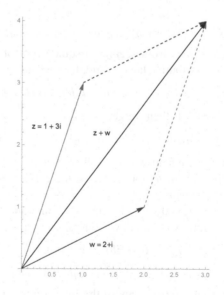

FIGURE 1. Addition of complex numbers

Complex conjugation helps us to understand reciprocals. For $z \neq 0$,

$$\frac{1}{z} = \frac{\overline{z}}{|z|^2}. \tag{6.1}$$

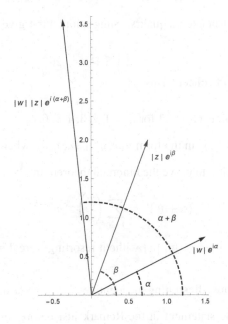

FIGURE 2. Multiplication of complex numbers

Hence the reciprocal of $x + iy$ is $\frac{x-iy}{x^2+y^2}$. In other words,

$$\frac{1}{x+iy} = \frac{x}{x^2+y^2} + i\frac{-y}{x^2+y^2}. \tag{6.2}$$

The author suggests remembering (6.1) rather than (6.2). In general it is easier to work with z and $\bar{z}$ than with x and y.

REMARK. Let us connect the definition (2) of complex multiplication with linear algebra. If we write $z = x + iy$ and $w = a + ib$, then of course $zw = (ax - by) + i(bx + ay)$. We can write this formula using matrices as

$$\begin{pmatrix} a & -b \\ b & a \end{pmatrix} \begin{pmatrix} x \\ y \end{pmatrix} = \begin{pmatrix} ax - by \\ bx + ay \end{pmatrix}. \tag{$*$}$$

The mapping $z \mapsto wz$ is linear from $\mathbb{C}$ to $\mathbb{C}$. By (*) it corresponds to a very special kind of linear map from $\mathbb{R}^2$ to $\mathbb{R}^2$. Multiplying by matrices of the form in (*) preserves angles between vectors.

EXERCISE 1.1. Suppose $a, b, c \in \mathbb{C}$ and $a \neq 0$. Prove the quadratic formula: the solutions of

$$az^2 + bz + c = 0$$

are given by

$$z = \frac{-b \pm \sqrt{b^2 - 4ac}}{2a}.$$

What properties of a field did you use? See [V] for a brilliant one-page paper.

EXERCISE 1.2. Given $w \in \mathbb{C}$, show that there is z with $z^2 = w$. Doing so is easy using the polar form discussed below. Show it by instead solving $(x + iy)^2 = a + ib$ for x, y.

EXERCISE 1.3. Show that $|z - w|^2 = |z|^2 - 2\text{Re}(z\overline{w}) + |w|^2$.

EXERCISE 1.4. Prove the triangle inequality. Suggestion: First give new names to $z - w$ and $z - \zeta$. The inequality is equivalent to

$$|z_1| - |z_2| \leq |z_1 - z_2|.$$

To prove it, square both sides and collect terms.

EXERCISE 1.5. Plot the points $(1 + i)^n$ for $n = 1, 2, 3, 4, 5, 6$.

EXERCISE 1.6. Write $(x + iy)^3$ in the form $u(x, y) + iv(x, y)$, where u, v are real.

EXERCISE 1.7. Use induction to prove the binomial theorem in $\mathbb{C}$:

$$(z + w)^n = \sum_{k=0}^{n} \binom{n}{k} z^k w^{n-k}.$$

EXERCISE 1.8. Prove that $|zw| = |z|\,|w|$ without resorting to real and imaginary parts. Suggestion: Square both sides and use $|z|^2 = z\overline{z}$.

EXERCISE 1.9. Draw pictures illustrating the identities $z + \overline{z} = 2x$ and $z - \overline{z} = 2iy$.

EXERCISE 1.10. Verify the statement in the Remark just before the Exercises. Thus, let M be the matrix $M = \begin{pmatrix} a & -b \\ b & a \end{pmatrix}$, with not both a, b zero. If v, w are vectors in $\mathbb{R}^2$, show that the angle between Mv and Mw is the same as the angle between v and w. (Section 1 of Chapter 3 recalls the formula for the angle between vectors.) The discussion around (10) below gives an instant proof using complex variables.

2. Limits

The distance function and the triangle inequality enable us to define limits and establish their usual properties. Once we have limits we can define convergent power series. Power series arise throughout the rest of this book. In the next section we define the complex exponential function via its power series and continue by developing the connection with trig. In Chapter 4 we make a systematic study of complex analytic functions, namely those functions given locally by convergent power series.

Definition 2.1. Let $\{z_n\}$ be a sequence of complex numbers. Let $L \in \mathbb{C}$. We say that $\{z_n\}$ converges to L, and write $\lim_{n \to \infty} z_n = L$ if, for each $\epsilon > 0$, there is an integer N such that

$$n \geq N \implies |z_n - L| < \epsilon.$$

As in real analysis, a sequence of complex numbers converges if and only if it is a Cauchy sequence. We discuss this concept in detail in Chapter 4. The definition of a limit says that a sequence converges to L if, no matter how small a ball about L we pick, the terms of the sequence are eventually in that ball. Figure 3 depicts a ball of radius ϵ about L. A sequence converges to L if, eventually, all the terms are close to L. The idea of a Cauchy sequence is that the terms are eventually all close to each other; no limit is specified.

Definition 2.2. Let $\{z_n\}$ be a sequence of complex numbers. We say that the infinite series $\sum_{n=0}^{\infty} z_n$ *converges* to L if

$$\lim_{N \to \infty} \sum_{n=0}^{N} z_n = L.$$

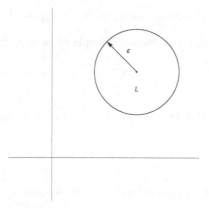

FIGURE 3. Ball of radius ϵ about L

The finite sum in Definition 2.2 is called the N-th **partial sum** of the series. In some situations it is possible to find the partial sums explicitly and then evaluate the limit. For a general series, doing so is impossible. Hence calculus books include many tests for deciding whether a series converges, without knowing the value of the limit. When the limit of the partial sums does not exist, one says that the series *diverges*.

EXAMPLE 2.1. Put $z_n = \frac{1}{(n+1)(n+2)}$. Then

$$\sum_{n=0}^{N} z_n = \sum_{n=0}^{N} \frac{1}{(n+1)(n+2)} = \sum_{n=0}^{N} \left(\frac{1}{n+1} - \frac{1}{n+2} \right) = 1 - \frac{1}{N+2}.$$

Hence $\sum_{n=0}^{\infty} z_n = \lim_{N\to\infty}(1 - \frac{1}{N+2}) = 1$. Here we can find the partial sum explicitly.

EXAMPLE 2.2. Put $z_n = \frac{1}{n}$. Then $\sum_{n=1}^{N} z_n$ is called the N-th harmonic number. There is no simpler formula for this sum. Since the partial sums grow without bound, the harmonic series $\sum_{n=1}^{\infty} \frac{1}{n}$ diverges. See Exercise 2.6.

We can apply many of the tests for convergence from the real case to the complex case, because of the following simple result.

Proposition 2.3 (Comparison test). *Assume $\sum_{n=0}^{\infty} |z_n|$ converges. Then $\sum_{n=0}^{\infty} z_n$ also converges.*

PROOF. We show that the partial sums $\{S_N\}$ of the series $\sum_n z_n$ form a Cauchy sequence. We know that the partial sums $\{T_N\}$ of the series $\sum_n |z_n|$ form a Cauchy sequence. We have

$$|S_N - S_M| = \Big| \sum_{M+1}^{N} z_n \Big| \leq \sum_{M+1}^{N} |z_n| = |T_N - T_M|.$$

Therefore S_N is Cauchy if T_N is Cauchy.   □

We next state a weak version of the *ratio test*; this version is adequate for showing that the exponential series converges. Since the test involves only absolute values, its proof in the real case implies the result in the complex case. See [R] for such a proof. We also mention the ratio test and the *root test* for power series in Section 2 of Chapter 5.

Proposition 2.4 (Ratio test). *Suppose z_n satisfies $\lim_{n \to \infty} \left| \frac{z_{n+1}}{z_n} \right| = L < 1$. Then $\sum_{n=0}^{\infty} z_n$ converges.*

The geometric series provides a fundamental example of an infinite series. First consider the identity

$$(1-z) \sum_{j=0}^{n} z^j = 1 - z^{n+1}. \tag{7}$$

If $z \neq 1$, we divide by $1 - z$ in (7) and obtain the *finite geometric series*

$$\sum_{j=0}^{n} z^j = \frac{1 - z^{n+1}}{1 - z}. \tag{FGS}$$

Let $n \to \infty$ in (FGS). The limit does not exist if $|z| \geq 1$. Assume $|z| < 1$. Then $\lim z^n = 0$ by Exercise 2.4. Hence

$$\lim_{n \to \infty} \frac{1 - z^{n+1}}{1-z} = \frac{1}{1-z}.$$

Hence we obtain the geometric series, one of the two series you must commit to memory. (The other is the series for the exponential function, defined in the next section.)

$$\sum_{n=0}^{\infty} z^n = \frac{1}{1-z}, \quad |z| < 1 \tag{GS}$$

EXERCISE 2.1. Verify (7) by induction. Show that (7) remains valid if we replace z by a linear map.

EXERCISE 2.2. Use the definition to prove that $\lim_{n \to \infty} z_n = 0$ if and only if $\lim_{n \to \infty} |z_n| = 0$.

EXERCISE 2.3. Assume that w_n is a bounded sequence (that is, there is a constant C such that $|w_n| \leq C$ for all n) and that $\lim_{n \to \infty} z_n = 0$. Prove that $\lim_{n \to \infty} w_n z_n = 0$. Observe that the conclusion holds even if $\lim w_n$ does not exist.

EXERCISE 2.4. Assume that $|z| < 1$. Prove that $\lim_{n \to \infty} z^n = 0$.

EXERCISE 2.5. Recall or look up the ratio test and the root test for power series in a real variable. Then state and prove their analogues for power series in a complex variable.

EXERCISE 2.6. Show that the sequence of partial sums of the harmonic series from Example 2.2 is unbounded.

EXERCISE 2.7. Use the geometric series and the comparison test to prove the ratio test.

EXERCISE 2.8. Assume $b_n > 0$ for all n and $\sum b_n$ converges. Assume for all n that

$$\left| \frac{a_{n+1}}{a_n} \right| \leq \frac{b_{n+1}}{b_n}.$$

Prove that $\sum a_n$ converges.

EXERCISE 2.9. Give an example of a sequence $\{z_n\}$ of complex numbers such that $\sum z_n$ converges but $\sum z_n^3$ diverges. Suggestion: First notice that $1 + w + w^2 = 0$ when $w \neq 1$ but $w^3 = 1$. Then try a series whose terms are $\frac{1}{n^{1/3}}$ and $\frac{w}{n^{1/3}}$ and $\frac{w^2}{n^{1/3}}$.

3. The exponential function and trig

Definition 3.1. For $z \in \mathbb{C}$, we define the complex exponential e^z by

$$\sum_{n=0}^{\infty} \frac{z^n}{n!} = e^z. \qquad (exp)$$

The series converges for all z.

The proof of convergence uses the ratio test and is left to the reader (Exercise 3.1).

Our next step is to recall and understand the polar coordinate representation of a complex number. Limits play a key but subtle role, because we have defined the exponential function as an infinite series.

Proposition 3.2 (Euler's formula). *For real t we have $e^{it} = \cos(t) + i \sin(t)$.*

PROOF. The proof of this formula depends on how one defines exponentials, cosine, and sine. Below we summarize the approach used here. $\qquad\qquad\square$

In our approach, the trig functions are *defined* in terms of the complex exponential, and Euler's formula holds for $t \in \mathbb{C}$.

Definition 3.3 (trig functions). For $z \in \mathbb{C}$,

$$\cos(z) = \frac{e^{iz} + e^{-iz}}{2}$$

$$\sin(z) = \frac{e^{iz} - e^{-iz}}{2i}.$$

The definitions have two desirable immediate consequences. For complex z, w,

$$\cos^2(z) + \sin^2(z) = 1. \qquad (8)$$

Furthermore, the identity $e^{iz} = \cos z + i \sin z$ holds for all *complex* z. Thus Euler's formula holds; when t is *real*, the definitions imply

$$\cos(t) = \operatorname{Re}(e^{it}) \qquad (8.1)$$

$$\sin(t) = \operatorname{Im}(e^{it}). \qquad (8.2)$$

This approach also leads to a definition of 2π, as the circumference of the unit circle. Now we can analyze the complete derivation of Euler's formula:

- We define $\mathbb{C}$ to be $\mathbb{R}^2$ with addition as in (1) and multiplication as in (2). Then $\mathbb{C}$ is a field.
- We interpret $|z - w|$ as the distance between z and w, allowing us to define and use limits.
- We define an infinite series to be the limit of its partial sums when this limit exists:

$$\sum_{n=0}^{\infty} z_n = \lim_{N \to \infty} \sum_{n=0}^{N} z_n.$$

- We define the complex exponential by

$$e^z = \sum_{n=0}^{\infty} \frac{z^n}{n!}.$$

This series converges for all z by the ratio test.

- Exponentiation converts addition into multiplication: $e^z e^w = e^{z+w}$. We formally verify this identity: We have $e^z = \sum_{n=0}^{\infty} \frac{z^n}{n!}$ and $e^w = \sum_{m=0}^{\infty} \frac{w^m}{m!}$. Therefore

$$e^z e^w = \sum_{n=0}^{\infty} \sum_{m=0}^{\infty} \frac{z^n w^m}{n! m!} = \sum_{n=0}^{\infty} \sum_{k=n}^{\infty} \frac{z^n w^{k-n}}{n!(k-n)!} = \sum_{k=0}^{\infty} \sum_{n=0}^{k} \frac{z^n w^{k-n}}{n!(k-n)!}$$

$$= \sum_{k=0}^{\infty} \frac{(z+w)^k}{k!} = e^{z+w}.$$

We have substituted $k = m + n$ in the double sum and eliminated m. Thus $n \leq k$. We omit the analysis needed to justify the interchange of order of summation.
- Complex conjugation is continuous. In particular $e^{\bar{z}} = \overline{e^z}$.
- For $t \in \mathbb{R}$ we have $\overline{e^{it}} = e^{\overline{it}} = e^{-it}$. Therefore

$$|e^{it}|^2 = e^{it} e^{-it} = e^0 = 1.$$

Hence e^{it} is a point on the unit circle. By (8.1) and (8.2) we obtain $e^{it} = \cos(t) + i\sin(t)$, where cosine and sine have their usual meanings from trig. We have fully justified Definition 3.3.

Corollary 3.4. $e^{ik\pi} = 1$ for all even integers k and $e^{im\pi} = -1$ for all odd integers m.

The logical development here is remarkable. The two identities $e^{z+w} = e^z e^w$ and $e^{\bar{z}} = \overline{e^z}$, together with the formulas (8.1) and (8.2), (which are *definitions*), imply all of trig! There are no vague, unwarranted uses of angles. We also note that $t \to e^{it}$ for $0 \leq t \leq 2\pi$ parametrizes the unit circle. This fact will be used many times in the sequel.

To introduce and fully appreciate the power of the polar coordinate representation of a complex number, we further develop our view of complex numbers as certain types of two-by-two matrices. Consider multiplying a complex number z by w; we obtain a linear function from $\mathbb{C}$ to itself: $z \mapsto wz$. We can regard this transformation also as a linear map from $\mathbb{R}^2$ to itself:

$$\begin{pmatrix} x \\ y \end{pmatrix} \mapsto \begin{pmatrix} ax - by \\ bx + ay \end{pmatrix} = \begin{pmatrix} a & -b \\ b & a \end{pmatrix} \begin{pmatrix} x \\ y \end{pmatrix}$$

In this manner we identify the complex number $a + ib$ with the matrix $\begin{pmatrix} a & -b \\ b & a \end{pmatrix}$.

This identification amounts to identifying a complex number w with the real linear transformation given by multiplication by w. The collection of these linear maps from $\mathbb{R}^2$ to $\mathbb{R}^2$ is a field. The multiplicative identity is the matrix

$$\begin{pmatrix} 1 & 0 \\ 0 & 1 \end{pmatrix}.$$

The additive identity is the matrix

$$\begin{pmatrix} 0 & 0 \\ 0 & 0 \end{pmatrix}.$$

Multiplying by i in $\mathbb{C}$ is the same as multiplying by J in $\mathbb{R}^2$, where

$$J = \begin{pmatrix} 0 & -1 \\ 1 & 0 \end{pmatrix}. \tag{9}$$

Adding complex numbers corresponds to adding these matrices, and multiplying complex numbers corresponds to multiplying these matrices. The reciprocal of a non-zero complex number corresponds to the matrix inverse. In mathematical language, the set of these matrices with the given operations of addition and multiplication form a field isomorphic to $\mathbb{C}$.

Note the following matrix identity, when a, b are not both zero:

$$\begin{pmatrix} a & -b \\ b & a \end{pmatrix} = \sqrt{a^2 + b^2} \begin{pmatrix} \frac{a}{\sqrt{a^2+b^2}} & -\frac{b}{\sqrt{a^2+b^2}} \\ \frac{b}{\sqrt{a^2+b^2}} & \frac{a}{\sqrt{a^2+b^2}} \end{pmatrix}. \tag{10}$$

This identity suggests the polar coordinate representation of a complex number. When $a + ib \neq 0$, the matrix on the right-hand side of (10) is a rotation matrix. Multiplying by $w = a + ib$ involves two geometric ideas. We rotate through an angle, and we stretch things by the factor $\sqrt{a^2 + b^2} = |w|$. We summarize:

$$(x, y) = (r\cos\theta, r\sin\theta) = \sqrt{x^2 + y^2}(\cos\theta, \sin\theta)$$
$$x + iy = z = |z|(\cos\theta + i\sin\theta) = |z|e^{i\theta}.$$

If also $w = |w|e^{i\phi}$, then we have

$$wz = |w||z|e^{i(\theta+\phi)}.$$

When $z \neq 0$ and we write $z = |z|e^{i\theta}$, we call θ the *argument* of z, the *phase* of z, or the *angle* of z. This angle is determined only up to multiples of 2π.

When $a^2 + b^2 = 1$, the matrix $\begin{pmatrix} a & -b \\ b & a \end{pmatrix}$ has a geometric meaning, as rotation through an angle θ with $\tan\theta = \frac{b}{a}$. Recall that the imaginary unit i can be identified with the matrix $J = \begin{pmatrix} 0 & -1 \\ 1 & 0 \end{pmatrix}$. Multiplying by i corresponds to a counterclockwise rotation by $\frac{\pi}{2}$ radians. This geometric interpretation helps in the analysis of alternating current.

Let us find the eigenvalues of J. We have $\det(J - \lambda I) = \lambda^2 + 1 = 0$. Thus the eigenvalues are $\lambda = \pm i$. The corresponding eigenspaces are easy to determine as well: E_i is spanned by $(i, 1)$ and E_{-i} is spanned by $(-i, 1)$. Multiplying by the matrix J on $\mathbb{R}^2$ corresponds to a counterclockwise rotation by an angle of $\frac{\pi}{2}$. Even though its entries are real, its eigenvalues are imaginary. This matrix J can be thought of as a square root of minus one and will be discussed in Chapter 2. See also Exercise 7.8 of Chapter 1.

We now show how to use the polar form of a complex number to solve certain equations. Let us consider *roots of unity*. See Figures 4 and 5. Let m be a positive integer. A complex number ω is called a primitive m-th root of unity if m is the smallest positive integer with $\omega^m = 1$. For example i is a primitive fourth root of unity and $\frac{-1+i\sqrt{3}}{2}$ is primitive cube root of unity.

EXAMPLE 3.1. Using polar coordinates, we find all solutions to $z^{16} = 1$ and $z^3 = 27$. Let $z = |z|e^{i\theta}$. Then $z^{16} = |z|^{16}e^{16i\theta} = 1e^{2\pi in}$. We obtain sixteen solutions; they are equally spaced on the unit circle. The angles are given by $\theta = \frac{n\pi}{8}$ for $n = 0, 1, \ldots, 15$. When $\omega^{16} = 1$, the finite geometric series gives

$$\sum_{j=0}^{15} \omega^j = 0.$$

If we regard the powers of ω as forces, we see that the sum of the forces is 0.

To solve $z^3 = 27$, put $z = |z|e^{i\theta}$. Then

$$z^3 = |z|^3 e^{3i\theta} = 27e^{2n\pi i}.$$

Thus $|z|^3 = 27$ and $3i\theta = 2n\pi i$. When $n = 0$ we have $\theta = 0$; when $n = 1$ we have $\theta = \frac{2\pi}{3}$; when $n = 2$ we have $\theta = \frac{4\pi}{3}$. It is instructive to write the complex cube roots of 1 as $\frac{-1\pm i\sqrt{3}}{2}$. The three cube roots of 27 are $z = 3$ and $z = \frac{3}{2}(-1 + i\sqrt{3})$ and $z = \frac{3}{2}(-1 - i\sqrt{3})$.

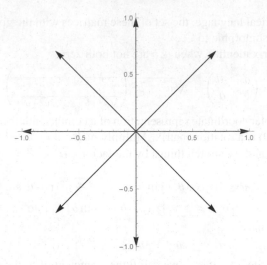

FIGURE 4. Eighth roots of unity

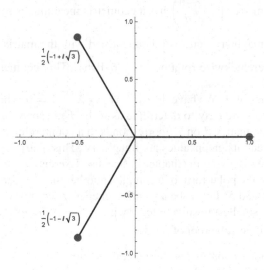

FIGURE 5. Cube roots of unity

We close this section with a simple observation about even and odd functions.

Definition 3.5. A function $f : \mathbb{C} \to \mathbb{C}$ is called **even** if $f(-z) = f(z)$ for all z. A function $f : \mathbb{C} \to \mathbb{C}$ is called **odd** if $f(-z) = -f(z)$ for all z.

Proposition 3.6. *Let $f : \mathbb{C} \to \mathbb{C}$ be an arbitrary function. Then there exists a unique way to write $f = A + B$ where A is even and B is odd.*

PROOF. Assume $f(z) = A(z) + B(z)$, with A even and B odd. We then have
$$f(-z) = A(-z) + B(-z) = A(z) - B(z).$$

These two equations for $f(z)$ and $f(-z)$ can be written

$$\begin{pmatrix} f(z) \\ f(-z) \end{pmatrix} = \begin{pmatrix} 1 & 1 \\ -1 & 1 \end{pmatrix} \begin{pmatrix} A(z) \\ B(z) \end{pmatrix}.$$

We solve for A and B to obtain

$$A(z) = \frac{1}{2}(f(z) + f(-z))$$

$$B(z) = \frac{1}{2}(f(z) - f(-z))$$

These formulas establish both the existence and uniqueness at the same time. $\qquad\square$

EXAMPLE 3.2. If $f(z) = e^z$, then $f(z) = \cosh(z) + \sinh(z)$ is the decomposition of f into the sum of an even function and an odd function. Notice that the series expansion for $\cosh(z)$ consists of the terms of even degree in the series expansion for e^z, and the series for $\sinh(z)$ consists of the terms of odd degree.

EXERCISE 3.1. Use the ratio test to prove that the series for e^z converges for all z.

EXERCISE 3.2. Assume $(x + iy)^{493} = 1$. What is $(x + iy)^{492}$? (the answer is very simple!)

EXERCISE 3.3. Find all sixth roots of unity, in the form $x + iy$.

EXERCISE 3.4. Let $p : \mathbb{C} \to \mathbb{C}$ be a function. Show that p can be written as a sum $A + B + C + D$ of four functions, where $A(iz) = A(z)$ and $B(iz) = iB(z)$ and $C(iz) = -C(z)$ and $D(iz) = -iD(z)$. What are A, B, C, D when p is a polynomial or power series?

EXERCISE 3.5 (Fourier matrix). Let ω be an n-th root of unity. Find the inverse of the n-by-n matrix defined by:

$$F = \begin{pmatrix} 1 & 1 & 1 & \cdot & 1 \\ 1 & \omega & \omega^2 & \cdot & \omega^{n-1} \\ 1 & \omega^2 & \omega^4 & \cdot & \omega^{2(n-1)} \\ \cdot & \cdot & \cdot & \cdot & \cdot \\ 1 & \omega^{n-1} & \omega^{2(n-1)} & \cdot & \omega^{(n-1)^2} \end{pmatrix}.$$

EXERCISE 3.6. Use complex variables to solve the coupled system of ODE given by

$$\begin{pmatrix} x' \\ y' \end{pmatrix} = \begin{pmatrix} a & -b \\ b & a \end{pmatrix} \begin{pmatrix} x \\ y \end{pmatrix}.$$

4. Subsets of $\mathbb{C}$

Let us consider circles, lines, and other geometric subsets of the complex plane. See Figure 6.

A *circle* is the set of z such that $|z - p| = r$. Here p is the center of the circle and the positive number r is its radius. The interior of the circle, an open disk, is defined by the inequality $|z - p| < r$, and the closed disk is defined by $|z - p| \leq r$. In many contexts, especially regarding integration, we will regard the circle as the image of the map $t \mapsto p + re^{it}$ for $0 \leq t \leq 2\pi$.

It is also useful to give a parametric equation for a line in $\mathbb{C}$. To specify a line, we need a point on the line and a direction. Suppose $w \neq 0$ gives the direction of a line containing p. Then the line is the set of z satisfying

$$z = p + tw \quad \text{for} \ t \in \mathbb{R}. \tag{11}$$

Since t is real, we can rewrite this equation by saying $\frac{z-p}{w}$ is real, or by writing

$$\frac{z - p}{w} = \frac{\bar{z} - \bar{p}}{\bar{w}}.$$

Putting $w = |w|e^{i\phi}$ and rewriting, we obtain the equation

$$z - p = e^{2i\phi}(\overline{z} - \overline{p}).$$

In the author's opinion, the parametric representation (11) is the most useful. One need not assume $|w| = 1$ when using the parametric form.

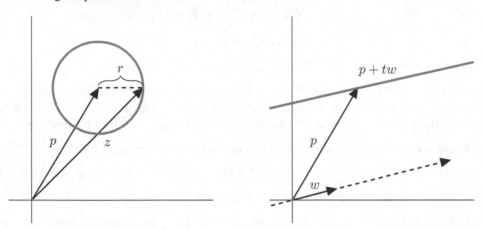

FIGURE 6. The circle $|z - p| = r$ and the line $z = p + tw$

Note also that the *line segment* between z and w can be represented parametrically by

$$t \mapsto z + t(w - z) = tw + (1 - t)z.$$

for $0 \le t \le 1$. A subset $K \subseteq \mathbb{C}$ is called **convex** if, whenever z and w are in K, then the line segment connecting them is also in K.

An *ellipse* is the set of points in a plane for which the sum of the distances to two given points is constant. These points are called the *foci* of the ellipse. Therefore the set of z for which

$$|z - p| + |z - q| = 1$$

defines an ellipse if it is not empty. In case $p = q$ the two foci correspond, and we obtain a circle whose center is p.

A *hyperbola* is the set of points in a plane for which the absolute difference of the distance between two fixed points is a positive constant, smaller than the distance between these points. These two points are assumed to be distinct; they are also called the foci. Therefore the set of z satisfying

$$|z - p| - |z - q| = \pm 1$$

is a hyperbola when $|p - q| > 1$.

It is possible and sometimes convenient to use complex variables when studying these objects. For example, setting the real and imaginary parts of z^2 equal to non-zero constants gives equations for orthogonal hyperbolas. See Chapter 4.

EXERCISE 4.1. Show that a rectangle (including its interior) is a convex set.

EXERCISE 4.2. Show that a disk is convex.

EXERCISE 4.3. Sketch the set of z for which both $1 < \text{Re}(z^2) < 2$ and $2 < \text{Im}(z^2) < 4$.

EXERCISE 4.4. Find a parametric equation $t \mapsto z(t)$ for one branch of the hyperbola $x^2 - y^2 = 1$.

EXERCISE 4.5. For a subset $K \subseteq \mathbb{C}$, define its translate $K + w$ to be the set of z such that $z = \zeta + w$ for some $\zeta \in K$. Show that $K + w$ is convex if K is convex.

EXERCISE 4.6. Find the equation of an ellipse with foci at $\pm i$ and passing through 1.

EXERCISE 4.7. Consider a line through $1 + i$ and -1. Where does it intersect the imaginary axis?

EXERCISE 4.8. Given distinct points w, ζ in $\mathbb{C}$, find the parametric equation of the perpendicular bisector of these points (the line through their midpoint and perpendicular to the segment connecting them).

5. Complex logarithms

We want to define complex logarithms. First we recall some background information, both from this book and from elementary calculus.

We have defined the complex exponential by

$$e^z = \sum_{n=0}^{\infty} \frac{z^n}{n!}.$$

This series converges for all complex numbers z. Note that $e^{z+2\pi i} = e^z$ for all z, and hence the exponential function is not injective. One must therefore be careful in trying to define an inverse.

Trig guides the development. Recall that

$$\cos(z) = \frac{e^{iz} + e^{-iz}}{2}$$

$$\sin(z) = \frac{e^{iz} - e^{-iz}}{2i}.$$

A certain power series computation justifies these definitions.

Suppose we had defined cosine and sine in some other fashion. These functions have their own Taylor series expansions, for t real:

$$\cos(t) = \sum_{n=0}^{\infty} (-1)^n \frac{t^{2n}}{(2n)!}$$

$$\sin(t) = \sum_{n=0}^{\infty} (-1)^n \frac{t^{2n+1}}{(2n+1)!}.$$

Note that $i^{2k} = (-1)^k$. These series therefore imply Euler's formula:

$$e^{it} = \sum_{n=0}^{\infty} \frac{(it)^n}{n!} = \sum_{n \text{ odd}}^{\infty} \frac{(it)^n}{n!} + \sum_{n \text{ even}}^{\infty} \frac{(it)^n}{n!} = i \sum_{k=0}^{\infty} (-1)^k \frac{z^{2k+1}}{(2k+1)!} + \sum_{k=0}^{\infty} (-1)^k \frac{z^{2k}}{(2k)!} = \cos(t) + i \sin(t).$$

We remind the reader, however, that precise definitions of cosine and sine are required to derive these series.

The map $t \to e^{it} = z(t)$ parametrizes the unit circle. This discussion suggests that trig and logs are intimately linked. (Of course twigs are just small logs.) Let us now approach logs as is done in calculus.

We recall the definition of the natural logarithm on $\mathbb{R}$.

Definition 5.1. For $x > 0$ we define $\ln(x)$ by

$$\ln(x) = \int_1^x \frac{1}{t} dt.$$

By changing variables (Exercise 5.1) one easily proves that $\ln(xy) = \ln(x) + \ln(y)$. See Figure 7. It follows that $\ln(x)$ tends to infinity as x does. Furthermore, by the fundamental theorem of calculus, $\ln$ is differentiable and its derivative at x is $\frac{1}{x}$. Hence $\ln$ is strictly increasing, and therefore has an inverse function exp. We write $\exp(y) = e^y$ as usual. Then $\exp(\ln(x)) = e^{\ln(x)} = x$ for $x > 0$.

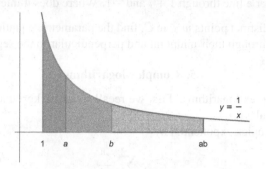

FIGURE 7. $\ln(ab) = \ln(a) + \ln(b)$

We wish to define log for complex numbers, but things become quite subtle. Assume $z = |z|e^{i\theta} \neq 0$. It is natural to presume that

$$\log(z) = \log(|z|e^{i\theta}) = \log(|z|) + i\theta.$$

One immediate problem is that the phase θ is determined only up to multiples of 2π.

Definition 5.2. For $z \neq 0$, we put $\log(z) = \ln(|z|) + i\theta$, where θ is any angle associated with z. Choose one value θ_0 of the phase. We interpret log as a "multi-valued function" whose values are parametrized by the integers:

$$\log(z) = \ln|z| + i\theta_0 + i2n\pi. \tag{12}$$

EXAMPLE 5.1. We compute some logarithms:

$$\log(-e) = \log(e) + i\theta = 1 + i(2n+1)\pi, \ \ n \in \mathbb{Z}.$$

For $x > 0$ we have

$$\log(x) = \ln(x) + 2n\pi i, \ \ n \in \mathbb{Z}.$$

Also, for $y > 0$

$$\log(iy) = \ln(|y|) + i(\frac{\pi}{2} + 2n\pi), \ \ n \in \mathbb{Z}.$$

REMARK. We warn the reader that the identity

$$\log(zw) \neq \log(z) + \log(w)$$

is false in general, even if we allow all possible values on both sides of the equation. We must therefore be careful when using the complex logarithm.

EXAMPLE 5.2. True or False? $\log(z^2) = 2\log(z)$. The answer is false, even if we consider all values. For example, put $z = i$. Then

$$\log(z^2) = \log(-1) = \log(1) + i(2n+1)\pi = i(2n+1)\pi$$
$$\log(z) = \log(i) = \log(|i|) + i(\frac{1}{2} + 2n)\pi = i(\frac{1}{2} + 2n)\pi$$

$$2 \log(z) = 2 \log(i) = i(4n+1)\pi.$$

Thus $\log(z^2)$ includes values not included in $2 \log(z)$.

Definition 5.3. For $z \neq 0$ and $\alpha \in \mathbb{C}$ we define

$$z^{\alpha} = e^{\alpha \log(z)}.$$

For most z and α there are infinitely many values.

EXAMPLE 5.3. All values of i^i are real!

$$i^i = e^{i \log(i)} = e^{i\left(i\left(\frac{\pi}{2}+2n\pi\right)\right)} = e^{-\frac{\pi}{2}-2n\pi}, \ n \in \mathbb{Z}.$$

Logs also arise when we consider functions such as *arc-cosine*. Recall that $\cos(z) = \frac{e^{iz}+e^{-iz}}{2}$. Let $w = \cos(z)$. Informally we write $z = \cos^{-1}(w)$. We solve $w = \cos(z)$ for z to get

$$e^{iz} - 2w + e^{-iz} = 0$$
$$(e^{iz})^2 - 2we^{iz} + 1 = 0.$$

We obtain a quadratic equation in e^{iz}, and hence

$$e^{iz} = w \pm \sqrt{w^2 - 1}.$$

Therefore

$$z = \frac{1}{i} \log(w \pm \sqrt{w^2 - 1}) = \cos^{-1} w.$$

EXAMPLE 5.4 (Hyperbolic functions). In our discussion of even and odd functions we defined:

$$\cosh(z) = \frac{e^z + e^{-z}}{2}$$

$$\sinh(z) = \frac{e^z - e^{-z}}{2}.$$

These formulas yield the identity

$$\cosh^2(z) - \sinh^2(z) = 1.$$

Just as cosine and sine are sometimes called *circular* functions, cosh and sinh are called *hyperbolic* functions. The map $t \mapsto \cosh(t) + i \sinh(t)$ parametrizes one branch of the hyperbola defined by $\mathrm{Re}(z^2) = x^2 - y^2 = 1$. The identities

$$\cos(z) = \frac{e^{iz} + e^{-iz}}{2} = \cosh(iz)$$

$$\sin(z) = \frac{e^{iz} - e^{-iz}}{2i} = \sinh(iz)$$

suggest a deeper connection we will not discuss.

The multiple values of the logarithm can be annoying. Often we choose a single *branch* of the logarithm. Doing so amounts to defining the phase θ in a precise fashion. For example, suppose $z \neq 0$. Put

$$\log(z) = \ln|z| + i\theta,$$

where $-\pi < \theta < \pi$. This function is well-defined. Another branch would be given by

$$g(z) = \log(z) = \ln|z| + i\theta,$$

where $0 < \theta < 2\pi$. Within a given problem one must be careful to choose only one branch, or contradictions can result. We return to this matter in Chapter 4.

As the inverse of exponentials, logs are closely connected to trig. We have defined cosine and sine already. The other trig functions are defined in terms of cosine and sine as usual. For example, we play a bit with the tangent function.

Definition 5.4 (Tangent function). For $\cos z \neq 0$, we define the tangent function by

$$\tan(z) = \frac{\sin z}{\cos z}.$$

To study the inverse-tangent function, we let $w = \tan z$ and try to solve for z:

$$w = \frac{\sin z}{\cos z} = \frac{\frac{e^{iz} - e^{-iz}}{2i}}{\frac{e^{iz} + e^{-iz}}{2}} = \frac{1}{i}\left(\frac{e^{iz} - e^{-iz}}{e^{iz} + e^{-iz}}\right),$$

To solve for z we compute

$$iw(e^{iz} + e^{-iz}) = e^{iz} - e^{-iz}$$

$$iw(e^{2iz} + 1) = e^{2iz} - 1$$

$$(iw - 1)e^{2iz} = -1 - iw$$

$$e^{2iz} = \frac{1 + iw}{1 - iw}$$

$$\tan^{-1}(w) = z = \frac{1}{2i}\log\left(\frac{1 + iw}{1 - iw}\right) = \frac{1}{2i}\log\left(\frac{i - w}{i + w}\right). \tag{13}$$

We have reduced the inverse-tangent function to the determination of all logs of the map $w \mapsto \frac{i-w}{i+w}$.

REMARK. In calculus the indefinite integral of $\frac{dx}{x^2+1}$ is the inverse-tangent. Formula (13) reveals what happens if we use partial fractions. See Exercise 5.5.

The mapping $w \mapsto \frac{i-w}{i+w}$ is an example of a *linear fractional transformation*, the subject of the next section. These functions arise in many applications, including networks, fluid flow, and electrostatics. As we shall see, the image of a line under a linear fractional transformation is always either a line or a circle.

EXERCISE 5.1. For x, y real, show that

$$\int_1^{xy} \frac{1}{t}dt = \int_1^x \frac{1}{t}dt + \int_1^y \frac{1}{t}dt.$$

EXERCISE 5.2. Show that the natural log function is unbounded on $\mathbb{R}$.

EXERCISE 5.3. Find all logarithms of $1 + \sqrt{3}i$.

EXERCISE 5.4. Put $w = \sec(z)$. Solve for z as a (multi-valued) function of w.

EXERCISE 5.5. Find $\int \frac{dx}{1+x^2}$ using partial fractions.

EXERCISE 5.6. True or false? The set of values of i^{2i} is the same as the set of values of $(-1)^i$.

EXERCISE 5.7. Find a w such that $\cos^{-1}(w) = -i$. Find a z such that $\cos(z) = -i$.

EXERCISE 5.8. Find all values of $\left|(1 + i)^{1+i}\right|$.

6. Linear fractional transformations

We develop the basic properties of linear fractional transformations.

Definition 6.1. Let $a, b, c, d \in \mathbb{C}$ and assume $ad - bc \neq 0$. Put $f(z) = \frac{az+b}{cz+d}$. The function $z \to f(z)$ is called a **linear fractional transformation**.

Without loss of generality, the assumption $ad - bc \neq 0$ can be replaced with $ad - bc = 1$. The reason: we do not change the map if we multiply the numerator and denominator by the same non-zero constant.

EXAMPLE 6.1. $f(z) = \frac{i-z}{i+z}$ and $w = f(z)$. Then $f(0) = 1$ and $f(\infty) = -1$ and $f(1) = i$. We claim that f maps the real line in the z-plane to the unit circle in the w-plane, as in Figure 8. For z real, we must show that $\left|\frac{i-z}{i+z}\right| = 1$. Equivalently, we must show $|i - z|^2 = |i + z|^2$. This equality follows from

$$|\alpha \pm \beta|^2 = |\alpha|^2 \pm 2\mathrm{Re}(\alpha\overline{\beta}) + |\beta|^2. \tag{14}$$

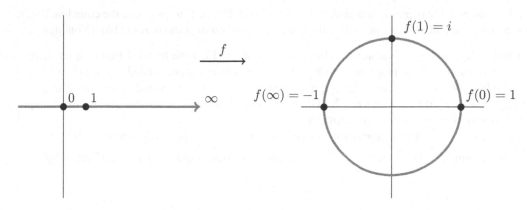

FIGURE 8. The real line gets mapped to the unit circle under $f(z) = \frac{i-z}{i+z}$.

The composition of linear fractional transformations is also a linear fractional transformation. We utilize this observation in the next section by reinterpreting linear fractional transformations as matrices.

REMARK. The identity

$$|\alpha + \beta|^2 + |\alpha - \beta|^2 = 2|\alpha|^2 + 2|\beta|^2, \tag{15}$$

obtained from adding the identities in (14), is called the *parallelogram law*.

When $c \neq 0$ in Definition 6.1, we notice that f is not defined when $z = -\frac{d}{c}$. It is natural in this case to say that $f(\frac{-d}{c}) = \infty$. This simple idea leads to the *Riemann sphere*; we add a point to the complex plane, called the **point at infinity**. We discuss this construction in Section 8. First we analyze the geometric properties of linear fractional transformations.

Let $\mathcal{S} = \{$lines and circles in $\mathbb{C}\}$ and $\mathcal{L} = \{$ linear fractional transformations $\}$.

Theorem 6.2. *Each $f \in \mathcal{L}$ maps $\mathcal{S}$ to itself.*

Before providing the proof, we make a useful observation about the collection $\mathcal{S}$.

Lemma 6.3. *For $A, C \in \mathbb{R}$ and $B \in \mathbb{C}$, we put*

$$h(z, \overline{z}) = A|z|^2 + Bz + \overline{B}\,\overline{z} + C. \tag{16}$$

Assume not all of A, B, C are zero. Then the zero-set of h is either empty, a single point, or an element of $\mathcal{S}$. Conversely, each element of $\mathcal{S}$ is the zero set of such a function h.

PROOF. First suppose $A = 0$. Then $h(z, \overline{z}) = 0$ if and only if $2\mathrm{Re}(Bz) = -C$. If $B = 0$, then the zero-set is empty, because $C \neq 0$. If $B \neq 0$, this zero-set is a line. Next suppose $A \neq 0$. After dividing by A and rewriting, we see that the zero-set of h is given by the equation

$$\left|z + \frac{\overline{B}}{A}\right|^2 = \frac{|B|^2}{A^2} - \frac{C}{A}. \tag{17}$$

The equation (17) defines a single point if its right-hand side is 0 and has no solutions if its right-hand side is negative. Otherwise (17) defines a circle.

Conversely, the circle given by $|z - p|^2 = r^2$ is the zero-set of some h as in (16); put $A = 1$, put $B = -\overline{p}$, and put $C = |p|^2 - r^2$. Consider a line. We may assume that it is given by the parametric equation $z = p + tv$ for $t \in \mathbb{R}$, and $v \neq 0$. Choose B such that Bv is purely imaginary. (See Exercise 6.5.) Then

$$Bz + \overline{B}\,\overline{z} = B(p + tv) + \overline{B}\,\overline{(p + tv)} = Bp + \overline{B}\,\overline{p}.$$

Since t is real and Bv is purely imaginary, $2\mathrm{Re}(tBv)$ is 0. Hence t drops out of the equation, the right-hand side is a constant, and we have described the line as the zero set of some h as in (16). (Note that $A = 0$.) □

PROOF. The proof of Theorem 6.2 has two main steps. First we factor f into simpler maps, and then we show that each of these maps preserves S. Three kinds of maps are needed: the translations $z \to z + \alpha$, the multiplications $z \to \beta z$, and the reciprocal map $z \to \frac{1}{z}$. The factorization into these maps T, M, R is precisely analogous to factoring a two-by-two matrix into elementary row-matrices. Rather than proving the factorization in this way, we prove it directly.

Let $f(z) = \frac{az+b}{cz+d}$. The factorization does not require the reciprocal map when $c = 0$.

- Assume $c = 0$. Then $f(z) = \frac{a}{d}z + \frac{b}{d}$. Thus f is the composition $f = TM$ given by

$$z \mapsto \frac{a}{d}z \mapsto \frac{a}{d}z + \frac{b}{d}.$$

- Assume $c \neq 0$. Then

$$z \mapsto cz \mapsto cz + d \mapsto \frac{1}{cz+d} \mapsto \frac{1}{cz+d} + \alpha = \frac{\alpha(cz+d)+1}{cz+d}$$

$$\mapsto \beta\frac{\alpha(cz+d)+1}{cz+d} = \frac{\beta\alpha cz + \beta\alpha d + \beta}{cz+d} = \frac{az+b}{cz+d}.$$

To make the last step valid, we must choose α and β such that both equations hold:

$$\beta\alpha c = a$$

$$\beta\alpha d + \beta = b.$$

Without loss of generality, we may assume that $ad - bc = 1$. Since $c \neq 0$, we solve both equations by putting $\alpha = -a$ and $\beta = \frac{-1}{c}$. We conclude that $f = MTRTM$ in this case.

We have completed the first main step. Before completing the second step, we comment on notation. If either of the translations is the identity map $z \to z + 0$, or either of the multiplications is the identity $z \to 1z$, we do not need to include them in the notation $MTRTM$. For example, if $f(z) = \frac{2}{z}$, we would write $f = MR$.

It is elementary to see that each translation or multiplication preserves S. To finish the proof, we must show that the reciprocal map does as well. Thus we need to show that the map $z \to \frac{1}{z}$ sends S to S. Our proof handles lines and circles at the same time. The converse part of the proposition tells that an object in S is given by the solution set of some equation

$$0 = A|z|^2 + 2\mathrm{Re}(Bz) + C.$$

Divide both sides by $|z|^2$ to get

$$0 = A + 2\mathrm{Re}\left(\frac{\overline{B}}{z}\right) + \frac{C}{|z|^2}.$$

Put $w = \frac{1}{z}$ to get

$$0 = C|w|^2 + 2\mathrm{Re}(\overline{B}w) + A.$$

We get an equation of the same form for w; by the proposition, the zero-set is either empty, a single point, or an element in $\mathcal{S}$. The first two possibilities are obviously impossible. $\qquad\square$

REMARK. One can prove Theorem 6.2 without factoring the linear fractional transformation, by plugging its formula into the equation $h = 0$ from Lemma 6.3. See Exercise 6.4. We use the factorization method for several reasons. It is geometrically appealing. It reinforces some ideas from linear algebra. Furthermore factorization into translations, rotations, and the reciprocal map has many uses in applied mathematics. See the next section for a glimpse at these ideas.

We next carefully study what happens to the circle $\{z : |z - z_0| = r\}$. See Figures 9 and 10. This circle passes through 0 if and only if $|z_0| = r$. We first assume that the circle does not pass through 0, and hence we can divide by $|z_0|^2 - r^2$. We will solve the equation $|\frac{1}{z} - w_0| = R$ for appropriate w_0 and R. After squaring and simplifying, this equation becomes

$$2\mathrm{Re}(zw_0) + (R^2 - |w_0|^2)|z|^2 = 1.$$

Assuming $R^2 - |w_0|^2 \neq 0$, we divide to get

$$\frac{2\mathrm{Re}(zw_0)}{R^2 - |w_0|^2} + |z|^2 = \frac{1}{R^2 - |w_0|^2}.$$

Compare with the equation $|z|^2 - 2\mathrm{Re}(z\overline{z_0}) + |z_0|^2 = r^2$ obtained by squaring $|z - z_0| = r$. To make the two equations identical, we require two conditions:

$$r^2 - |z_0|^2 = \frac{1}{R^2 - |w_0|^2} \tag{18.1}$$

$$\frac{w_0}{R^2 - |w_0|^2} = -\overline{z}_0. \tag{18.2}$$

Using (18.1) in (18.2) determines w_0:

$$w_0 = -\overline{z}_0(R^2 - |w_0|^2) = \frac{-\overline{z}_0}{(r^2 - |z_0|^2)}.$$

Once w_0 is known, (18.1) determines R. Thus, if the domain circle does not go through 0, its image under the reciprocal map is a circle of radius R and center w_0.

If the domain circle does go through 0, then we must have $0 = |z|^2 - 2\mathrm{Re}(z\overline{z}_0)$, which can be rewritten for $z \neq 0$ as $2\mathrm{Re}(\frac{z_0}{z}) = 1$. Put $w = \frac{1}{z}$; we obtain a linear equation in w and the image is a line.

The reciprocal map sends the unit circle to itself, while sending the interior of the unit circle to its exterior. Consider a circle whose image under the reciprocal map is also a circle. Does the interior go to the interior or the exterior? Since the map $z \mapsto \frac{1}{z}$ exchanges 0 and ∞, the answer depends on whether 0 lies in the interior of the domain circle. If 0 is in the exterior of the domain circle, then the function $z \mapsto \frac{1}{z}$ is continuous on the closed disk and hence bounded there. In this case, the image of the open disk in the domain must be the interior in the image space. If 0 is in the interior of the domain disk, then the image must be unbounded and hence must be the exterior.

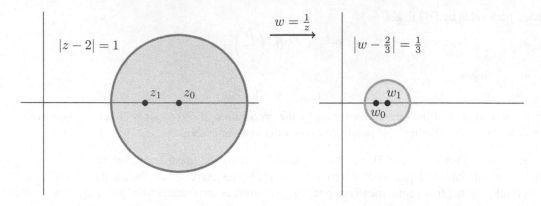

FIGURE 9. Interior gets mapped to interior

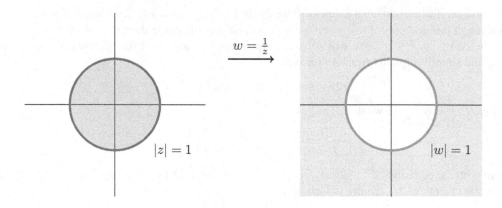

FIGURE 10. Interior gets mapped to exterior

REMARK. When working on the Riemann sphere, one can regard lines in $\mathbb{C}$ as circles on the Riemann sphere that pass through the point at infinity. This idea can be useful in elementary Euclidean geometry, as Example 6.2 shows.

EXAMPLE 6.2 (Inversion in a circle). Let S denote the unit circle. For each point a in the unit disk, we may consider the point $\frac{1}{\overline{a}}$. This point is called the *inversion* or *reflection* of a across the circle. By definition, the inversion of 0 is infinity. Many geometric properties about circle inversion follow from properties of linear fractional transformations. We mention one of these properties here. Let ω be an arbitrary point on S. Assume that a circle C containing both a and ω is orthogonal to S at the points of intersection. Then the reflected point $\frac{1}{\overline{a}}$ also lies on C. We outline a simple proof and suggest a computational proof in the exercises.

EXERCISE 6.1. Why is (15) called the parallelogram law? Prove it using geometry.

EXERCISE 6.2. Given a with $|a| < 1$, show that the linear fractional transformation

$$z \mapsto \frac{a - z}{1 - \overline{a}z}$$

maps the unit circle to itself and maps a to the origin. Show that the image of the circle C from Example 6.2 must be a line. (You must use orthogonality.) Since the reflected point $\frac{1}{a}$ goes to infinity, it must be on this line. Explain why the general result follows.

EXERCISE 6.3. Prove the result about inversion (Example 6.2) by filling in the following steps. First show, without loss of generality, that we may take $\omega = 1$. Assume that C is given by $|z - p| = r$. Hence we have both $|1 - p| = r$ and $|a - p| = r$. We also know that C is orthogonal to the unit circle at the point 1. Show that this property means that $p = 1 + it$ for some real t. Show that $|t| = r$. The conclusion amounts to showing that

$$|\frac{1}{a} - p|^2 = r^2.$$

Verify this conclusion by writing $a = x + iy$ and computing the needed squared distance.

EXERCISE 6.4. Let h be as in Lemma 6.3. Let $w = f(z)$ define a linear fractional transformation. Show by direct computation that w satisfies a quadratic equation of the same form.

EXERCISE 6.5. In the proof of Lemma 6.3, we were given $v \neq 0$ and we chose B to make Bv purely imaginary. Check that we can always do so. (There are several possible immediate proofs.)

EXERCISE 6.6. What happens if you compose the map in Exercise 6.2 with itself?

7. Linear fractional transformations and matrices

Consider first the map $z \to \frac{az+b}{cz+d} = f(z)$, where a, b, c, d are complex constants. We assume that at least one of c and d is not zero. If both a and b are 0, then f is the constant map 0. More generally, if $\frac{a}{c} = \frac{b}{d} = \lambda$, then $f(z) = \lambda$ for all z, and again f is a constant map. The condition $\frac{a}{c} = \frac{b}{d}$ is the same as $ad - bc = 0$. We do not wish to regard constant maps as linear fractional transformations. Hence we assume $\Delta = ad - bc \neq 0$. In fact, we may assume that $\Delta = 1$, as follows. We divide each of the constants a, b, c, d by $\sqrt{\Delta}$, obtaining new constants a', b', c', d'. We have $\frac{a'z+b'}{c'z+d'} = \frac{az+b}{cz+d}$ for all z, but now $a'd' - b'c' = 1$.

We therefore define a linear fractional transformation to be a map

$$z \to f(z) = \frac{az + b}{cz + d},$$

where $ad - bc = 1$. We notice that replacing each of the four constants by its negative yields the same map. We can identify the map with the matrix F, defined by

$$F = \begin{pmatrix} a & b \\ c & d \end{pmatrix},$$

and having determinant 1. Again, both F and $-F$ yield the same map. The analogy between maps and matrices is deep and worth developing.

Given $f(z) = \frac{az+b}{cz+d}$, and $g(z) = \frac{Az+B}{Cz+D}$, we have

$$(g \circ f)(z) = g(f(z)) = \frac{A\frac{az+b}{cz+d} + B}{C\frac{az+b}{cz+d} + D} = \frac{(Aa + Bc)z + (Ab + Bd)}{(Ca + Dc)z + (Cb + Dd)}.$$

Notice that

$$\begin{pmatrix} a & b \\ c & d \end{pmatrix} \begin{pmatrix} A & B \\ C & D \end{pmatrix} = \begin{pmatrix} Aa + Bc & Ab + Bd \\ Ca + Dc & Cb + Dd \end{pmatrix}.$$

Thus composition of linear fractional transformations is equivalent to multiplying matrices. In algebraic language, the group of linear fractional transformations under composition is isomorphic to the group of two-by-two complex matrices M with determinant 1, where M and $-M$ are identified. The identity function

$f(z) = z$ corresponds to the identity matrix I. The inverse of a linear fractional transformation corresponds to the inverse matrix. This correspondence is useful in engineering and in optics, as we now indicate.

In electrical engineering it is common to introduce ABCD matrices for two-port networks. Let V_1 denote the voltage drop across the first port, and let I_1 denote the current into the first port. Similarly, define V_2 and I_2. One analyzes complicated networks by multiplying matrices!

The conventions are a bit strange, but these matrices are defined by

$$\begin{pmatrix} V_1 \\ I_1 \end{pmatrix} = \begin{pmatrix} A & B \\ C & D \end{pmatrix} \begin{pmatrix} V_2 \\ -I_2 \end{pmatrix}. \tag{19}$$

The minus sign in (19) can be omitted if we regard the current I_2 as directed out of the port. We give several examples of ABCD matrices.

The ABCD matrix for a **resistor**:

$$\begin{pmatrix} 1 & R \\ 0 & 1 \end{pmatrix}.$$

The ABCD matrix for a **shunt resistor**:

$$\begin{pmatrix} 1 & 0 \\ \frac{1}{R} & 1 \end{pmatrix}.$$

The ABCD matrix for a **series conductor**:

$$\begin{pmatrix} 1 & \frac{1}{G} \\ 0 & 1 \end{pmatrix}.$$

The ABCD matrix for a **transmission Line**:

$$\begin{pmatrix} \cosh(\gamma l) & -Z_0 \sinh(\gamma l) \\ -\frac{1}{Z_0} \sinh(\gamma l) & \cosh(\gamma l) \end{pmatrix}.$$

Here Z_0 is the *characteristic impedance*, γ is the *propagation constant*, and l is the *length* of the transmission line.

When cascading networks, one simply multiplies their ABCD matrices. Note that all these matrices have determinant 1. Hence any product of them also does. We can identify a two-by-two matrix of determinant 1 with a linear fractional transformation, and the matrix product then corresponds to composition. Hence linear fractional transformations provide insight into transmission lines.

A similar approach applies in the design of certain optical systems. Consider parallel planes of reference, called the input and output planes. The optical axis is perpendicular to these planes. A light ray hits the input plane. The variables are a pair (x_1, θ_1). Here x_1 is the distance between the optical axis and where the light ray hits the input plane, and θ_1 is the angle between the light ray and the optical axis. The **ray transfer matrix** M is the two-by-two matrix such that

$$\mathbf{M} \begin{pmatrix} x_1 \\ \theta_1 \end{pmatrix} = \begin{pmatrix} x_2 \\ \theta_2 \end{pmatrix}.$$

Here (x_2, θ_2) are the variables when the ray hits the output plane. When the planes are located within the same medium, the determinant of M equals 1. More generally, indices of refraction arise. As with transmission lines, one breaks down a complicated optical system into a product of ray transfer matrices.

EXAMPLE 7.1. The ray transfer matrix for free space is given by

$$S = \begin{pmatrix} 1 & d \\ 0 & 1 \end{pmatrix}.$$

Here d is the distance between the input and output planes; the approximation $\tan(\theta_1) = \theta_1$ is tacitly assumed here.

The ray transfer matrix of a thin lens is given by

$$L = \begin{pmatrix} 1 & 0 \\ \frac{-1}{f} & 1 \end{pmatrix}.$$

Here f is the focal length of the lens; it is assumed that f is much greater than the thickness of the lens.

EXERCISE 7.1. For S, L as in Example 7.1, compute SL and LS. They are not equal. Can you interpret this result? Compute the eigenvalues of LS. If we assume that the system is *stable*, then these eigenvalues both lie on the unit circle. Can you interpret this result?

EXERCISE 7.2. Suppose a mirror has curvature R. Show that the ray transfer matrix for reflection is

$$\begin{pmatrix} 1 & 0 \\ \frac{-2}{R} & 1 \end{pmatrix}.$$

EXERCISE 7.3. If you know what it means for a network to be **lossless**, show that the diagonal entries of the ABCD matrix of a lossless network are real and the off diagonal entries are purely imaginary.

EXERCISE 7.4. Diagonalize the matrix A, given by

$$A = \begin{pmatrix} \cosh(z) & \sinh(z) \\ \sinh(z) & \cosh(z) \end{pmatrix}.$$

Find a simple formula for A^n.

8. The Riemann sphere

We have seen that it is sometimes useful to add a point at infinity to the complex number system $\mathbb{C}$. We consider this idea in this section.

First we discuss limits. Let f be a function. We can consider the limit of $f(z)$ as z tends to infinity (limit **at** infinity) or we can consider the limit of $f(z)$ to be infinity as z tends to some w (limit **of** infinity).

Definition 8.1. Limits **of** infinity. Let $\{z_n\}$ be a sequence in $\mathbb{C}$. We say that $\lim_{n\to\infty} z_n = \infty$ if $\lim_{n\to\infty}(\frac{1}{z_n}) = 0$. Similarly, when f is a function, we say that $\lim_{z\to L} f(z) = \infty$ if $\lim_{z\to L} \frac{1}{f(z)} = 0$.

Limits **at** infinity. Let f be a function, defined for $|z| > R$ for some R. We say that $\lim_{z\to\infty} f(z) = L$, if $\lim_{w\to 0} f(\frac{1}{w}) = L$.

Sometimes it is easier to use definitions of limits involving infinity that are analogous to the usual $\epsilon - \delta$ definition. For limits of infinity, we have

$$\lim_{z\to w} f(z) = \infty$$

if, for all $R > 0$, there is a $\delta > 0$ such that $|z - w| < \delta$ implies $|f(z)| > R$. Thus, by making z close enough to w, we can make $|f(z)|$ as large as we wish.

For limits at infinity, we have

$$\lim_{z\to\infty} f(z) = L$$

if, for all $\epsilon > 0$, there is an R such that $|z| > R$ implies $|f(z) - L| < \epsilon$. Thus, by making $|z|$ sufficiently large, we can make $f(z)$ as close as we wish to L.

One can combine both types of limits.

EXAMPLE 8.1. Here are some examples of limits involving ∞.

- $\lim_{z\to i} \frac{z+i}{z-i} = \infty$.

- For n a negative integer, $\lim_{z \to \infty} z^n = 0$.
- $\lim_{z \to \infty} e^z$ does not exist. We have $e^z = e^x e^{iy}$. When $y = 0$ and $|x|$ is large, $|e^z|$ is large. When $x = 0$ and $|y|$ is large, $|e^z| = 1$.
- When p is a non-constant polynomial, $\lim_{z \to \infty} p(z) = \infty$.
- Let $\frac{p}{q}$ be a rational function. To find $\lim_{z \to \infty} \frac{p(z)}{q(z)}$, we proceed as in calculus. If the degree of p is less than the degree of q, the limit is 0. If the degree of p exceeds the degree of q, the limit is ∞. If p and q have the same degree d, then the limit is the ratio of the coefficients of degree d.

These examples of limits suggest that all points z for which $|z|$ is large are somehow close. To realize this idea geometrically we introduce the Riemann sphere.

Let S denote the unit sphere in $\mathbb{R}^3$. Thus $S = \{(x_1, x_2, x_3) : x_1^2 + x_2^2 + x_3^2 = 1\}$. We call the point $(0, 0, 1)$ the *north pole* and the point $(0, 0, -1)$ the *south pole*. Given a point $p \in S$, other than the north pole, we consider the line through the north pole and p. The line intersects the plane defined by $x_3 = 0$ in a unique point, called the *stereographic projection* of p onto the plane. If we regard this plane as $\mathbb{C}$, then stereographic projection is a one-to-one correspondence between S minus the north pole and $\mathbb{C}$. The north pole corresponds to ∞. In this manner we think of S as $\mathbb{C} \cup \infty$.

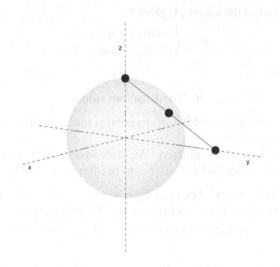

FIGURE 11. Stereographic projection

Algebraic geometry offers another way to think of this situation. We have two copies of $\mathbb{C}$. One, called U, is in one-to-one correspondence with S minus the north pole. The other, called V, is in one-to-one correspondence with S minus the south pole. We pass from U to V by taking reciprocals.

A linear fractional transformation $\frac{az+b}{cz+d}$ can be regarded as a map from S to itself, with $f(\infty) = \frac{a}{c}$ and $f(\frac{-d}{c}) = \infty$. In this way of thinking, all points on S are geometrically the same. Furthermore, each line in $\mathbb{C}$ is simply a circle on S that passes through the point at infinity. We gain new insight into Theorem 6.2.

We close this section by giving explicit formulas for passing between $\mathbb{C}$ and the Riemann sphere. Let (A, B, C) be a point on the unit sphere in $\mathbb{R}^3$, other than the north pole $(0, 0, 1)$. Put $p(A, B, C) = \frac{A+iB}{1-C}$, the stereographic projection of (A, B, C) onto $\mathbb{C}$. Thus $p(A, B, C)$ is the point where the line through the north pole and (A, B, C) intersects the plane in $\mathbb{R}^3$ defined by $x_3 = 0$. We put $p(0, 0, 1) = \infty$. Suppose

$z = x + iy \in \mathbb{C}$. We define $q : \mathbb{C} \to S$ by

$$q(z) = \left(\frac{2x}{|z|^2 + 1}, \frac{2y}{|z|^2 + 1}, \frac{|z|^2 - 1}{|z|^2 + 1} \right). \tag{20}$$

Proposition 8.2. *With p stereographic projection, and q as in (20), $p \circ q$ and $q \circ p$ are the identity maps.*

PROOF. First we note that a simple computation gives

$$\left(\frac{2x}{|z|^2 + 1} \right)^2 + \left(\frac{2y}{|z|^2 + 1} \right)^2 + \left(\frac{|z|^2 - 1}{|z|^2 + 1} \right)^2 = 1.$$

Thus $q(z)$ is on S, and hence $p(q(z))$ is defined. Next note that

$$1 - \frac{|z|^2 - 1}{|z|^2 + 1} = \frac{2}{|z|^2 + 1},$$

the $1 - C$ term in the definition of p. Then we compute $p \circ q$ to obtain

$$p(q(z)) = x + iy = z.$$

To verify the other direction we first note when (A, B, C) is on S that

$$A^2 + B^2 + (1 - C)^2 = 2 - 2C.$$

Using this formula, and multiplying through by $(1 - C)^2$ at the right place, we find that

$$q(p(A, B, C)) = q(\frac{A + iB}{1 - C}) = \frac{\left(2A(1 - C), 2B(1 - C), A^2 + B^2 - (1 - C)^2 \right)}{A^2 + B^2 + (1 - C)^2}$$

$$= \frac{1}{2 - 2C} \left(2A(1 - C), 2B(1 - B), C(2 - 2C) \right) = (A, B, C).$$

$\square$

EXERCISE 8.1. Verify the claims in Example 8.1.

EXERCISE 8.2. True or false? $\lim_{z \to \infty} e^{z^2} = \infty$.

EXERCISE 8.3. Let R be positive and s complex. Assume $\operatorname{Re}(s) > 0$. Prove that

$$\lim_{R \to \infty} e^{-Rs} = 0.$$

What happens if $\operatorname{Re}(s) = 0$? What if $\operatorname{Re}(s) < 0$? These facts are useful in understanding the Laplace transform.

EXERCISE 8.4. Determine $\lim_{n \to \infty} (1 + \frac{z}{n})^n$.

CHAPTER 3

Vector analysis

In this chapter we review some ideas from multi-variable calculus. We particularly emphasize line and surface integrals and their interpretations as work and flux. We develop the ideas in terms of both vector fields and differential forms, including the major theorems of vector analysis. We discuss Maxwell's equations from both points of view and our use of differential forms goes well beyond a typical multi-variable calculus course.

1. Euclidean geometry

The reader is surely familiar with the notions of Euclidean dot product (or inner product) and length in $\mathbb{R}^n$. In Chapter 6 we discuss inner products in the setting of Hilbert spaces. Here we recall basic facts about Euclidean geometry.

For $p = (p_1, \ldots, p_n)$ and $q = (q_1, \ldots, q_n)$ we have

$$p \cdot q = \sum_{j=1}^{n} p_j q_j.$$

$$||p||^2 = p \cdot p = \sum_{j=1}^{n} p_j^2.$$

The scalar $p \cdot q$ is called the *dot product* or *inner product* of p and q. The non-negative number $||p||$ is called the (Euclidean) *length*, *magnitude*, or *norm* of p. The formula

$$p \cdot q = ||p|| \, ||q|| \, \cos(\theta) \tag{1}$$

relates the dot product to the angle between two vectors. The Cauchy–Schwarz inequality, stated and proved (in more generality) in Theorem 1.3 of Chapter 6, states that

$$|p \cdot q| \leq ||p|| \, ||q||$$

when p and q are vectors in an arbitrary inner product space. One can then define angles using (1).

In three dimensions we have the notion of cross product of two vectors:

$$p \times q = \det \begin{pmatrix} \mathbf{i} & \mathbf{j} & \mathbf{k} \\ p_1 & p_2 & p_3 \\ q_1 & q_2 & q_3 \end{pmatrix} = (p_2 q_3 - p_3 q_2, p_3 q_1 - p_1 q_3, p_1 q_2 - p_2 q_1). \tag{2}$$

Here $\mathbf{i} = (1, 0, 0)$, $\mathbf{j} = (0, 1, 0)$, and $\mathbf{k} = (0, 0, 1)$ form the standard orthonormal basis of $\mathbb{R}^3$. Note in particular that $p \times q = 0$ whenever p is a scalar multiple of q, because of the alternating property of the determinant.

The cross product of two vectors in $\mathbb{R}^3$ is a vector in $\mathbb{R}^3$. Its direction is perpendicular to both p and q (determined by the right-hand rule) and its length is given by

$$||p \times q|| = ||p|| \, ||q|| \, \sin(\theta).$$

The dot product is symmetric in its two inputs: $p \cdot q = q \cdot p$. The cross product is skew-symmetric: $p \times q = -(q \times p)$. As a consequence, $p \times p = 0$ for each p. By contrast, unless $p = 0$, we have $p \cdot p > 0$. The dot product and cross product are related by the formula

$$||p \times q||^2 + |p \cdot q|^2 = ||p||^2 \, ||q||^2.$$

The **Pythagorean theorem** holds: $||v + w||^2 = ||v||^2 + ||w||^2$ if and only if $v \cdot w = 0$; in other words, if and only if v is perpendicular to w. Mathematicians generally say that v and w are **orthogonal** when $v \cdot w = 0$. The Euclidean distance between v and w is defined by $||v - w||$. The Cauchy–Schwarz inequality implies the **triangle inequality** $||v + w|| \leq ||v|| + ||w||$.

When vectors depend on time, we can use the product rule for differentiation to compute derivatives of dot products and cross products:

$$\frac{d}{dt}\big(p(t) \cdot q(t)\big) = p'(t) \cdot q(t) + p(t) \cdot q'(t).$$

$$\frac{d}{dt}\big(p(t) \times q(t)\big) = p'(t) \times q(t) + p(t) \times q'(t).$$

Both the dot product and the cross product arise throughout physics. The formulas for derivatives have several interesting corollaries. For example, suppose a particle moves in a sphere of constant radius r. Let $\gamma(t)$ denote its position at time t. Then its velocity $\gamma'(t)$ is perpendicular to its position $\gamma(t)$. The proof is easy: if $r^2 = ||\gamma(t)||^2 = \gamma(t) \cdot \gamma(t)$, then differentiating gives

$$0 = \gamma'(t) \cdot \gamma(t) + \gamma(t) \cdot \gamma'(t) = 2\gamma(t) \cdot \gamma'(t).$$

Conversely, suppose that $\gamma(t)$ is orthogonal to $\gamma'(t)$ for all t. Then $||\gamma(t)||^2$ is a constant, and the particle travels in a sphere.

We pause to illustrate uses of the dot and cross products in physics. These ideas become more highly developed later in the chapter.

In this paragraph we denote vector quantities in boldface for consistency with physics notation. A constant force $\mathbf{F}$ applied to a particle causes the particle to be moved through a displacement vector $\mathbf{d}$. Both $\mathbf{F}$ and $\mathbf{d}$ are vectors. Their dot product $\mathbf{F} \cdot \mathbf{d}$ defines the *work* done. For an example of the cross product, consider a moving particle whose position in $\mathbb{R}^3$ at time t is denoted by $\mathbf{r}(t)$. Its linear momentum is denoted by $\mathbf{p}$. The *angular momentum* $\mathbf{L}$ is defined by the cross product $\mathbf{r} \times \mathbf{p}$. By Newton's second law, the force $\mathbf{F}$ on the particle satisfies

$$\mathbf{F} = \frac{d\mathbf{p}}{dt}. \tag{3}$$

Taking the cross product of both sides of (3) with the position vector $\mathbf{r}$ gives the torque τ:

$$\tau = \mathbf{r} \times \mathbf{F} = \mathbf{r} \times \frac{d\mathbf{p}}{dt}.$$

Differentiating the angular momentum gives

$$\frac{d\mathbf{L}}{dt} = \mathbf{v} \times \mathbf{p} + \mathbf{r} \times \frac{d\mathbf{p}}{dt} = \mathbf{r} \times \frac{d\mathbf{p}}{dt} = \tau.$$

The term $\mathbf{v} \times \mathbf{p}$ is 0, because the momentum $\mathbf{p}$ is a multiple (by the mass) of the velocity $\mathbf{v}$. We conclude that $\tau = \frac{d\mathbf{L}}{dt}$, the rotational analogue of Newton's second law.

We return to the mathematical development. A subset of $\mathbb{R}^n$ is **open** if, for each $p \in \Omega$, there is an $\epsilon > 0$ such that $q \in \Omega$ whenever $||p - q|| < \epsilon$. In other words, for each p there is a ball about p contained in Ω. A subset of $\mathbb{R}$ is **closed** if its complement is open. A subset $K \subseteq \mathbb{R}^n$ is **bounded** if there is a positive number C such that $||p|| \leq C$ for all $p \in K$.

We can characterize closed sets using limits of sequences.

Definition 1.1. Let $\{p_m\}$ be a sequence in $\mathbb{R}^n$. Then $\{p_m\}$ converges to v if, for every $\epsilon > 0$, there is an integer M such that $m \geq M$ implies $||p_m - v|| < \epsilon$.

The following simple result (Exercise 1.6) characterizes closed sets in Euclidean space.

Proposition 1.2. *A subset K of $\mathbb{R}^n$ is closed if and only if, whenever $\{p_m\}$ is a sequence in K and $\{p_m\}$ converges to v, then $v \in K$.*

If $g : \mathbb{R}^n \to \mathbb{R}$ is a function and c is a scalar, then $\{x : g(x) = c\}$ is called a **level set** of g. Level sets help us understand the geometry of objects in dimensions too high to visualize. Level sets arise in almost every quantitative subject. An *equipotential surface* is a level set of the potential energy function. An *isotherm* is a level set of the temperature function. In economics, an *indifference curve* is a level set of the utility function. In topography, *contour lines* are level sets of the height function.

EXERCISE 1.1. Verify the formula $||p \times q||^2 = |p \cdot q|^2 + ||p||^2 \, ||q||^2$, by expanding both sides in coordinates. Verify the formula assuming that the cross product and dot product are *defined* in terms of the angle between p and q.

EXERCISE 1.2. Put $f(x,y) = (y - x^2)(y - 4x^2)$. Graph the level set $\{f = 0\}$. Put $+$ or $-$ on each region of the plane on which f is positive or negative. For a, b not both zero, consider the line $t \to (at, bt)$. Show that the restriction of f to each of these lines has a minimum of 0 at $t = 0$, but that f itself is negative somewhere in every neighborhood of $(0, 0)$. Explain!

EXERCISE 1.3. Let u, v, w be vectors in $\mathbb{R}^3$. Verify the following identities:

$$u \times (v \times w) = (u \cdot w) \, v - (u \cdot v) \, w.$$

$$u \times (v \times w) + w \times (u \times v) + v \times (w \times u) = 0.$$

$$u \times (v \times w) - (u \times v) \times w = (w \times u) \times v.$$

The third identity shows that we must be careful to put in parentheses when we take a triple cross product. Comment: Proving these identities requires little computation. Take advantage of linearity to prove the first identity. Then use the first to prove the second, and use the second to prove the third. The second identity is called the *Jacobi identity*; its generalization to arbitrary Lie algebras is of fundamental significance in both mathematics and physics.

EXERCISE 1.4. Let P be a plane in $\mathbb{R}^3$, containing q. Assume that u, v are linearly independent vectors, and $q + u$ and $q + w$ lie in P. Give the defining equation of the plane, in the form $\mathbf{n} \cdot (x - q) = 0$.

EXERCISE 1.5. For $x, \mathbf{n} \in \mathbb{R}^n$ and $\mathbf{n} \neq 0$, the equations $\mathbf{n} \cdot x = c_1$ and $\mathbf{n} \cdot x = c_2$ define parallel hyperplanes. What is the distance between them?

EXERCISE 1.6. Prove both implications in Proposition 1.2. Hint: Assuming the complement of K is open, the sequence property is easy to check. The other direction is a bit harder. Assuming the sequence property holds, we must prove that the complement is open. To do so, consider an arbitrary point q in the complement and balls of radius $\frac{1}{n}$ about q. If all these balls intersect K, then by choosing a point in each such ball we construct a sequence that violates the sequence property. Therefore, for n large enough, each such ball is in the complement, and hence the complement is open.

2. Differentiable functions

Let Ω be an open set in $\mathbb{R}^n$. Given $f : \Omega \to \mathbb{R}$, we wish to define the phrase "f is differentiable at p". The definition is similar to the one variable case, but some care is needed.

Definition 2.1. Suppose $f : \Omega \to \mathbb{R}$ and $p \in \Omega$. We say that f is **differentiable** at p if there is a vector $(\nabla f)(p)$ such that
$$f(p + h) = f(p) + (\nabla f)(p) \cdot h + \text{error},$$
where $\lim_{\|h\| \to 0} \frac{\text{error}}{\|h\|} = 0$.

In Definition 2.1, the error is a function of h. The function is differentiable when the error is much smaller than $\|h\|$. Below we give some examples.

Definition 2.2 (Directional derivative). Assume that Ω is an open subset of $\mathbb{R}^n$, that $f : \Omega \to \mathbb{R}$ is a function, that $p \in \Omega$, and finally that v is a non-zero vector. We define the directional derivative $\frac{\partial f}{\partial v}(p)$ by the limit:
$$\lim_{t \to 0} \frac{f(p + tv) - f(p)}{t}.$$

The idea behind the definition of the directional derivative is quite simple: the function $t \to p + tv$ defines a line in $\mathbb{R}^n$; we are restricting f to this line and taking the ordinary derivative of f along this line. This derivative measures the (instantaneous) rate of change of f at p if we move in the direction v. Partial derivatives are special cases, namely, directional derivatives in the coordinate directions. When all the first partial derivatives of a function are themselves continuous, the function is differentiable in the sense of Definition 2.1. In this case, it is easy to compute directional derivatives, as we describe next. See, however, Exercise 2.14 for what can happen without the hypothesis of continuous partials.

The vector $\nabla f(p)$ is called the gradient of f at p. When all the first-order partials exist at p, we must have the formula
$$\nabla f(p) = \left(\frac{\partial f}{\partial x_1}(p), \ldots, \frac{\partial f}{\partial x_n}(p) \right).$$

Unless these partials are continuous, however, this vector of partial derivatives might not satisfy Definition 2.1. The gradient, as defined in Definition 2.1, has many uses in calculus, several of which we now recall.

- The gradient vector $\nabla f(p)$ is perpendicular to the level set defined by the equation $f(x) = f(p)$, and it points in the direction in which f is increasing most rapidly.
- The expression $\nabla f(p) \cdot v$ gives the directional derivative of f in the direction v at p.
- The *tangent plane* at p to the level surface $\{f = f(p)\}$ in $\mathbb{R}^3$ is defined by the equation
$$(\mathbf{x} - p) \cdot \nabla f(p) = 0.$$

    In higher dimensions this equation defines the tangent *hyperplane*.
- Let f be differentiable on $\mathbb{R}^n$ and let $t \to \gamma(t)$ be a curve in $\mathbb{R}^n$. The **Chain Rule** holds:
$$\frac{d}{dt} f(\gamma(t)) = \nabla f(\gamma(t)) \cdot \gamma'(t).$$

- Points at which the gradient vector is 0 are called *critical points* of f.

REMARK. The term *temperature gradient* is common. It means, of course, the gradient vector of the temperature function. Wikipedia says "A temperature gradient is a physical quantity that describes in which direction and at what rate the temperature changes the most rapidly around a particular location." This sentence is essentially correct; one needs to interpret the phrase *physical quantity* as *vector field*.

The following result is proved in most advanced calculus books. Let Ω be an open ball in $\mathbb{R}^n$. Suppose $f : \Omega \to \mathbb{R}$ is differentiable and $\nabla f = 0$ on that ball. Then f is a constant.

Partial derivatives are directional derivatives in the coordinate directions. If f is differentiable at p, then all the partial derivatives $\frac{\partial f}{\partial x_j}$ exist at p and

$$\nabla f(p) = \left(\frac{\partial f}{\partial x_1}(p), \cdots, \frac{\partial f}{\partial x_n}(p) \right).$$

Conversely, even if all the partial derivatives exist at p, it need not be true that f is differentiable there. But, if all the partial derivatives of f are continuous on an open set, then f is differentiable there. See Exercise 2.14 for what can happen when partials fail to be continuous.

EXAMPLE 2.1. Define $f : \mathbb{R}^2 \to \mathbb{R}$ by $f(0,0) = 0$ and

$$f(x,y) = \frac{xy}{x^2 + y^2}$$

otherwise. Then both partials of f exist at $(0,0)$ but f is not even continuous there. See Exercise 2.1.

EXAMPLE 2.2. Define f by $f(x,y) = x^3 - y^3$. Put $p = (2,1)$. Then $f(p) = 7$. We compute an approximate value of $f(2.01, .99)$ using the gradient.

$$f(2.01, 0.99) \cong f(2,1) + \nabla f(2,1) \cdot (0.01, -0.01) = 7 + (12, -3) \cdot (.01, -.01) = 7.15.$$

The exact value is 7.1503.

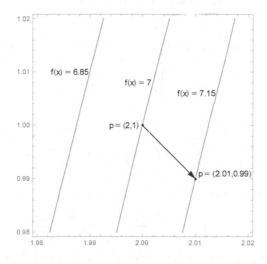

FIGURE 1. Use of the gradient

EXAMPLE 2.3. On the real line, put $f(x) = x^k$ for k a positive integer. Then $f'(x) = kx^{k-1}$. One way to see this basic fact is to use the binomial expansion:

$$f(x+h) = (x+h)^k = x^k + kx^{k-1}h + \frac{k(k-1)}{2}h^2 + \cdots = f(x) + kx^{k-1}h + \text{error}.$$

Definition 2.1 holds because the terms in the error are all divisible by h^2.

EXAMPLE 2.4. Many physics and engineering courses ask students to memorize some specific approximations. For example, $\sin(h)$ is approximately h when h is small. Also $\frac{1}{1-h}$ is approximately $1 + h$ and $\sqrt{1+h}$ is approximately $1 + \frac{h}{2}$ when h is small. These approximations are nothing more than restating the definition of the derivative! The error in the approximation is identical with the error function used in Definition 2.1.

Often in this book we assume that functions are **continuously differentiable**; all partial derivatives are continuous functions. Even on the real line, there exist differentiable functions that are not continuously differentiable.

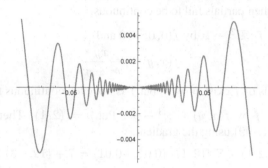

FIGURE 2. The graph of $x^2 \sin(\frac{1}{x})$

EXAMPLE 2.5. Here is a simple example of a differentiable function whose derivative is not continuous.

$$f(x) = \begin{cases} x^2 \sin(\frac{1}{x}) & \text{if } x \neq 0 \\ 0 & \text{if } x = 0 \end{cases}$$

For $x \neq 0$,

$$f'(x) = 2x \sin\left(\frac{1}{x}\right) + x^2 \cos\left(\frac{1}{x}\right)\left(-\frac{1}{x^2}\right) = 2x \sin\left(\frac{1}{x}\right) - \cos\left(\frac{1}{x}\right). \tag{4}$$

For $x = 0$, we use the limit quotient definition to see that $f'(0) = 0$:

$$f'(0) = \lim_{h \to 0} \frac{f(h) - f(0)}{h} = \lim_{h \to 0} \frac{h^2 \sin(\frac{1}{h})}{h} = 0.$$

In the language of Definition 2.1, the error $E(h)$ here is $h^2 \sin(\frac{1}{h})$, and $\frac{E(h)}{h}$ tends to zero. The limit in (4) of $f'(x)$ as x tends to 0 does not exist, but $f'(0) = 0$. In summary, f is differentiable at all points, and f' is not continuous at 0.

We will be observing many examples of **orthogonal trajectories** in $\mathbb{R}^2$. We will be given differentiable functions u and v with the property that the level sets of u are perpendicular to the level sets of v at each point p. It follows that $\nabla u(p) \cdot \nabla v(p) = 0$ at each point.

EXAMPLE 2.6. For $x > 0$, put

$$u(x,y) = \ln\left((x^2 + y^2)^{\frac{1}{2}}\right)$$

$$v(x,y) = \tan^{-1}\left(\frac{y}{x}\right).$$

Put $z = x + iy$. We note that $(u + iv)(x, y) = \ln(|z|) + i\theta = \log(z)$. The level sets of these functions form orthogonal trajectories; for $c \geq 0$, the level set $u = c$ is a circle centered at the origin. The level sets of v are lines through the origin. See Figure 3.

$$\ln((x^2 + y^2)^{\frac{1}{2}}) = c_1 \Rightarrow x^2 + y^2 = e^{2c_1}$$

$$\tan^{-1}(\frac{y}{x}) = c_2 \Rightarrow y = \tan(c_2)x.$$

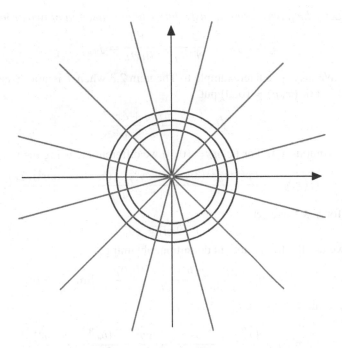

FIGURE 3. Level sets for real and imaginary parts of $\log(z)$.

EXAMPLE 2.7. For $z \in \mathbb{C}$, recall that $z^2 = (x + iy)^2 = x^2 - y^2 + i(2xy)$. Put

$$u(x,y) = x^2 - y^2$$

$$v(x,y) = 2xy.$$

Then $\nabla u = (2x, -2y)$ and $\nabla v = (2y, 2x)$. Thus $\nabla u \cdot \nabla v = 0$ at every point. The point $(x, y) = (0, 0)$ is special; both gradients equal the zero vector there, and the level sets are not perpendicular. The complex derivative $2z$ of z^2 is 0 at $z = 0$, and hence 0 is a **critical point**.

EXAMPLE 2.8. Consider z^3. We have $z^3 = (u + iv)(x, y)$, where

$$u(x,y) = x^3 - 3xy^2$$

$$v(x,y) = 3x^2y - y^3.$$

Here $\nabla u = (3x^2 - 3y^2, -6xy)$ and $\nabla v = (6xy, 3x^2 - 3y^2)$. Thus $\nabla u \cdot \nabla v = 0$ at every point. Again, away from the origin, we obtain orthogonal trajectories.

Later we will glimpse why such orthogonal trajectories arise in science and how complex variable theory, via the Cauchy–Riemann equations, explains what is happening. We mention now a fundamental result concerning the equality of mixed partial derivatives. This result helps us understand conservation laws in physics, and it will be used several times in this chapter. We also use the notation of subscripts for partial derivatives for the first time; this notation is efficient and convenient. Thus f_y means the partial derivative of f with respect to y, and f_{yx} means that we first differentiate with respect to y and then with respect to x. The notation $\frac{\partial^2 f}{\partial x \partial y}$ also means that we first differentiate with respect to y. Fortunately, for twice continuously differentiable functions, the order of differentiation does not matter.

Theorem 2.3. *Let f be a twice continuously differentiable function of two or more variables. Then*

$$f_{yx} = \frac{\partial^2 f}{\partial x \partial y} = \frac{\partial^2 f}{\partial y \partial x} = f_{xy}.$$

EXAMPLE 2.9. We give a counterexample to Theorem 2.3 when f is not twice continuously differentiable. Put $f(0,0) = 0$. For $(x, y) \neq (0, 0)$ put

$$f(x, y) = \frac{bx^3 y + cxy^3}{x^2 + y^2}.$$

This function f is continuous at 0. For $(x, y) \neq (0, 0)$ we differentiate using the rules of calculus:

$$f_x(x, y) = \frac{(x^2 + y^2)(3bx^2 y + cy^3) - (bx^3 y + cxy^3)(2x)}{(x^2 + y^2)^2}.$$

Evaluating at $(0, y)$ for $y \neq 0$ we get

$$f_x(0, y) = cy.$$

For $(x, y) = (0, 0)$ we use the limit quotient definition, obtaining:

$$f_x(0, 0) = \lim_{h \to 0} \frac{f(h, 0) - f(0, 0)}{h} = \lim_{h \to 0} \frac{0}{h} = 0.$$

By symmetry or similar computation we get

$$f_y(x, y) = \frac{(x^2 + y^2)(bx^3 + 3cxy^2) - (bx^3 y + cxy^3)(2y)}{(x^2 + y^2)^2}.$$

Evaluating as above we obtain

$$f_y(x, 0) = bx.$$

Again using the limit quotient definition we obtain

$$f_y(0, 0) = \lim_{h \to 0} \frac{f(0, h) - f(0, 0)}{h} = \lim_{h \to 0} \frac{0}{h} = 0.$$

To find the mixed second partial derivatives f_{xy} and f_{yx}, we use the limit quotient definition on the above formulas to get:

$$f_{xy}(0, 0) = \lim_{h \to 0} \frac{f_x(0, h) - f_x(0, 0)}{h} = \lim_{h \to 0} \frac{ch}{h} = c.$$

$$f_{yx}(0, 0) = \lim_{h \to 0} \frac{f_y(h, 0) - f_y(0, 0)}{h} = \lim_{h \to 0} \frac{bh}{h} = b.$$

When $c \neq b$ we see that $f_{xy}(0, 0) \neq f_{yx}(0, 0)$.

EXERCISE 2.1. Verify the statements in Example 2.1.

EXERCISE 2.2. Define a function f by $f(0,0) = 0$ and otherwise by

$$f(x,y) = \frac{x^2 y}{x^4 + y^2}.$$

Show that the limit of f is 0 along every line approaching $(0,0)$ but that the limit of f is *not* 0 if we approach along the parabola $y = x^2$. Explain! Draw some level sets of f near the origin. See also Exercise 2.12.

EXERCISE 2.3. Graph several level sets of u and v in Example 2.8.

EXERCISE 2.4. Assume $f : \mathbb{R}^n \to \mathbb{R}$ is differentiable, and $f(x) \geq f(p)$ for all x near p. Prove that $\nabla f(p) = 0$. Suggestion: use Definition 2.1 directly.

EXERCISE 2.5. Assume $f : \mathbb{R}^3 \to \mathbb{R}$ is differentiable, $f(1,2,3) = 17$, and $\nabla f(1,2,3) = (4,9,3)$. Find approximately $f(1.1, 1.9, 3.2)$.

EXERCISE 2.6. For x close to 0, one often uses x as an approximation for both $\sin(x)$ and $\tan(x)$. Which is more accurate? Where ... denotes higher-order terms, show that

$$\sin(x) = x - \frac{x^3}{6} + \ldots$$

$$\tan(x) = x + \frac{x^3}{3} + \ldots$$

EXERCISE 2.7. Put $f(x,y,z) = x^2 + y^2 - z^2$. Describe the level sets $\{f = c\}$ for $c < 0$, for $c = 0$, and for $c > 0$.

EXERCISE 2.8. Again put $f(x,y,z) = x^2 + y^2 - z^2$. Show that $(0,0,0)$ is the only critical point of f. How is this information relevant to the previous exercise?

EXERCISE 2.9. Find the critical points of the function in Exercise 1.2.

EXERCISE 2.10. Suppose f is a differentiable function on $\mathbb{R}$ with exactly one critical point, and f has a local minimum there. Show that f has a global minimum there.

EXERCISE 2.11. True or false? Suppose $f(x,y)$ is a polynomial in two variables, and f has exactly one critical point. Suppose that f has a local minimum there; then f has a global minimum there.

EXERCISE 2.12. For positive numbers a, b, c, d define a function f by $f(0,0) = 0$ and otherwise by

$$f(x,y) = \frac{|x|^a |y|^b}{|x|^{2c} + |y|^{2d}}.$$

Give a necessary and sufficient condition on a, b, c, d such that f is continuous at $(0,0)$.

EXERCISE 2.13. Assume that $f : \mathbb{R} \to \mathbb{R}$ satisfies

$$|f(x)| \leq Cx^2$$

for all x and some constant C. Prove that f is differentiable at 0. Find such an f that is discontinuous at every other x.

EXERCISE 2.14. Put $f(x,y) = (x^3 + y^3)^{\frac{1}{3}}$. Use the limit quotient definition to find both partial derivatives $\frac{\partial f}{\partial x}(0,0)$ and $\frac{\partial f}{\partial y}(0,0)$. Compute these partials at other points using the rules of calculus. Show that these partials are not continuous at $(0,0)$. Given $v \neq 0$, compute the directional derivative $\frac{\partial f}{\partial v}(0,0)$ and show that the formula

$$\frac{\partial f}{\partial v}(p) = \nabla f(p) \cdot v$$

fails at the origin for most v.

EXERCISE 2.15. Put $f(x,y) = \frac{x^2 y^8}{x^4 + y^4}$ for $(x,y) \neq (0,0)$. Put $f(0,0) = 0$. Either directly or by using Exercise 2.12, show that f is continuous at $(0,0)$. Use Definition 2.1 to show that f is differentiable at $(0,0)$. Note first, if f is differentiable there, then $\nabla f(0,0)$ must be $(0,0)$. Then show that the error term goes to 0 even after dividing by $|h|$.

EXERCISE 2.16. Find $f_{xy}(0,0)$ if $f(0,0) = 0$ and $f(x,y) = \frac{x^4}{x^2 + y^2}$ otherwise.

EXERCISE 2.17. Let f be the function in Example 2.9. Let $t \to (\cos(\theta)t, \sin(\theta)t)$ be a line through the origin. Find the restriction of f to this line.

3. Line integrals and work

A **vector field** on an open set $\Omega \subseteq \mathbb{R}^n$ is a function $\mathbf{E} : \Omega \to \mathbb{R}^n$. At each point p, we think of $\mathbf{E}(p)$ as a vector placed at p. The gradient provides one basic example. This use of the word *field* has nothing to do with the algebraic notion of a field from Definition 5.1 of Chapter 1.

A (parametrized) curve in $\mathbb{R}^n$ is a continuous function from an interval $[a,b]$ in $\mathbb{R}$ to $\mathbb{R}^n$. We usually assume that this function is at least piecewise differentiable. We often identify a curve γ with its image set. Thus, for example, we regard the unit circle as a curve in the plane. The curve is officially the function $t \to (\cos(t), \sin(t))$ defined for $0 \leq t \leq 2\pi$.

A curve is **simple** if its image does not cross itself and **closed** if $\gamma(a) = \gamma(b)$. More precisely, a curve is simple if $\gamma(s) \neq \gamma(t)$ for $s \neq t$, with the possible exception that $\gamma(a) = \gamma(b)$ is allowed, in which case γ is closed. If γ is a simple closed curve, then we say that γ is **positively oriented** if the interior of the image of γ lies on the left as t increases from a to b. Thus we travel in a counterclockwise sense. When γ is a simple closed curve in $\mathbb{R}^2$ or in $\mathbb{C}$, we write $int(\gamma)$ for the open, bounded region determined by the image of γ. Henceforth we assume all curves are piecewise-smooth; intuitively we mean that the curve consists of finitely many smooth pieces. For example, rectangles and triangles are (images of) piecewise-smooth curves.

Let γ be a curve in $\mathbb{R}^n$ and let $\mathbf{E}$ be a vector field. The line integral $\int_\gamma \mathbf{E} \cdot d\mathbf{l}$ measures the *work* done by the force $\mathbf{E}$ on a particle moving along γ. For example, if γ is a line segment, and $\mathbf{E}$ is tangent to γ, then the work is positive if the particle moves in the direction $\mathbf{E}$ points. The sign can be confusing; one can regard the force field as given and measure the work done by an *external agent* against the force. The work is then multiplied by -1.

Consider a vector field in $\mathbb{R}^2$. If $\mathbf{E} = (P, Q)$ and $d\mathbf{l} = (dx, dy)$, then

$$\mathbf{E} \cdot d\mathbf{l} = P dx + Q dy$$

$$\int_\gamma \mathbf{E} \cdot d\mathbf{l} = \int_\gamma P dx + Q dy.$$

The expression $P dx + Q dy$ is called a differential 1-form. We start using differential forms now; Section 8 offers a detailed discussion. We evaluate line integrals by *pulling back*. Assume $\gamma : [a,b] \to \mathbb{R}^2$ is smooth. If $\gamma(t) = (x(t), y(t))$, then we replace each x with $x(t)$, we replace dx with $x'(t)dt$, and we do the analogous thing with y. We then obtain an ordinary integral over $[a,b]$. When γ has finitely many smooth pieces, we parametrize each piece separately and add the results.

EXAMPLE 3.1. Put $\mathbf{E} = (x^2, xy)$. Consider a triangle with vertices at $(0,0)$, $(1,0)$, and $(1,2)$. Let μ be the bottom and right side of the triangle, and let η be the left side of the triangle. Then

$$\int_\mu \mathbf{E} \cdot d\mathbf{l} = \int_\mu x^2 dx + xy dy = \int_0^1 t^2 dt + \int_0^2 t dt = \frac{1}{3} + 2 = \frac{7}{3}.$$

$$\int_\eta x^2 dx + xy\,dy = \int_0^1 (t^2 + t(2t)2)\,dt = \int_0^1 5t^2 dt = \frac{5}{3}.$$

The work done by $\mathbf{E}$ in going from $(0,0)$ to $(1,2)$ depends on which path we take. Figure 4 illustrates the situation, although we used the vector field $\frac{\mathbf{E}}{4}$ instead of $\mathbf{E}$ to keep the arrows at a manageable length. Notice that $\mathbf{E}$ is tangent to η.

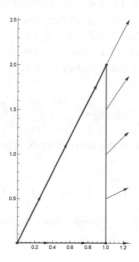

FIGURE 4. Line integrals and work

Green's theorem allows us to evaluate certain line integrals around closed curves. If γ is a simple closed curve in $\mathbb{R}^2$, then it surrounds an open set, (imprecisely) called the *interior* of γ and written $int(\gamma)$.

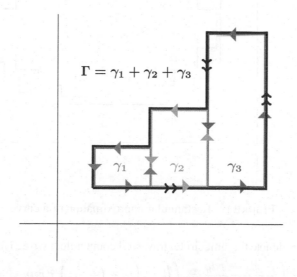

FIGURE 5. Curve whose sides are parallel to one of the axes

Theorem 3.1 (Green's theorem). *Let γ be a piecewise-smooth, simple, positively oriented closed curve in $\mathbb{R}^2$. Suppose P and Q are smooth on and inside γ. Then*

$$\int_\gamma \mathbf{E} \cdot d\mathbf{l} = \int_\gamma Pdx + Qdy = \iint_{int(\gamma)} \left(\frac{\partial Q}{\partial x} - \frac{\partial P}{\partial y} \right) dx\,dy. \tag{5}$$

PROOF. We prove Green's theorem only for a rectangle. The reader should then observe that the result follows for any simple closed curve consisting of finitely many sides, each parallel to one of the axes. See Figure 5. By taking limits, the formula holds for rather general curves, including piecewise-smooth curves. Figure 6 illustrates how to approximate an ellipse by a closed curve all of whose sides are parallel to an axis.

We therefore consider a rectangle whose vertices are at (a_1, b_1), (a_2, b_1), (a_2, b_2), and (a_1, b_2). We begin with the right-hand side of (5).

$$J = \int_{b_1}^{b_2} \int_{a_1}^{a_2} (\frac{\partial Q}{\partial x} - \frac{\partial P}{\partial y})dx\,dy = \int_{b_1}^{b_2} \int_{a_1}^{a_2} \frac{\partial Q}{\partial x} dxdy - \int_{b_1}^{b_2} \int_{a_1}^{a_2} \frac{\partial P}{\partial y} dxdy$$

We interchange the order of integration in the second term. Then we integrate each term directly. By the fundamental theorem of calculus, we get

$$J = \int_{b_1}^{b_2} \big(Q(a_2, y) - Q(a_1, y) \big) dy - \int_{a_1}^{a_2} \big(P(x, b_2) - P(x, b_1) \big) dx. \tag{6}$$

The four terms in (6) correspond to the line integrals along the four sides of the rectangle. $\square$

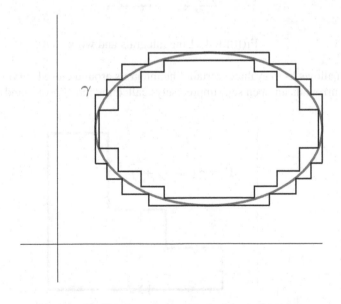

FIGURE 6. Rectangular approximation of a curve

EXAMPLE 3.2. Let γ denote the unit circle, traversed counterclockwise. Then

$$\int_\gamma \frac{-ydx + xdy}{2} = \iint_{int(\gamma)} \left(\frac{1}{2} - \left(-\frac{1}{2} \right) \right) dx\,dy = \pi.$$

More generally, $\int_\gamma \frac{-ydx+xdy}{2}$ measures the area of the interior of γ (when γ is a simple closed curve).

Most of our examples have been in $\mathbb{R}^2$. We therefore pause to note the formula for line integrals in $\mathbb{R}^3$. There $dl = (dx, dy, dz)$. If $\mathbf{E} = (E_x, E_y, E_z)$, then

$$\int_\gamma \mathbf{E} \cdot dl = \int_\gamma \mathbf{E} \cdot \mathbf{T} dl = \int_\gamma E_x dx + E_y dy + E_z dz.$$

Definition 3.2. A vector field $\mathbf{E}$ is called **conservative** if for all γ, the integral $\int_\gamma \mathbf{E} \cdot dl$ depends only on the starting and ending point of γ. We also say that $\int_\gamma \mathbf{E} \cdot dl$ is **path independent**.

When $\mathbf{E}$ is conservative, the work done in going from p to q is independent of the path taken. In any dimension, when $\mathbf{E}$ is the gradient of a function, it is a conservative field. The function f is called the **potential** and $\int_\gamma \mathbf{E} \cdot dl$ is the change in potential energy. For example, if $\mathbf{E} = (\frac{\partial f}{\partial x}, \frac{\partial f}{\partial y}, \frac{\partial f}{\partial z}) = \nabla f$, then

$$\int_\gamma \mathbf{E} \cdot dl = \int_\gamma \frac{\partial f}{\partial x} dx + \frac{\partial f}{\partial y} dy + \frac{\partial f}{\partial z} dz = f(\gamma(b)) - f(\gamma(a)).$$

Mathematicians simply write

$$\int_\gamma df = f(\gamma(b)) - f(\gamma(a)). \tag{7}$$

EXAMPLE 3.3. Given $\mathbf{E} = (ye^{xy}, xe^{xy})$, find $\int_\gamma \mathbf{E} \cdot dl$. Let γ be an arbitrary curve starting at $(0,0)$ and ending at $(2,4)$. Here $\mathbf{E} = \nabla f$, where $f(x,y) = e^{xy}$. Hence

$$\int_\gamma \mathbf{E} \cdot dl = f(2,4) - f(0,0) = e^8 - 1.$$

Examples of conservative fields include the gravitational field due to a fixed mass and the electric field due to a fixed charge. In each case, the vector field is a constant times the gradient of the reciprocal of the distance to the mass or charge. Because of the division by 0, however, one must be careful. Mathematicians understand these ideas using differential forms. See Section 8.

The vector field in Example 3.1 is not conservative. We revisit that example now that we know Green's theorem.

EXAMPLE 3.4. Let γ be the curve whose image is the triangle with vertices at $(0,0)$, $(1,0)$, and $(1,2)$.

$$\int_\gamma x^2 dx + xy dy = \iint_\Omega (y - 0) \, dx \, dy = \int_0^1 \int_0^{2x} y \, dy \, dx = \int_0^1 2x^2 dx = \frac{2}{3}.$$

Notice that $\frac{2}{3} = \frac{7}{3} - \frac{5}{3}$, and hence this result agrees with Example 3.1.

EXAMPLE 3.5. Let γ be the unit circle, positively oriented. For $(x,y) \neq (0,0)$ define $\mathbf{E}$ by

$$\mathbf{E} = (\frac{-y}{x^2 + y^2}, \frac{x}{x^2 + y^2}).$$

We evaluate

$$\int_\gamma \mathbf{E} \cdot dl = \int_\gamma \frac{-y}{x^2 + y^2} dx + \frac{x}{x^2 + y^2} dy = \int_\gamma \omega$$

by putting $(x,y) = (\cos(t), \sin(t))$. We get

$$\int_\gamma \mathbf{E} \cdot dl = \int_0^{2\pi} (\sin^2 t + \cos^2 t) dt = 2\pi.$$

For $(x,y) \neq (0,0)$, we have $\frac{\partial P}{\partial y} = \frac{y^2 - x^2}{(x^2+y^2)^2}$ and $\frac{\partial Q}{\partial x} = \frac{y^2 - x^2}{(x^2+y^2)^2}$. But P, Q are not differentiable at $(0,0)$, which is in Ω. Thus Green's theorem does not apply. When expressed in complex variable notation, this example will be fundamental in the next chapter. In particular, we mention the following. The line integral of this form ω around any closed curve not surrounding the origin is 0, and the line integral of ω around any closed curve surrounding the origin equals 2π times the number of times the curve winds around the origin.

EXERCISE 3.1. Graph the set $\{(x,y) : x^3 + y^3 = 3xy\}$. Use Green's theorem to find the area enclosed by the loop. Set $y = tx$ to find a parametrization of the loop, and note that $dy = x\,dt + t\,dx$.

EXERCISE 3.2. Use (7) to compute

$$\int_\gamma x\,dx + y\,dy + z\,dz$$

if γ is a path from the origin to (a, b, c). Why is the integral path-independent?

EXERCISE 3.3. Let γ denote the unit circle, traversed counterclockwise. Compute the following line integrals:

$$\int_\gamma x\,dx + y\,dy.$$

$$\int_\gamma -y\,dx + x\,dy.$$

$$\int_\gamma y e^{xy}dx + x e^{xy}dy.$$

$$\int_\gamma y e^{xy}dx + (x + x e^{xy})dy.$$

EXERCISE 3.4. Determine the intersection of the surfaces defined by $2x+y+z = 3$ and $x^2 - y - z = 0$.

EXERCISE 3.5. Find parametric equations for the curve defined by the intersection of the surfaces given by $2x + y + 4z = 4$ and $y = x^2 + z^2$.

4. Surface integrals and flux

Let $\mathbf{E}$ be a vector field, and let S be a surface in $\mathbb{R}^3$. The **flux** of $\mathbf{E}$ across S is the rate of flow of $\mathbf{E}$ in a direction perpendicular to S. We use an integral, defined in (8) below, as the definition of flux.

We remind the reader how to compute surface integrals such as

$$\iint_S \mathbf{E} \cdot \mathbf{n}\,dS = \iint_S \mathbf{E} \cdot d\mathbf{S}. \tag{8}$$

One often writes $\mathbf{E} = (E_x, E_y, E_z)$ in terms of its components. Although $\mathbf{E}$ is an arbitrary smooth vector field for most of what we do, it is particularly useful to think of it as the electric field.

In formulas such as (8) we can regard the vector surface area form $d\mathbf{S}$ as follows:

$$d\mathbf{S} = (dy \wedge dz, dz \wedge dx, dx \wedge dy).$$

It then follows that

$$\iint_S \mathbf{E} \cdot \mathbf{n}\,dS = \iint_S E_x dy \wedge dz + E_y dz \wedge dx + E_z dx \wedge dy. \tag{8.1}$$

The expressions $dy \wedge dz$, $dz \wedge dx$, and $dx \wedge dy$ are what mathematicians call differential 2-forms. Because of orientation issues we have $dy \wedge dx = -dx \wedge dy$ and so on. Once we have all the 2-forms properly oriented, we drop the $\wedge$ symbol when doing integrals.

Just as with line integrals, we can compute either with vector fields or with differential forms. We remind the reader how to compute explicit examples. Let us assume that the surface S is parametrized by a function Φ as follows. Let Ω be an open set in $\mathbb{R}^2$, with variables (u, v). Assume $\Phi : \Omega \to \mathbb{R}^3$ is a smooth injective function. To compute a surface integral, we express $\mathbf{n}dS$ in terms of the derivatives of Φ as follows:

$$\mathbf{n}dS = d\mathbf{S} = \frac{\partial \Phi}{\partial u} \times \frac{\partial \Phi}{\partial v} du dv \tag{9.1}$$

$$dS = ||\frac{\partial \Phi}{\partial u} \times \frac{\partial \Phi}{\partial v}|| \, du \, dv. \tag{9.2}$$

Most calculus books provide intuitive explanations for these formulas. Mathematicians usually use differential forms to compute surface integrals; the method is virtually the same but the notation differs somewhat. See Example 4.2.

We compare work and flux. The line integral

$$\int_\gamma \mathbf{E} \cdot \mathbf{T} dl = \int_\gamma \mathbf{E} \cdot d\mathbf{l} = \int_\gamma P dx + Q dy + R dz,$$

measures the work done by the field $\mathbf{E}$ in traveling along the curve γ. We put $\mathbf{E} = (P, Q, R)$ and $d\mathbf{l} = (dx, dy, dz)$.

The surface integral

$$\int_S \mathbf{E} \cdot \mathbf{n} dS = \int_S \mathbf{E} \cdot d\mathbf{S} = \int_S P dy \wedge dz + Q dz \wedge dx + R dx \wedge dy$$

represents the flux of $\mathbf{E}$ across the surface S. The wedge product has the property $dy \wedge dz = -dz \wedge dy$, and similarly for the other 2-forms.

To continue the analogy with line integrals, we require an analogue for surfaces of piecewise-smooth positively oriented curve. We define orientation only when the surface bounds a region. The convention is then to make the normal vector $\mathbf{n}$ point outwards, in which case we say that S is *oriented by its outer normal*. See Figure 8 in Section 5 for the orientation of a rectangular box. In Figure 7 below, showing a right circular cone, the vector $\mathbf{n}$ points downward from the bottom. It points outward from the side of the cone. It is not defined at the vertex nor along the circle bounding the bottom; the cone is not smooth at these points. Thus the cone is a piecewise-smooth surface.

We illustrate how to compute surface integrals with several examples. First we find the surface area of a cone.

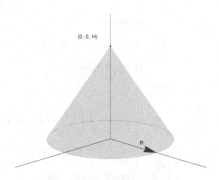

FIGURE 7. Right circular cone

EXAMPLE 4.1. Figure 7 shows a cone C with vertex at $(0, 0, H)$ in $\mathbb{R}^3$. It is a right circular cone; its base is determined by the circle given by $x^2 + y^2 = R^2$.

$$\iint_C dS = \iint_{\text{side}} dS + \iint_{\text{bottom}} dS$$

The integral over the bottom equals plus or minus the area πR^2, depending on orientation. In finding the surface area, we use πR^2. We compute the integral over the side. The axis of symmetry suggests that we use cylindrical coordinates (r, θ, z). Here $z = H - \frac{H}{R}r$, and we have

$$(x, y, z) = \left(r\cos\theta, r\sin\theta, H(1 - \frac{r}{R})\right) = \Phi(r, \theta).$$

Here $\Phi : \{0 \leq r \leq R, 0 \leq \theta \leq 2\pi\} \to \mathbb{R}^3$; its image is the side of the cone. Following the recipe for computing dS, we get

$$\frac{\partial \Phi}{\partial r} = \left(\cos\theta, \sin\theta, -\frac{H}{R}\right)$$

$$\frac{\partial \Phi}{\partial \theta} = (-r\sin\theta, r\cos\theta, 0)$$

$$d\mathbf{S} = (\frac{\partial \Phi}{\partial r} \times \frac{\partial \Phi}{\partial \theta})dr d\theta = \left(\frac{H}{R}r\cos\theta, \frac{H}{R}r\sin\theta, r\right) dr\, d\theta.$$

Therefore,

$$dS = r\sqrt{(\frac{H}{R})^2 + 1}dr\, d\theta = \frac{r}{R}\sqrt{H^2 + R^2}\, dr\, d\theta$$

The surface area of the side equals

$$\int_0^{2\pi} \int_0^R \frac{r}{R}\sqrt{H^2 + R^2}dr d\theta = \pi R\sqrt{H^2 + R^2}.$$

The surface area of the cone is therefore $\pi R^2 + \pi R\sqrt{H^2 + R^2}$.

We next compute a flux integral using (9.1) and also by using differential forms. The two methods are equivalent, but good students should be able to do both.

EXAMPLE 4.2. Put $\mathbf{E} = (x, y, z)$ in $\mathbb{R}^3$. Let the surface S be the paraboloid defined by $z = R^2 - x^2 - y^2$ for $z \geq 0$ and the disk in the plane $z = 0$ for $x^2 + y^2 \leq R^2$ defining the bottom. Thus the surface consists of two pieces.

In cylindrical coordinates, the paraboloid is given by $z = R^2 - r^2$. First we use formula (9.1). For the bottom, $\mathbf{n}$ points downward. We parametrize the bottom by $(r, \theta) \to (r\cos\theta, r\sin\theta, 0)$. We have

$$\mathbf{E} = (r\cos\theta, r\sin\theta, 0)$$

$$\mathbf{n} = (0, 0, -1)$$

and thus $\mathbf{E} \cdot \mathbf{n} = 0$ on the bottom. Therefore

$$\iint_{\text{bottom}} \mathbf{E} \cdot \mathbf{n}\, dS = \iint_{\text{bottom}} 0\, dS = 0. \tag{10}$$

Next we do the side. Put $\Phi(r, \theta) = (r\cos\theta, r\sin\theta, R^2 - r^2)$. Therefore

$$\frac{\partial \Phi}{\partial r} = (\cos\theta, \sin\theta, -2r),$$

$$\frac{\partial \Phi}{\partial \theta} = (-r\sin\theta, r\cos\theta, 0),$$

$$\frac{\partial \Phi}{\partial r} \times \frac{\partial \Phi}{\partial \theta} = (2r^2 \cos\theta, 2r^2 \sin\theta, r)$$

We obtain

$$\mathbf{E} \cdot d\mathbf{S} = (r\cos\theta, r\sin\theta, R^2 - r^2) \cdot (2r^2 \cos\theta, 2r^2 \sin\theta, r)\, dr\, d\theta = (r^3 + rR^2)dr d\theta.$$

Therefore

$$\iint_{\text{side}} \mathbf{E} \cdot d\mathbf{S} = \int_0^{2\pi} \int_0^R 2r^3 \cos^2\theta + 2r^3 \sin^2\theta + r(R^2 - r^2)dr\, d\theta$$

$$= \int_0^{2\pi} \int_0^R (r^3 + rR^2)dr d\theta = \frac{3}{2}\pi R^4. \tag{11}$$

Next we compute the same integral using differential forms. If necessary, the reader can look ahead to Section 8. First we compute the side. We need to express $dy \wedge dz$, $dz \wedge dx$, and $dx \wedge dy$ in terms of the variables (r, θ). We have

$$x = r\cos(\theta) \implies dx = \cos(\theta)dr - r\sin(\theta)d\theta.$$

$$y = r\sin(\theta) \implies dy = \sin(\theta)dr + r\cos(\theta)d\theta.$$

$$z = R^2 - r^2 \implies dz = -2rdr.$$

Computing the wedge products, and using $dr \wedge dr = 0$, we get

$$dy \wedge dz = 2r^2 \cos(\theta)dr \wedge d\theta$$

$$dz \wedge dx = 2r^2 \sin(\theta)dr \wedge d\theta$$

$$dx \wedge dy = rdr \wedge d\theta.$$

We then can compute $Pdy \wedge dz + Qdx \wedge dz + Rdx \wedge dy$, where (in this problem),

$$(P, Q, R) = (x, y, z) = (r\cos(\theta), r\sin(\theta), R^2 - r^2).$$

Finally we obtain the integral

$$\int_0^{2\pi} \int_0^R \left(2r^3 \cos^2(\theta) + 2r^3 \sin^2(\theta) + (R^2 - r^2)r\right) dr \wedge d\theta,$$

the same integral as in (11).

To compute the integral over the bottom is easy. Since $z = 0$, also $dz = 0$. Therefore $dz \wedge dx = 0$ and $dy \wedge dz = 0$. Also $zdx \wedge dy = 0$ because $z = 0$. Hence, as above, the integral over the bottom is 0.

REMARK. Orientation issues can be subtle, even on the real line. For example, everyone agrees that $\int_a^b dx = b - a$, even when $b < a$. The reason is that dx is a 1-form. Suppose, however, that we regard the interval $[a, b]$ as a curve in the plane. If we compute $\int_\gamma dl$, then we expect to get the length of the curve, which is $|b - a|$. Similar issues arise for surfaces. When we parametrize a surface with a map

$$(u, v) \mapsto (x(u, v), y(u, v), z(u, v))$$

we are presuming that u precedes v in the orientation of $\mathbb{R}^2$.

EXERCISE 4.1. Define Φ by $\Phi(u, v) = (u^2, \sqrt{2}uv, v^2)$. Find the surface area of the image of the unit disk under Φ. Take multiplicity into account.

EXERCISE 4.2. Let C be the piecewise smooth surface defined as follows. It consists of the cylinder C_R given by $0 \leq z \leq H$ and $x^2 + y^2 = R^2$, together with the top (a disk in the plane $z = H$) and the bottom (a disk in the plane $z = 0$). Find the flux integrals

$$\iint_C (x^2, y, z) \cdot d\mathbf{S}$$

$$\iint_{C_R} (x^2, y, 2z) \cdot d\mathbf{S}.$$

EXERCISE 4.3. A Gaussian surface M is a piecewise-smooth, oriented closed surface in $\mathbb{R}^3$ that encloses the origin. Evaluate the flux integral and explain its significance. What would the answer be if the origin is outside M?

$$\iint_M \frac{x\, dy\, dz + y\, dz\, dx + z\, dx\, dy}{(x^2 + y^2 + z^2)^{\frac{3}{2}}}.$$

EXERCISE 4.4. Find a parametrization of the hyperboloid of one sheet defined by $x^2 + y^2 = 1 + z^2$. Compute $d\mathbf{S}$. Suggestion: Use θ and z.

EXERCISE 4.5. You wish to coat a surface S with a thin layer of silver. Since silver is expensive, you need to compute the surface area before submitting your budget. Sketch S and find its area. Here S is the subset of $\mathbb{R}^3$ such that $x, y \geq 0$, also $x^2 + y^2 = 25$, and $0 \leq z \leq 10 - x - y$. All units are in feet.

EXERCISE 4.6. Let Δ denote the unit disk in $\mathbf{R}^2$. Find the surface areas of each of the three given surfaces.

- The subset of the plane defined by $z = 1$ lying above Δ.
- The subset of the plane defined by $z = x + y$ lying above Δ.
- The subset of the paraboloid defined by $z = R^2 - x^2 - y^2$ lying above Δ.

EXERCISE 4.7. Assume $-1 \leq a \leq b \leq 1$. Show that the surface area of the part of the unit sphere between the planes $z = a$ and $z = b$ depends only upon the difference $b - a$. Check your answer when $a = -1$ and $b = 1$. (It is easiest to use spherical coordinates.)

5. The divergence theorem

Green's theorem allowed us to compute a line integral around a closed curve by doing a related double integral over the interior of the curve. In a similar fashion, the divergence theorem (or Gauss's theorem), allows us to compute surface integrals over surfaces that bound a region by doing a related triple integral over this region. The divergence theorem holds in all dimensions, and its two-dimensional version implies Green's theorem.

First we recall the notion of divergence of a vector field. This term does not mean that a sequence or series is diverging! But the use of the word is consistent with ideas about how much of something is leaving a given set. It is also convenient to define the curl of a vector field at this time.

Put $\mathbf{E} = (P, Q, R)$. Then the divergence of E is the scalar

$$\operatorname{div}(\mathbf{E}) = \nabla \cdot \mathbf{E} = \frac{\partial P}{\partial x} + \frac{\partial Q}{\partial y} + \frac{\partial R}{\partial z}.$$

This quantity has units flux per volume.

The curl of $\mathbf{E}$ is a vector field. It measures rotation (or circulation) about an axis.

$$\text{curl}(\mathbf{E}) = \nabla \times \mathbf{E} = \det \begin{pmatrix} i & j & k \\ \frac{\partial}{\partial x} & \frac{\partial}{\partial y} & \frac{\partial}{\partial z} \\ P & Q & R \end{pmatrix} = \left(\frac{\partial R}{\partial y} - \frac{\partial Q}{\partial z}, \frac{\partial P}{\partial z} - \frac{\partial R}{\partial x}, \frac{\partial Q}{\partial x} - \frac{\partial P}{\partial y} \right).$$

REMARK. As written, $\text{curl}(\mathbf{E})$ makes sense only if $\mathbf{E}$ is a vector field on a subset of $\mathbb{R}^3$. We identify $\mathbf{E} = (P, Q)$ with $(P, Q, 0)$ to compute its curl. We get $\text{curl}(\mathbf{E}) = (0, 0, Q_x - P_y)$, where we have used subscripts to denote partial derivatives. Thus the $\mathbf{k}$ component of the curl is the expression arising in Green's theorem.

The divergence is defined in all dimensions:

$$\text{div}(\mathbf{E}) = \sum_{j=1}^{n} \frac{\partial E_j}{\partial x_j}$$

if $\mathbf{E} = (E_1, E_2, \cdots, E_n)$. Furthermore, although we will not prove it here, the divergence can be computed in this way no matter what coordinates are used.

For completeness, we mention standard physical interpretations of curl and divergence and a standard mnemonic device for remembering which is which.

Definition 5.1. A vector field in $\mathbb{R}^3$ is called **irrotational** or **curl-free** if its curl is the vector 0. A vector field in $\mathbb{R}^n$ is called **incompressible** or **divergence-free** if its divergence is the scalar 0.

REMARK. Both curl and cross begin with the letter c, and $\text{curl}(\mathbf{E}) = \nabla \times \mathbf{E}$. Both divergence and dot begin with the letter d, and $\text{div}(\mathbf{E}) = \nabla \cdot \mathbf{E}$.

The following major result, called the *divergence theorem* or *Gauss's theorem*, provides an analogue of Green's theorem for flux integrals. It arises in Maxwell's equations and it tells us how to perform integration by parts in higher dimensions. We state it only in three dimensions, but the analogue of formula (12) holds in all dimensions.

Theorem 5.2. *Let S be a positively oriented piecewise-smooth surface in $\mathbb{R}^3$ bounding a region Ω. Suppose $\mathbf{E} = (P, Q, R)$ is continuously differentiable on and inside S. Then*

$$\iint_S \mathbf{E} \cdot d\mathbf{S} = \iiint_\Omega \text{div}(\mathbf{E}) dV. \tag{12}$$

PROOF. We merely sketch one possible proof. First one proves the result for a rectangular box. See Figure 8 below. As in the proof of Green's theorem, the result follows for objects built from such rectangular boxes. Then one approximates a nice surface by such objects. $\square$

The left-hand side of (12) measures the flux of $\mathbf{E}$ across S. Since it equals the right-hand side, the divergence measures flux per volume.

EXAMPLE 5.1. We compute Example 4.2 using the divergence theorem. The surface S is defined by $z = R^2 - r^2$ on the top, and $z = 0$ on the bottom. Since $\mathbf{E} = (x, y, z)$, we get $\text{div}(\mathbf{E}) = 3$. Hence

$$\iint_S \mathbf{E} \cdot d\mathbf{S} = \iiint_\Omega 3 dV = 3 \int_0^R \int_0^{R^2 - r^2} \int_0^{2\pi} r d\theta dz dr = \frac{3}{2} \pi R^4$$

One of the most important and illuminating examples concerns the electric field due to a point charge, which we place at the origin. For an appropriate constant a, depending on the units used, we have

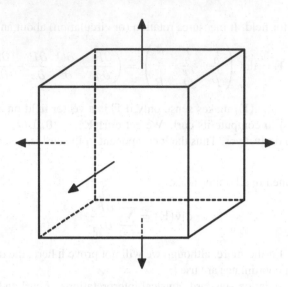

FIGURE 8. Orientation of a rectangular box

$$\mathbf{E} = a \, \frac{(x, y, z)}{(x^2 + y^2 + z^2)^{\frac{3}{2}}}. \tag{13}$$

To better understand this field, put $r(x, y, z) = \sqrt{x^2 + y^2 + z^2}$, the distance to $(0, 0, 0)$. The function $\frac{1}{r}$ gives the electric potential away from the origin. Then

$$\nabla\left(-\frac{1}{r}\right) = \frac{1}{2(x^2 + y^2 + z^2)^{\frac{3}{2}}}(2x, 2y, 2z) = \frac{(x, y, z)}{(x^2 + y^2 + z^2)^{\frac{3}{2}}}$$

Then the flux across a sphere of radius R about the origin is

$$\iint_{S_R} \mathbf{E} \cdot \mathbf{n} \, d\mathbf{S} = a \iint_{S_R} \frac{(x, y, z)}{(x^2 + y^2 + z^2)^{\frac{3}{2}}} \cdot \frac{(x, y, z)}{(x^2 + y^2 + z^2)^{\frac{1}{2}}} \, dS$$

$$= a \iint_{S_R} \frac{1}{R^2} \, dS = \frac{a}{R^2} 4\pi R^2 = 4\pi a = \frac{q}{\epsilon_0}. \tag{14}$$

In (14) we have introduced the charge q and the permittivity constant ϵ_0. Thus the constant a has the value $\frac{q}{4\pi\epsilon_0}$. Note also that the value of this flux integral is independent of R. The divergence theorem does not apply to $\mathbf{E}$, because $\frac{1}{x^2 + y^2 + z^2}$ is not differentiable at $(0, 0, 0)$. Nonetheless, the divergence theorem is highly relevant here. It explains the notion of Gaussian surface and leads to two of the four Maxwell equations for the electromagnetic field.

Let $\mathbf{E}$ represent the electric field and $\mathbf{B}$ represent the magnetic field. The first two of Maxwell's equations involve the divergences of $\mathbf{E}$ and $\mathbf{B}$. We discuss Maxwell's equations in Sections 7 and 8.

- div$(\mathbf{E}) = \frac{\rho}{\epsilon_0}$, where ρ is the charge density, and ϵ_0 is the permittivity constant.
- div$(\mathbf{B}) = 0$. Thus no magnetic monopole exists.

To describe the other pair of Maxwell equations, we need to discuss curl and the classical Stokes' theorem, the subject of the next section.

Many quantities from physics can be computed using multiple integrals. For example, consider an object S in $\mathbb{R}^3$ and an axis of rotation. Assume that the mass density of the object is δ; thus

$$\iiint_S \delta(x, y, z)\, dx\, dy\, dz = M.$$

The *moment of inertia* of S about the axis l is the integral

$$\iiint_S (\text{dist}(x, y, z), l)^2 \delta(x, y, z)\, dx\, dy\, dz.$$

Its units are mass times distance squared.

For example, the moment of inertia of a circular cylinder of radius R, uniform mass density, and total mass M, about its axis of symmetry, is $\frac{1}{2}MR^2$. Here we are assuming constant mass density. Thus $\delta = \frac{M}{V}$. Not surprisingly, cylindrical coordinates make this computation easy. Cylindrical coordinates also work well in the subsequent exercises. In a certain funny way, the sphere turns out to be the average of the cone and the cylinder.

The parallel axis theorem for moments of inertia appears in most physics books. We leave its proof to the reader, but we provide a generous hint. The idea of the proof arises several times in this book. For example, in Chapter 6 we discuss squared norms of vectors and Bessel's inequality, and in Chapter 7 we give two formulas for the variance of a random variable. Both proofs rely on the same idea as the proof of the following theorem.

Theorem 5.3 (Parallel axis theorem). *Let I_0 denote the moment of inertia of a solid S of mass M, rotating about an axis l_0 through the center of mass of S. Let I_1 denote the moment of inertia of the same solid rotating about a line parallel to l_0 and at distance d to l_0. Then*

$$I_1 = I_0 + Md^2.$$

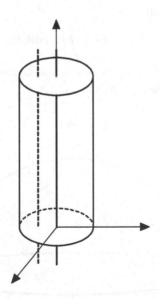

FIGURE 9. Parallel axis theorem

EXERCISE 5.1. A circular cylinder has height a and radius z. What food does its volume suggest?

EXERCISE 5.2. The volume of a sphere of radius r is $\frac{4}{3}\pi r^3$ and the surface area is $4\pi r^2$, the derivative with respect to r of the volume. Using the divergence theorem or otherwise, explain why.

EXERCISE 5.3. Prove the parallel axis theorem. Start by assuming that the center of mass is located at the origin, and l_0 is the z-axis.

EXERCISE 5.4. Show using a triple integral that the moment of inertia of a sphere of mass M and radius R, about an axis through the center, is $\frac{2}{5}MR^2$. Assume the mass density is constant.

EXERCISE 5.5. Show using a triple integral that the moment of inertia of a circular cone of mass M and radius R, about an axis through the center, is $\frac{3}{10}MR^2$. Assume the mass density is constant.

6. The classical Stokes' theorem

Theorem 6.1. *Let S be a positively oriented smooth surface whose boundary γ is a piecewise-smooth curve. Assume that the vector field $\mathbf{E}$ is continuously differentiable on S and on γ. Then*

$$\iint_S \operatorname{curl}(\mathbf{E}) \cdot d\mathbf{S} = \oint_\gamma \mathbf{E} \cdot d\mathbf{l}.$$

Thus the work done in traveling around a closed loop is the same as the flux of the curl across a surface whose boundary is this loop. It follows that

$$\iint_S \operatorname{curl}(\mathbf{E}) \cdot d\mathbf{S}$$

depends only on the boundary of S, but not on S itself. This idea is analogous to work and conservative vector fields.

Consider an oriented surface S formed from surfaces S_1 and S_2 with the same boundary γ. Picture a paraboloid lying over a disk, as in Example 4.2.

$$\iint_{S_1 \cup S_2} \operatorname{curl}(\mathbf{E}) \cdot d\mathbf{S} = \iint_{S_1} \operatorname{curl}(\mathbf{E}) \cdot d\mathbf{S} + \iint_{S_2} \operatorname{curl}(\mathbf{E}) \cdot d\mathbf{S} = \oint_\gamma \mathbf{E} \cdot d\mathbf{S} - \oint_\gamma \mathbf{E} \cdot d\mathbf{S} = 0.$$

Thus $\iint_{S_1} \operatorname{curl}(\mathbf{E}) \cdot d\mathbf{S} = -\iint_{S_2} \operatorname{curl}(\mathbf{E}) \cdot d\mathbf{S}$. The minus sign indicates the orientation. If we orient S_1 and S_2 in the same way, then the minus sign goes away.

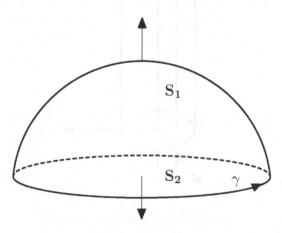

FIGURE 10. Classical Stokes' theorem

For clarity, we perform the analogous computation for work. Suppose γ is a simple closed curve, written as $\gamma_1 + \gamma_2$ to denote the sum of two curves. Then

$$0 = \int_\gamma \nabla f \cdot d\mathbf{l} = \int_{\gamma_1} \nabla f \cdot d\mathbf{l} + \int_{\gamma_2} \nabla f \cdot d\mathbf{l}.$$

Therefore

$$\int_{\gamma_1} \nabla f \cdot d\mathbf{l} = -\int_{\gamma_2} \nabla f \cdot d\mathbf{l}.$$

We pause to summarize some of the formulas from vector analysis, to introduce the triple scalar product, and to discuss the relationships among the three major theorems.

$$\mathrm{curl}(\mathbf{E}) = \nabla \times \mathbf{E}.$$
$$\mathrm{div}(\mathrm{curl}(\mathbf{E})) = \nabla \cdot (\nabla \times \mathbf{E})$$

Given vectors $\mathbf{V_1}, \mathbf{V_2}, \mathbf{V_3}$, their triple scalar product is given by

$$\mathbf{V_1} \cdot (\mathbf{V_2} \times \mathbf{V_3}) = \det \begin{pmatrix} \mathbf{V_1} \\ \mathbf{V_2} \\ \mathbf{V_3} \end{pmatrix}.$$

Its value is therefore plus or minus the value of the box they span. The formal analogue:

$$\nabla \cdot (\nabla \times \mathbf{E}) = \det \begin{pmatrix} \frac{\partial}{\partial x} & \frac{\partial}{\partial y} & \frac{\partial}{\partial z} \\ \frac{\partial}{\partial x} & \frac{\partial}{\partial y} & \frac{\partial}{\partial z} \\ P & Q & R \end{pmatrix}$$

should therefore, because of the repeated row, be 0. Computing the determinant does give zero, as will be noted in Proposition 7.3, because of the equality of mixed partial derivatives.

Green's theorem is a special case of both Stokes' theorem and the divergence theorem. Let $\mathbf{E}$ be a vector field in $\mathbb{R}^2$ with no component in the z direction. Thus $\mathbf{E} = (P, Q, 0)$, and

$$\oint_\gamma \mathbf{E} \cdot d\mathbf{l} = \int P dx + Q dy = \iint_{int(\gamma)} (\frac{\partial Q}{\partial x} - \frac{\partial P}{\partial y}) dx\, dy.$$

Think of $\mathbb{R}^2$ as a subset of $\mathbb{R}^3$; we then have

$$\mathrm{curl}(\mathbf{E}) = \det \begin{pmatrix} \mathbf{i} & \mathbf{j} & \mathbf{k} \\ \frac{\partial}{\partial x} & \frac{\partial}{\partial y} & \frac{\partial}{\partial z} \\ P & Q & R \end{pmatrix} = (0, 0, \frac{\partial Q}{\partial x} - \frac{\partial P}{\partial y}) = (0, 0, Q_x - P_y).$$

Also, $\mathbf{n} = (0, 0, 1)$. Therefore $\mathrm{curl}(\mathbf{E}) \cdot \mathbf{n} = Q_x - P_y$. Putting these formulas together gives the formula in Green's theorem:

$$\oint_\gamma P dx + Q dy = \oint \mathbf{E} \cdot d\mathbf{l} = \iint \mathrm{curl}(\mathbf{E}) \cdot \mathbf{n}\, dS = \iint (Q_x - P_y) dS$$

To obtain Green's theorem from the divergence theorem, we first need to state the divergence theorem in the plane. Assume γ is a positively oriented simple closed curve. Assume $\mathbf{E} = (P, Q, 0)$ is smooth on and inside γ. The vector field $\mathbf{F} = (Q, -P, 0)$ is a clockwise rotation of $\mathbf{E}$ by $\frac{\pi}{2}$. We can also rotate the unit normal $\mathbf{n}$ by $\frac{\pi}{2}$ to get the unit tangent $\mathbf{T}$; then $\mathbf{F} \cdot \mathbf{n} = \mathbf{E} \cdot \mathbf{T}$. The divergence theorem implies that

$$\int_\gamma P dx + Q dy = \int_\gamma \mathbf{E} \cdot \mathbf{T} dl = \int_\gamma \mathbf{F} \cdot \mathbf{n} dl = \int_{int(\gamma)} \mathrm{div}(\mathbf{F}) dx dy = \int_{int(\gamma)} (Q_x - P_y) dx dy.$$

In the next chapter, we will do line integrals in $\mathbb{R}^2$ using complex variable notation. We give an example here to illustrate the simplification that results.

EXAMPLE 6.1. Let γ be the (positively oriented) unit circle. We compute $\int_\gamma \frac{dz}{z}$.

• Using real variables: since $dz = dx + idy$ and $z = x + iy$,

$$\frac{dz}{z} = \frac{dx + idy}{x + iy} = \frac{x}{x^2 + y^2}dx + \frac{y}{x^2 + y^2}dy + i\frac{xdy - ydx}{x^2 + y^2}.$$

Next put $x = \cos(t)$ and $y = \sin(t)$ to get

$$\int_\gamma \frac{dz}{z} = \int_0^{2\pi}(\cos(t)(-\sin(t)) + \sin(t)\cos(t))dt + i\int_0^{2\pi} dt = 0 + 2\pi i = 2\pi i.$$

• Using complex variables directly: since $z(t) = e^{it}$ and $0 \le t \le 2\pi$, we have $dz = ie^{it}dt$. Then

$$\int_\gamma \frac{dz}{z} = \int_0^{2\pi} \frac{ie^{it}}{e^{it}}dt = \int_0^{2\pi} idt = 2\pi i.$$

EXERCISE 6.1. For n a positive integer, use complex variables to find $\int_\gamma z^n\, dz$ where γ is the positively oriented unit circle. When $n = 2$ and $n = 3$, write your result in terms of real line integrals.

EXERCISE 6.2. Let $t \mapsto \gamma(t)$ be a smooth curve in $\mathbb{R}^3$. Its velocity vector $v(t)$ is $\gamma'(t)$ and its acceleration vector $a(t)$ is $\gamma''(t)$. Its speed is $||v(t)||$. Let $\mathbf{T}$ denote the unit tangent vector and let $\mathbf{N}$ denote the unit normal vector; by definition $\mathbf{N}$ is orthogonal to $\mathbf{T}$ and lies in the plane spanned by v and a. Both $\mathbf{T}$ and $\mathbf{N}$ depend on time. Let κ denote curvature. Give a physical explanation of the formula

$$a = \frac{d}{dt}||v||\,\mathbf{T} + \kappa||v||^2\,\mathbf{N}. \tag{$*$}$$

EXERCISE 6.3. Let $\gamma(t) = (a\cos(t), a\sin(t), bt)$ define a cylindrical helix. Compute the quantities from the previous exercise. Verify that the formula for curvature agrees with your intuition in the limiting cases when a tends to 0 or when b tends to 0.

EXERCISE 6.4. Use $(*)$ from Exercise 6.2 to prove that $\kappa = \frac{||v \times a||}{||v||^3}$.

EXERCISE 6.5. Let f be a twice differentiable function on $\mathbb{R}$. The curvature of the curve defined by $y = f(x)$ is given by the formula

$$\kappa = \frac{|f''|}{(1 + (f')^2)^{\frac{3}{2}}}.$$

Prove this formula using Exercise 6.4.

7. A quick look at Maxwell's equations

Electromagnetics describes the behavior of charged particles in the presence of electric and magnetic fields $\mathbf{E}$ and $\mathbf{B}$. The Lorentz force law gives the force on a moving charge. The four Maxwell equations describe the fields themselves.

Lorentz force Law. A particle of charge q moves through an electromagnetic field with velocity $\mathbf{v}$. The force $\mathbf{F}$ upon the particle is given by

$$\mathbf{F} = q(\mathbf{E} + \mathbf{v} \times \mathbf{B}).$$

Charge density. The charge density is a scalar defined by $\rho = \frac{\text{charge}}{\text{volume}}$. Hence the total charge q in a region Ω is given by an integral:

$$\iiint_\Omega \rho dV = q.$$

Current density. The current density $\mathbf{J}$ is a vector field whose magnitude at a point measures the current per area, and whose direction is the direction of the flow of the current. Thus the total current $\mathcal{I}$ across a surface S is the flux of $\mathbf{J}$ across S. Hence, when S encloses a region Ω, we have

$$\mathcal{I} = \iint_S \mathbf{J} \cdot \mathbf{n} dS = \iiint_\Omega \operatorname{div}(\mathbf{J}) dV.$$

Theorem 7.1 (Vacuum Maxwell's equations). *The electric and magnetic fields satisfy:*

- *Gauss's Law:* $\operatorname{div}(\mathbf{E}) = \frac{\rho}{\epsilon_0}$.
- *No magnetic monopole:* $\operatorname{div}(\mathbf{B}) = 0$.
- *Faraday's law:* $\operatorname{curl}(\mathbf{E}) = -\frac{\partial \mathbf{B}}{\partial t}$.
- *Ampere's law:* $\operatorname{curl}(\mathbf{B}) = \mu_0 \mathbf{J} + \mu_0 \epsilon_0 \frac{\partial \mathbf{E}}{\partial t}$.

REMARK. Both constants are small. ϵ_0 is approximately 8.854×10^{-12} farads/meter and $\mu_0 = 4\pi \times 10^{-7}$ henries/ meter. Their product is the reciprocal of the speed of light squared. The speed of light c is very close to 3×10^8 meters/second.

REMARK. We have expressed the fields $\mathbf{E}$ and $\mathbf{B}$ in terms of SI units. Maxwell's equations appear a bit differently in Gaussian units. For example, Gauss's law becomes $\operatorname{div}(\mathbf{E}) = 4\pi\rho$. The Lorentz force law becomes

$$\mathbf{F} = q\left(\mathbf{E} + \frac{\mathbf{v} \times \mathbf{B}}{c}\right).$$

Consider a surface M with boundary ∂M. The theorems of vector analysis then combine with Maxwell's equations to describe how work and flux are related:

$$\int_{\partial M} \mathbf{E} \cdot dl = \iint_M \operatorname{curl}(\mathbf{E}) \cdot d\mathbf{S} = \iint_M -\frac{\partial \mathbf{B}}{\partial t} \cdot \mathbf{n} dS = -\frac{\partial}{\partial t} \iint_M \mathbf{B} \cdot \mathbf{n} dS.$$

Thus a changing magnetic field gives rise to a current. The analogue for $\mathbf{B}$ is

$$\int_{\partial M} \mathbf{B} \cdot dl = \iint_M \operatorname{curl}(\mathbf{B}) \cdot \mathbf{n} dS = \iint_M \left(\mu_0 \mathbf{J} + \mu_0 \epsilon_0 \frac{\partial \mathbf{E}}{\partial t}\right) \cdot \mathbf{n} dS$$

$$= \mu_0 \mathcal{I} + \mu_0 \epsilon_0 \frac{\partial}{\partial t} \iint \mathbf{E} \cdot \mathbf{n} dS.$$

The original version of Ampere's law did not have the term $\mu_0 \epsilon_0 \frac{\partial \mathbf{E}}{\partial t}$. Maxwell discovered that conservation of charge forced such a term to appear. Today we realize this correction term must appear because mixed partial derivative operators commute: $\frac{\partial}{\partial x}\frac{\partial}{\partial y} = \frac{\partial}{\partial y}\frac{\partial}{\partial x}$.

A **continuity equation** typically expresses the time derivative of the density of a quantity in terms of the divergence of the flux of the quantity. For charge this equation is

$$0 = \operatorname{div}(\mathbf{J}) + \frac{\partial \rho}{\partial t}. \tag{15}$$

Here $\mathbf{J}$ is the current density and ρ is the charge density. We will verify that (15) is consistent with Ampere's law. Start with Ampere's law as stated above, and use $0 = \operatorname{div}(\operatorname{curl}(\mathbf{B}))$. We obtain

$$0 = \operatorname{div}\left(\mu_0 \mathbf{J} + \mu_0 \epsilon_0 \frac{\partial \mathbf{E}}{\partial t}\right) = \mu_0 \operatorname{div}(\mathbf{J}) + \mu_0 \epsilon_0 \operatorname{div}\left(\frac{\partial \mathbf{E}}{\partial t}\right).$$

After dividing by μ_0 and using Gauss's law $\rho = \epsilon_0 \operatorname{div}(\mathbf{E})$, we obtain

$$0 = \operatorname{div}(\mathbf{J}) + \frac{\partial \rho}{\partial t},$$

the continuity equation. This computation explains why Maxwell added the second term to Ampere's law.

The reason that $\text{div}(\text{curl}(\mathbf{B})) = 0$ is simply the equality $f_{xy} = f_{yx}$ for twice continuously differentiable functions f. This explanation appears throughout modern physics; we will glimpse several examples as we proceed. First we note the analogue for functions.

Proposition 7.2. *Let f be a smooth function on $\mathbb{R}^3$. Then $\text{curl}(\nabla f) = \mathbf{0}$.*

PROOF. Using subscripts to denote partial derivatives, we have $\nabla f = (f_x, f_y, f_z)$. Since

$$\text{curl}(P, Q, R) = (R_y - Q_z, P_z - R_x, Q_x - P_y)$$

we obtain, using Theorem 2.2,

$$\text{curl}(\nabla f) = (f_{zy} - f_{yz}, f_{xz} - f_{zx}, f_{yx} - f_{xy}) = \mathbf{0}.$$

Therefore a gradient field is both conservative and irrotational.   □

Proposition 7.3. *Let $\mathbf{F}$ be a smooth vector field in $\mathbb{R}^3$. Then*

$$\text{div}(\text{curl}(\mathbf{F})) = 0.$$

PROOF. Write $\mathbf{B} = (P, Q, R)$. We obtain

$$\text{curl}(\mathbf{B}) = (R_y - Q_z, P_z - R_x, Q_x - P_y).$$

Using Theorem 2.2 we get

$$\text{div}(\text{curl}(\mathbf{B})) = R_{yx} - Q_{zx} + P_{zy} - R_{xy} + Q_{xz} - P_{yz} = 0.$$

Therefore the curl of a vector field is incompressible.   □

Mathematicians view both of these propositions as versions of $d^2 = 0$. Here d is called the **exterior derivative**. It is an operator on differential forms. For interested readers we include a section on differential forms and the exterior derivative d. We then express Maxwell's equations using differential forms.

8. Differential forms

Nearly every calculus student has wondered what dx really means. Modern mathematics gives a clear but deep answer, which is best understood in higher dimensions. Furthermore the answer gives a profound new perspective to vector analysis, Maxwell's equations, and similar topics. References for this material include [Dar, D3, F, HH] and the classic [Sp].

Let Ω be an open set in $\mathbb{R}^n$, and suppose $f : \Omega \to \mathbb{R}$ is a smooth function. We will define df, using directional derivatives. In particular, this definition applies to the coordinate functions x_j. Hence we get a definition of dx_j and the formula

$$df = \sum_{j=1}^{n} \frac{\partial f}{\partial x_j} \, dx_j \qquad (16)$$

is an easy proposition; furthermore the right-hand side will be rigorously defined. (Many calculus books define *the total differential* of f by the right-hand side of (16), but do not define dx_j precisely.)

Given a point $p \in \Omega$, a scalar t, and a direction $\mathbf{v}$, we know that

$$f(p + t\mathbf{v}) = f(p) + t \, (\nabla f)(p) \cdot \mathbf{v} + E,$$

where E is an error:

$$\lim_{t \to 0} \frac{E(t)}{t} = 0.$$

Consider the machine (function) that assigns to the pair $(p, \mathbf{v})$ the number

$$(\nabla f)(p) \cdot \mathbf{v} = \frac{\partial f}{\partial \mathbf{v}}(p) = \lim_{t \to 0} \frac{f(p + t\mathbf{v}) - f(p)}{t}.$$

We write df for this machine; df first seeks a point p as an input, then seeks a direction $\mathbf{v}$, and finally computes the directional derivative of f at p in the direction $\mathbf{v}$. By definition,

$$df(p)(\mathbf{v}) = (\nabla f)(p) \cdot \mathbf{v} = \frac{\partial f}{\partial \mathbf{v}}(p).$$

Thus $df(p)$ is itself a function; its domain is $\mathbb{R}^n$ and the function $\mathbf{v} \to df(p)(\mathbf{v})$ is a linear functional. Thus the gradient and the differential provide the same information, but they are not the same objects! We next define the **dual space** of a (finite-dimensional) vector space; doing so clarifies the distinction between $df(p)$ and $(\nabla f)(p)$.

Definition 8.1. Let V be a finite-dimensional real vector space. Its dual space, written V^*, is the collection of linear maps $L : V \to \mathbb{R}$. When V is a finite-dimensional complex vector space, its dual space is the collection of linear maps $L : V \to \mathbb{C}$.

REMARK. When V is an n-dimensional real vector space, its dual space V^* is also n-dimensional.

We verify the statement in this remark using bases. We may assume that V is $\mathbb{R}^n$, with standard basis e_j. Define δ_j to be the linear functional with $\delta_j(e_k)$ equal to 1 for $j = k$ and equal to 0 otherwise. We claim that $\{\delta_j\}$ is a basis for the dual space. These functionals are obviously linearly independent. We check that they span V^*. Let L be an arbitrary linear functional. We claim that

$$L = \sum_{j=1}^{n} L(e_j)\delta_j.$$

To prove the claim, we show that both sides give the same result on an arbitrary vector v. We have

$$L(v) = L(\sum v_k e_k) = \sum v_k L(e_k).$$

By linearity and the definition of the δ_j we also have

$$\left(\sum L(e_j)\delta_j\right)(v) = \left(\sum L(e_j)\delta_j\right)\left(\sum v_k e_k\right) = \sum L(e_k)v_k = L(v).$$

Thus the claim holds and the δ_j span the dual space. We call $\{\delta_j\}$ the **dual basis** to $\{e_j\}$.

REMARK. The distinction between $\mathbb{R}^n$ and its dual may seem a bit pedantic. In fact, the distinction is crucial. We give two examples. In economics, one distinguishes between the space of quantities and the space of prices. These spaces differ because the units differ. They are actually dual to each other. In vector analysis, we have glimpsed the difference between a vector field $\mathbf{E}$ and a 1-form $\mathbf{e}$. At a point p, the object $\mathbf{E}(p)$ is a vector, and the object $\mathbf{e}(p)$ is in the dual space. Such objects are often called **covectors**.

Next we establish formula (16) for the total differential. The proof mimics the proof that the δ_j form a basis for the dual space, because $df(p)$ is dual to $\nabla f(p)$.

Proposition 8.2. *Let f be a smooth function. Then*

$$df = \sum \frac{\partial f}{\partial x_j} dx_j.$$

PROOF. The result holds because both sides gives the same result when applied to each $\mathbf{v}$. To see why, we first note for each p and $\mathbf{v}$, that

$$dx_j(p)(\mathbf{v}) = v_j,$$

by definition of the differential of the coordinate function x_j. Therefore

$$\left(\sum \frac{\partial f}{\partial x_j}(p)dx_j \right)(\mathbf{v}) = \sum \frac{\partial f}{\partial x_j}(p)v_j = (\nabla f)(p) \cdot \mathbf{v} = df(p)(\mathbf{v}).$$

$\square$

Definition 8.3. A differential 1-form ω on Ω is a smooth function $\omega : \Omega \to (\mathbb{R}^n)^*$. Thus the map $\mathbf{v} \to \omega(p)(\mathbf{v})$ is a linear functional.

Thus we can write

$$\omega = \sum A_j dx_j$$

in terms of the differentials dx_j. Here the A_j are smooth scalar-valued functions. A differential 1-form ω is therefore a function whose value at p lies in the dual space to the copy of $\mathbb{R}^n$ associated with p.

The value of a differential 1-form ω at a point p is a covector. We could also think of ω as requiring both p and $\mathbf{v}$ as inputs. First we evaluate at p; then, given a vector $\mathbf{v}$, we evaluate $\omega(p)(\mathbf{v})$ to get a scalar. If we regard the vectors at p as a vector space (called the *tangent space* to $\mathbb{R}^n$ at p), then $\omega(p)$ is a linear functional on this vector space. We say that $\omega(p)$ is in the *cotangent space*.

One of the nice features of differential 1-forms is that we can integrate them along piecewise-smooth curves. As in Section 3, the line integral $\int_\gamma \omega$ computes work. From this perspective, as we will see later, it is sometimes useful to regard vector fields such as $\mathbf{E}$ or $\mathbf{B}$ as 1-forms.

Given two 1-forms ω and η, we can create a new kind of object, called a 2-form, by taking their alternating product $\omega \wedge \eta$. We have

$$\omega \wedge \eta = -\eta \wedge \omega.$$

At a point p, the 2-form $\omega \wedge \eta$ seeks two vectors as inputs, and

$$(\omega \wedge \eta)(u, v) = \omega(u)\eta(v) - \omega(v)\eta(u).$$

We can integrate 2-forms over piecewise-smooth surfaces. We computed flux integrals in Section 4 using 2-forms; see equation (8.1) for example.

Differential k-forms make sense for each integer k with $0 \le k \le n$. A 0-form f on an open set Ω is simply a smooth function; we can evaluate f at p. A 1-form ω, after being evaluated at p, seeks one vector as an input. A k-form, after being evaluated at p, requires k vectors as inputs. It depends on these vectors in an alternating multi-linear fashion just as the determinant of a k-by-k matrix does.

We can integrate k-forms over k-dimensional objects. Line integrals, such as work, are integrals of 1-forms along curves. Surface integrals, such as flux, integrate 2-forms over 2-dimensional surfaces. An n-form can be integrated over an n-dimensional object, such as an open set in n-dimensional space. What happens when $k = 0$? A 0-form is a function. A zero-dimensional object is a point, and integration becomes evaluation.

Forms of different degrees are connected by the **exterior derivative** d. For f a function, we have already defined the 1-form df. For a 1-form $\omega = \sum A_j dx_j$, we define a 2-form $d\omega$ by

$$d\omega = d\left(\sum A_j dx_j \right) = \sum dA_j \wedge dx_j = \sum_{j,k} \frac{\partial A_j}{\partial x_k} dx_k \wedge dx_j$$

$$= \sum_{k<j} \left(\frac{\partial A_j}{\partial x_k} - \frac{\partial A_k}{\partial x_j} \right) dx_k \wedge dx_j. \tag{17}$$

There is a more invariant way of defining d, which the author prefers, using Theorem 8.4. Notice the alternating or wedge product in (17).

Given 1-forms $v_1, v_2, \ldots, v_k$ we create a k-form

$$v_1 \wedge v_2 \wedge \cdots \wedge v_k.$$

This product is linear in each vector, and multiplies by -1 if we switch two of the slots. Not every k-form is the wedge product of 1-forms. Nor is every $k + 1$ form $d\omega$ for some k-form ω.

EXAMPLE 8.1. In $\mathbb{R}^4$, consider the 2-form $\omega = dx_1 \wedge dx_2 + dx_3 \wedge dx_4$. Then

$$\omega \wedge \omega = dx_1 \wedge dx_2 \wedge dx_3 \wedge dx_4 + dx_3 \wedge dx_4 \wedge dx_1 \wedge dx_2 = 2dx_1 \wedge dx_2 \wedge dx_3 \wedge dx_4 \neq 0.$$

Note that the two terms do not cancel. Hence ω is not the wedge product of two 1-forms.

We write down the formula for d on a 2-form in $\mathbb{R}^3$.

$$d(Pdy \wedge dz + Qdz \wedge dx + Rdx \wedge dy) = (P_x + Q_y + R_z)dx \wedge dy \wedge dz.$$

This formula suggests a connection between the exterior derivative and the divergence. See Proposition 8.6. We state without proof (although it is not hard) an invariant way of defining the exterior derivative.

Theorem 8.4. *There is a unique operator d taking k-forms to $(k + 1)$-forms satisfying the following properties:*

- *On functions, df agrees with our definition above.*
- *(linearity) If ω and η are k-forms, then $d(\omega + \eta) = d(\omega) + d(\eta)$.*
- *(product rule) If ω is a k-form, then*

$$d(\omega \wedge \eta) = d\omega \wedge \eta + (-1)^k \omega \wedge d\eta.$$

- $d^2 = 0$.

Differential forms and the exterior derivative operator d simplify many of the ideas in vector analysis. There is a second operator $*$, called the Hodge star operator. It takes k-forms to $(n-k)$-forms and $** = \pm 1$. On a k-form in $\mathbb{R}^n$, we have $** = (-1)^{k(n-k)}$. If one thinks of a k-form ω as describing a k-dimensional plane, then $*\omega$ describes the orthogonal $(n - k)$-dimensional plane. The volume form (commonly written dV, but not really d of something!) can be regarded as $*(1)$. We also have $1 = *(dV)$. The operator $*$ is linear. Thus $*(\omega + \eta) = *\omega + *\eta$ and $*(f\omega) = f(*\omega)$ when f is a function and ω is a differential form.

In $\mathbb{R}^2$, we have $*dx = dy$ and $*dy = -dx$. If we regard $\mathbb{R}^2$ as $\mathbb{C}$ and work with the complex differentials, then we have

$$*dz = *(dx + idy) = dy - idx = -i(dx + idy) = -idz$$

$$*d\overline{z} = *(dx - idy) = dy + idx = i(dx - idy) = id\overline{z}.$$

Thus dz and $d\overline{z}$ are eigenvectors for the $*$ operator.

REMARK. The alert reader will realize that $*$ depends on the metric used. To illustrate this point, we compute $*$ in polar coordinates in two dimensions. Notice that $dr = \frac{xdx+ydy}{r}$ and $d\theta = \frac{-ydx+xdy}{r^2}$. Then

- We have $*(1) = rdrd\theta$ and $*(rdrd\theta) = 1$.
- We have $*(dr) = rd\theta$ and $*(d\theta) = \frac{-dr}{r}$.

In a general curved space, one defines $*$ as follows. Assume $\omega_1, \cdots, \omega_n$ are a positively-oriented orthonormal basis of 1-forms. Then

$$*(\omega_1 \wedge \cdots \wedge \omega_k) = \omega_{k+1} \wedge \cdots \wedge \omega_n.$$

Combining d and $*$ in the right way gives divergence and curl. We will derive and use these expressions when we express Maxwell's equations in a new way.

The modern Stokes' theorem includes Green's theorem, the divergence theorem, the classical Stokes' theorem, and the fundamental theorem of calculus all in one tidy formula:

Theorem 8.5 (Modern Stokes' theorem). *Let Ω be a smooth k-dimensional object with smooth oriented boundary $\partial\Omega$. Let ω be a smooth $(k-1)$-form. Then*

$$\int_\Omega dw = \int_{\partial\Omega} w. \tag{18}$$

See [HH] for a proof. On the right-hand side of (18) we are integrating a $(k-1)$-form around the boundary of Ω. This boundary is a $(k-1)$-dimensional object.

REMARK. When S is the boundary of a region Ω, and $d\eta$ is an exact form, Stokes' theorem and $d^2 = 0$ imply

$$\int_S \omega + d\eta = \int_\Omega d\omega + d(d\eta) = \int_\Omega d\omega = \int_S \omega.$$

Thus exact forms do not contribute to such integrals.

Stokes' theorem lies behind de Rham's theorem, which relates integrals of forms to topology. See [Fr] for an introduction to these ideas. We mention that there is a natural equivalence relation on closed forms: closed differential k-forms lie in the same class if they differ by an exact form. Rather than pursuing this beautiful topic, we simply relate the modern Stokes' theorem to Green's theorem, the divergence theorem, and the classical Stokes' theorem.

Consider a 1-form $Pdx + Qdy$ in $\mathbb{R}^2$. We have

$$d(Pdx + Qdy) = (Q_x - P_y)dx \wedge dy.$$

The formula in Green's theorem becomes

$$\int_\Omega (Q_x - P_y)dx \wedge dy = \int_\Omega d(Pdx + Qdy) = \int_{\partial\Omega} Pdx + Qdy.$$

Similarly we can derive the divergence theorem and the classical Stokes' theorem. Let $\mathbf{E} = (E_1, \ldots, E_n)$ be a vector field defined on an open set in $\mathbb{R}^n$. Its divergence is defined by

$$\operatorname{div}(\mathbf{E}) = \sum_{j=1}^n \frac{\partial E_j}{\partial x_j}.$$

It is remarkable that the divergence is independent of the coordinate system used. Given $\mathbf{E}$, we consider the *dual* 1-form $\mathbf{e}$ given by

$$\mathbf{e} = \sum_{j=1}^n E_j dx_j.$$

Then $\operatorname{div}(\mathbf{E}) = *d * \mathbf{e}$. Since $\mathbf{e}$ is a 1-form, $*\mathbf{e}$ is an $(n-1)$-form, and $d * \mathbf{e}$ is an n-form. Finally, one more $*$ gives us a 0-form, in other words, a function. Thus the modern Stokes' theorem says, under appropriate hypotheses, that

$$\int_\Omega \operatorname{div}(\mathbf{E})dV = \int_\Omega *d * \mathbf{e}\, dV = \int_\Omega d * \mathbf{e} = \int_{\partial\Omega} *\mathbf{e}.$$

In 3-space, suppose $\mathbf{E} = (P, Q, R)$. Then $\mathbf{e} = Pdx + Qdy + Rdz$. We then get

$$*\mathbf{e} = Pdy \wedge dz + Qdz \wedge dx + Rdx \wedge dy.$$

Thus formula (12) from Theorem 5.1 follows from the modern Stokes' theorem.

In three dimensions we can define the curl of a vector field by

$$\text{curl}(\mathbf{E}) = *d\mathbf{e}.$$

Notice that $\mathbf{e}$ is a 1-form. Hence, in three dimensions, $d\mathbf{e}$ is a 2-form, and therefore $*d\mathbf{e}$ is a 1-form. If we wish, we may identify it with a vector field. If $\mathbf{E} = (P, Q, R)$, then

$$\text{curl}(\mathbf{E}) = (R_y - Q_z, P_z - R_x, Q_x - P_y).$$

Using the notation of differential forms, we have

$$\text{curl}(\mathbf{e}) = *d\mathbf{e} = (R_y - Q_z)dx + (P_z - R_x)dy + (Q_x - P_y)dz.$$

Therefore the classical Stokes' theorem simply says

$$\int_\Omega \text{curl}(\mathbf{E}) \cdot dS = \int_\Omega *d\mathbf{e} \cdot dS = \int_\Omega d\mathbf{e} = \int_{\partial\Omega} \mathbf{e}.$$

The rule $d^2 = 0$ restates Theorem 2.2 on the equality of mixed partial derivatives. Notice that Propositions 7.2 and 7.3 become automatic. We mention only briefly the notions of closed and exact differential forms. A k-form α is called **exact** if there is a $(k-1)$-form ω with $d\omega = \alpha$. The form α is called **closed** if $d\alpha = 0$. Since $d^2 = 0$, we see that exact implies closed. The converse is false; if a form is closed in a region without holes, however, then it is exact there. In the next chapter we will see that $\frac{dz}{z}$ is closed in the complex plane minus the origin, but it is not exact. The electric field due to a point charge in 3-space provides another example. The corresponding 1-form is closed but not exact in $\mathbb{R}^3$ minus the point where the charge is located.

The following simple result summarizes some of our computations while providing a dictionary between the languages of vector analysis and differential forms.

Proposition 8.6. *Let $\mathbf{E}$ be a smooth vector field in an open set in $\mathbb{R}^3$, and let $\mathbf{e}$ be the corresponding dual 1-form. Let f be a smooth function. Then, since $d^2 = 0$ and $**$ is the identity on 2-forms,*

- $\text{div}(\mathbf{E}) = *d * \mathbf{e}$
- $\text{curl}(\mathbf{E}) = *d\mathbf{e}$
- $\text{div curl}(\mathbf{E}) = *d * (*d\mathbf{e}) = *d^2\mathbf{e} = 0.$
- $\text{curl}(\nabla f) = *d(df) = 0.$

Maxwell's equations have a remarkably simple expression in the language of differential forms. This expression even makes sense when considering relativity.

Rather than considering the electric and magnetic fields separately, we consider a combination of the two. First we regard these fields $\mathbf{E}$ and $\mathbf{B}$ as 1-forms $\mathbf{e}$ and $\mathbf{b}$. Let $*_n$ denote the Hodge star operator in n-space; here n can be both 3 and 4. Let c denote the speed of light. Put

$$\mathbf{F} = \mathbf{e} \wedge c\,dt + *_3\mathbf{b}.$$

Then $\mathbf{F}$ is a 2-form in $\mathbb{R}^4$. The four Maxwell equations become two equations, one finding $d\mathbf{F}$ and the other finding $d *_4 \mathbf{F}$. In this formulation, we need to introduce a 3-form, called the current 3-form, which incorporates both the current 1-form $\mathbf{J}$ and the charge density ρ. The current 3-form $\mathcal{J}$ is defined by

$$\mathcal{J} = \rho\, c\, dx \wedge dy \wedge dz + c\, dt \wedge *_3\mathbf{J}.$$

Now we can restate Maxwell's equations as

$$d\mathbf{F} = 0.$$
$$d(*_4\mathbf{F}) = \mathcal{J}.$$

We can simplify even further. In nice regions, $d\mathbf{F} = 0$ implies that $\mathbf{F} = d\mathbf{A}$ for some 1-form $\mathbf{A}$. This 1-form $\mathbf{A}$ is called the *potential*. Using A, we can write Maxwell's equations as

$$d *_4 d\mathbf{A} = \mathcal{J}.$$

The operator $d *_4 d$ is of second-order and somewhat analogous to the Laplace operator on functions. This approach is used in the branch of physics called *gauge theory*, where things are much more general, such as the Yang-Mills equations.

EXERCISE 8.1. Verify that each of the formulas from Propositions 7.2 and 7.3 says simply that $d^2 = 0$.

EXERCISE 8.2. Suppose $z = f(x, y)$ defines a surface M in $\mathbb{R}^3$. In other words, $M = \{(x, y, f(x, y)\}$ for (x, y) lying in an open set Ω in $\mathbb{R}^2$. Find and prove a formula for the surface area of M in terms of $\nabla(f)$.

EXERCISE 8.3. Put $\omega = (z - 1)dx \wedge dy - xdy \wedge dz$. Show that $d\omega = 0$. Find a 1-form u with $du = \omega$.

EXERCISE 8.4. Define an $(n - 1)$-form in the complement of the origin in $\mathbb{R}^n$ by the formula

$$\omega_n = \frac{\sum_{j=1}^n (-1)^{j+1} x_j dx_1 \wedge \cdots \wedge dx_{j-1} \wedge dx_{j+1} \cdots \wedge dx_n}{||x||^{\frac{n}{2}}}.$$

Note that dx_j is omitted in the numerator.

- Prove that ω_n is closed but not exact.
- Find $\int_S \omega_n$ if S is the unit sphere.
- What is the physical significance of w_3?
- The volume of a ball of radius R in $\mathbb{R}^n$ is $\lambda_n R^n$ for some constant λ_n. Prove that the surface area of the sphere of radius R is $n\lambda_n R^{n-1}$. Suggestion: Use ω_n.
- Compute λ_n in the previous item.

EXERCISE 8.5. For a smooth function f, express $*d * df$ in terms of derivatives of f. Use the remark after Theorem 8.4 to find $*d * df$ in polar coordinates in $\mathbb{R}^2$.

EXERCISE 8.6. Let ω be the 2-form defined in $\mathbb{R}^{2n}$ by

$$\omega = dx_1 \wedge dx_2 + dx_3 \wedge dx_4 + \cdots + dx_{2n-1} \wedge dx_{2n}.$$

Compute the wedge product of ω with itself n times. (The answer is not 0.)

9. Inverse and implicit functions

Perhaps the main point of calculus is *linearization*. We work with curved objects by approximating them with straight objects. This idea underlies both the theory of differentiation and the theory of integration. One differentiates to locally linearize a problem, and one integrates to recover the global behavior.

In this section we discuss the inverse and implicit function theorems for continuously differentiable functions. The results are essentially the same for infinitely differentiable functions. Especially for implicit functions, the easiest way to understand these theorems uses differential forms.

Consider an open set Ω in $\mathbb{R}^n \times \mathbb{R}^k$ and a continuously differentiable function $f : \Omega \to \mathbb{R}^k$. We write the points in Ω as pairs (x, y), where $x \in \mathbb{R}^n$ and $y \in \mathbb{R}^k$. Suppose that $f(x_0, y_0) = 0$. The implicit function theorem provides circumstances in which we can solve the equation $f(x, y) = 0$ for y as a function of x.

EXAMPLE 9.1. Put $f(x, y) = y + xe^{xy}$ for $x, y \in \mathbb{R}$. We cannot solve the equation $f(x, y) = 0$ for y as an explicit elementary function of x. Suppose, however, we wish to understand how y depends on x near

$(0,0)$. We would like to say that, near $(0,0)$, we can write $f(x,y) = 0$ if and only if $y = g(x)$ for some nice function g. We would also like to know the Taylor expansion of g. Recall that

$$e^{xy} = 1 + xy + \frac{x^2 y^2}{2} + \cdots$$

Replacing e^{xy} by 1, which is approximately correct near $(0,0)$, we obtain the approximate equation $y = -x$. Replacing e^{xy} by $1+xy$, we obtain $y = \frac{-x}{1+x^2}$. Proceeding in this way, we can approximate g. See Figure 11. One problem in this approach is that we obtain higher-order polynomial equations to solve. In this example, implicit differentiation gives

$$\frac{dy}{dx} = \frac{-e^{xy}(1+xy)}{1+x^2 e^{xy}}. \tag{19}$$

We have an explicit formula (19) for the derivative of the implicit function, but not for the function itself.

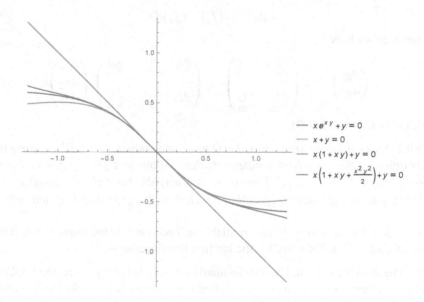

$$\begin{aligned}
&— \quad xe^{xy} + y = 0 \\
&— \quad x + y = 0 \\
&— \quad x(1+xy) + y = 0 \\
&— \quad x\left(1 + xy + \frac{x^2 y^2}{2}\right) + y = 0
\end{aligned}$$

FIGURE 11. Some implicit functions

EXAMPLE 9.2. Consider the equation $x^2 + y^2 = 1$ for a circle in the plane. Near the point $(1,0)$ on the circle, we cannot solve the equation for y, because $y = \pm\sqrt{1-x^2}$, and we do not know which branch to take. Note, by implicit differentiation, that $\frac{dy}{dx} = \frac{-x}{y}$ here, as long as $y \neq 0$. This example suggests that we must require a hypothesis on the derivative with respect to y if we want to solve for y.

EXAMPLE 9.3. Consider a pair of non-linear equations in five variables. We write the information as $f(x,y) = 0$. The author thinks of it this way, but the reader might prefer regarding this equation as an abbreviation for the pair of equations

$$f_1(x_1, x_2, x_3, y_1, y_2) = 0 \tag{20.1}$$

$$f_2(x_1, x_2, x_3, y_1, y_2) = 0. \tag{20.2}$$

Taking exterior derivatives of the equation $f(x, y) = 0$, we obtain

$$f_x dx + f_y dy = \frac{\partial f}{\partial x} dx + \frac{\partial f}{\partial y} dy = 0. \tag{21}$$

In (21), $\frac{\partial f}{\partial x}$ and $\frac{\partial f}{\partial y}$ are matrices, and dx and dy are column vectors of differentials. The reader might prefer to write this information as the pair of equations:

$$\frac{\partial f_1}{\partial x_1} dx_1 + \frac{\partial f_1}{\partial x_2} dx_2 + \frac{\partial f_1}{\partial x_3} dx_3 + \frac{\partial f_1}{\partial y_1} dy_1 + \frac{\partial f_1}{\partial y_2} dy_2 = 0 \tag{22.1}$$

$$\frac{\partial f_2}{\partial x_1} dx_1 + \frac{\partial f_2}{\partial x_2} dx_2 + \frac{\partial f_2}{\partial x_3} dx_3 + \frac{\partial f_2}{\partial y_1} dy_1 + \frac{\partial f_2}{\partial y_2} dy_2 = 0. \tag{22.2}$$

Suppose that the matrix $\frac{\partial f}{\partial y}$ is invertible at some point. Then we can solve the system of equations (22) for the differentials dy_j in terms of the differentials dx_j. In the abbreviated notation, we have

$$dy = -(f_y)^{-1}(f_x)dx. \tag{23}$$

In the longer notation we have

$$\begin{pmatrix} dy_1 \\ dy_2 \end{pmatrix} = - \begin{pmatrix} \frac{\partial f_1}{\partial y_1} & \frac{\partial f_1}{\partial y_2} \\ \frac{\partial f_2}{\partial y_1} & \frac{\partial f_2}{\partial y_2} \end{pmatrix}^{-1} \begin{pmatrix} \frac{\partial f_1}{\partial x_1} & \frac{\partial f_1}{\partial x_2} & \frac{\partial f_1}{\partial x_3} \\ \frac{\partial f_2}{\partial x_1} & \frac{\partial f_2}{\partial x_2} & \frac{\partial f_2}{\partial x_3} \end{pmatrix} \begin{pmatrix} dx_1 \\ dx_2 \\ dx_3 \end{pmatrix}.$$

Henceforth we use notation as in (23).

Theorem 9.1 (Implicit function theorem). *Let Ω be an open subset of $\mathbb{R}^n \times \mathbb{R}^k$. Assume that $f : \Omega \to \mathbb{R}^k$ is continuously differentiable (or infinitely differentiable). Assume that $f(x_0, y_0) = 0$. Assume further that the matrix $\frac{\partial f}{\partial y}$ is invertible at (x_0, y_0). Then there is a neighborhood U of (x_0, y_0) and a continuously differentiable (or infinitely differentiable) function g such that $f(x, y) = 0$ on U if and only if $y = g(x)$.*

REMARK. When f is continuously differentiable in Theorem 9.1, the implicit function also is. When f is infinitely differentiable in Theorem 9.1, the implicit function also is.

REMARK. Theorem 9.1 is local. Even if the matrix is invertible everywhere, the implicit function need not be defined everywhere. A common error in applied mathematics is to invoke the theorem globally when it holds only locally. A similar warning applies for the inverse function theorem. We give simple examples in Exercises 9.2 and 9.3.

REMARK. One reason for using the notation as in (23) is that this notation works in infinite-dimensional settings. The analogues of Theorems 9.1 and 9.2 hold as well.

Consider the equation $f(x, y) = 0$. Here x lives in n-space, y lives in k space, and 0 lives in k-space. Assume that f is continuously differentiable. We linearize by writing

$$f_x dx + f_y dy = 0,$$

where f_x and f_y are the matrices of derivatives as defined in Example 9.3. The implicit theorem says, in more technical language, the following thing. We are given the smooth system of equations $f(x, y) = 0$ in the $n + k$ variables (x, y). We know that $f(x_0, y_0) = 0$ holds at a single point (x_0, y_0) and that we can solve the equation (21) for the differentials dy at that point. Then, there is a nice function g such that, near this point, we have $f(x, y) = 0$ if and only if $y = g(x)$. Furthermore the derivatives of g are obtained by implicit differentiation of the equation $f(x, y) = 0$, using equations (21) or (23).

Recall that a function $f : A \to B$ is **injective** if $f(a) = f(t)$ implies $a = t$. An injective function has an inverse: for y in the image of f, we define $f^{-1}(y)$ to be the unique a with $f(a) = y$. A function $f : A \to B$ is **surjective** (or **onto**) if the image of A under f is all of B. Suppose f is both injective and surjective. By surjectivity, for each $b \in B$ there is an $a \in A$ with $f(a) = b$. By injectivity, this a is unique. This surjectivity guarantees existence of a solution and injectivity guarantees uniqueness of the solution.

We next state the inverse function theorem. The implicit function theorem is essentially equivalent to the inverse function theorem, in the sense that each can be easily derived from the other.

Theorem 9.2 (Inverse function theorem). *Let Ω be an open subset of $\mathbb{R}^n$. Assume that $f : \Omega \to \mathbb{R}^n$ is continuously differentiable (or infinitely differentiable). Assume that $f(x_0) = 0$ and that the matrix $\frac{\partial f}{\partial x}$ is invertible at x_0. Then there is a neighborhood U of x_0 on which f is injective. The inverse function is continuously differentiable (or infinitely differentiable).*

We refer to [HH] or [Sp] for proofs of these theorems.

EXERCISE 9.1. Consider the implicit equation $xe^y + 2ye^x = 0$. At what points can we locally solve for y as a function of x? What is $\frac{dy}{dx}$ at these points?

EXERCISE 9.2. Assume $F : \mathbb{R}^2 \to \mathbb{R}^2$ is given by

$$F(x_1, x_2) = (e^{x_1} \cos(x_2), e^{x_1} \sin(x_2)).$$

Show that the matrix $\frac{\partial F}{\partial x}$ is invertible everywhere, but that the function f is only locally injective. Write F as a single function of one complex variable.

EXERCISE 9.3. Assume $F : \mathbb{R}^2 \to \mathbb{R}^2$ is given by

$$F(x_1, x_2) = (x_1^2 - x_2^2, 2x_1 x_2).$$

At what points is F locally injective? Express F as a function of a single complex variable.

EXERCISE 9.4. First observe, if A is the matrix $\begin{pmatrix} 1 & 2 \\ 3 & 1 \end{pmatrix}$, then $A^2 = \begin{pmatrix} 7 & 4 \\ 6 & 7 \end{pmatrix}$. Can we find square roots of matrices near A^2 that are near A? Suggestion: Think of the entries as elements of $\mathbb{R}^4$, compute appropriate derivatives, and apply the inverse function theorem.

EXERCISE 9.5. Let $r : \mathbb{R}^n \to \mathbb{R}$ be smooth with $r(0) = 0$. Assume $dr(0) \neq 0$. Show that there are coordinates $(x_1, \ldots, x_n)$ and a smooth function g, defined near 0, such that (near 0) we can write $r(x) = 0$ if and only if $x_n = g(x_1, \ldots, x_{n-1})$.

EXERCISE 9.6. The implicit equation $x = W(x)e^{W(x)}$ arises throughout science. (See [CGHJK] and the Wikipedia page on the Lambert W-function.) Show that $W' = \frac{W}{x(1+W)}$ for $x \neq 0$ and $x \neq \frac{-1}{e}$. Define K by

$$K(x) = x \left(W(x) + \frac{1}{W(x)} - 1 \right).$$

Show that $K' = W$. Then show that $\int_0^e W(x)dx = e - 1$.

CHAPTER 4

Complex analysis

We return to the complex number system $\mathbb{C}$. We introduce complex differentiation and integration. Our reward is a beautiful and deep theory with applications throughout science.

1. Open subsets of $\mathbb{C}$

A subset of $\mathbb{C}$ is called **open** if, for each $p \in \Omega$, there is a ball about p also contained in Ω. In other words, Ω is open if, for each point p in Ω, all points sufficiently close to p are also in Ω. Recall that $|z - p|$ equals the distance between z and p, and $B_r(p) = \{z : |z - p| < r\}$ is the open ball of radius r about p. The triangle inequality asserts, for all $z, w, \zeta \in \mathbb{C}$, that

$$|z - w| \leq |z - \zeta| + |\zeta - w|. \tag{1}$$

In words, the distance between z and w is at most the sum of the distances from z to ζ and w to ζ. This inequality is used when verifying statements about limits. We give two standard examples; the reader should recognize these examples as proofs that the sum of two limits is the limit of the sum, and that the product of two limits is the limit of the product. After the examples you are asked to use the triangle inequality to verify that an open ball is an open set.

EXAMPLE 1.1. Suppose the side-lengths of a rectangle are x_0 and y_0. We measure them, with some small error. How accurately must we measure the sides to guarantee that the perimeter is within ϵ units of the correct answer? Call our measurements x and y. Since the perimeter is given by $2(x_0 + y_0)$, we want

$$|2(x + y) - 2(x_0 + y_0)| < \epsilon,$$

or equivalently

$$|(x - x_0) + (y - y_0)| < \frac{\epsilon}{2}.$$

By the triangle inequality, $|(x - x_0) + (y - y_0)| \leq |x - x_0| + |y - y_0|$. Hence we succeed if we make each measurement to within $\frac{\epsilon}{4}$. Here the errors in measurement add.

EXAMPLE 1.2. Again suppose the side-lengths of a rectangle are x_0 and y_0, but we wish to measure the area to within ϵ square units. We want

$$|xy - x_0 y_0| < \epsilon.$$

Using the triangle inequality we have

$$|xy - x_0 y_0| = |x_0 y_0 - xy_0 + xy_0 - xy| \leq |y_0| \, |x - x_0| + |x| \, |y - y_0|.$$

To ensure that the right-hand side is at most ϵ, we need to work a bit harder. One way to proceed is to use some bound for y_0 and x. Assume each of these numbers is at most B. Then

$$|x_0 y_0 - xy| \leq |y_0| \, |x - x_0| + |x| \, |y - y_0| \leq B(|x - x_0| + |y - y_0|) < \epsilon,$$

whenever both measurements are within $\frac{\epsilon}{2B}$. Here, the accuracy of the measurements required depends on the actual side lengths! See Figure 1.

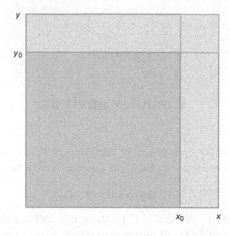

FIGURE 1. Area and the limit of products

We clarify the notion of open subset of $\mathbb{C}$ by giving some examples.

EXAMPLE 1.3. Open sets.

- The empty set $\varnothing$ is open.
- All of $\mathbb{C}$ is open.
- $\mathbb{R}$ is not an open subset of $\mathbb{C}$.
- The open first quadrant Q is open; here $Q = \{x + iy : x > 0 \text{ and } y > 0\}$.
- The open ball $B_r(p)$ is an open set.
- An arbitrary union of open sets is open. In particular, open sets do not need to be *connected*.
- An open interval on the real line is not open in $\mathbb{C}$, but it is open as a subset of $\mathbb{R}$.

We do not give a precise definition of connected set, but we do give a standard characterization of connected open set. Most readers should use the intuitive idea that a set is connected if it has one piece.

Definition 1.1. An open set Ω is connected if, whenever $p, q \in \Omega$, there is a polygonal path (with finitely many edges) from p to q that lies in Ω.

EXERCISE 1.1. Use (1) to prove rigorously that an open ball is an open set.

EXERCISE 1.2. Show that an arbitrary union of open sets is open. Show that a finite intersection of open sets is open. Give an example of a countable intersection of open sets that is not open.

EXERCISE 1.3. Let Ω be the complement of the non-negative real axis. Show that Ω is open.

2. Complex differentiation

In each situation, the derivative of a function at a point provides a linear approximation for that function. To define the derivative of a function f at a point p, one must therefore know the values of f near p. In other words, to define derivatives (in any context), one works with open sets. Let Ω be an open subset of $\mathbb{C}$.

Definition 2.1. A function $f : \Omega \to \mathbb{C}$ is **complex analytic** on Ω if, for all $p \in \Omega$, the limit $\lim_{h \to 0} \frac{f(p+h) - f(p)}{h}$ exists.

Thus f is complex analytic on Ω if it is complex differentiable at each point of Ω. We write $f'(p)$ for this limit and call it the derivative of f. The existence of this limit has far-reaching consequences. A function f is complex analytic at p if, for $|h|$ small, $f(p+h)$ is approximately equal to $f(p)+f'(p)h$. Thus, infinitesimally, f changes by multiplying h by $f'(p)$. The reader should compare with the definition of the gradient from Chapter 3.

EXAMPLE 2.1. Complex analytic functions.

- Put $f(z) = z^n$ for $n \in \mathbb{N}$. Then f is complex analytic at every point, and $f'(z) = nz^{n-1}$.
- Put $f(z) = \frac{1}{z}$. Then f is complex analytic at every point of the open set $\mathbb{C} - \{0\}$, with $f'(z) = \frac{-1}{z^2}$.
- Put $f(z) = e^z$, defined in Chapter 2 via its power series. Then f is complex analytic everywhere and $f' = f$.
- If $\sum_{n=0}^{\infty} a_n(z-p)^n$ converges for $|z-p| < R$, then this series defines a complex analytic function there. Its derivative at each z is given by $\sum_{n=0}^{\infty} na_n(z-p)^{n-1}$, which also converges for $|z-p| < R$.

The complex analytic functions on an open set form a vector space, because sums and constant multiples of analytic functions are analytic. By the product rule for derivatives, it is also true that the product of analytic functions is analytic. If f is analytic and not zero, then $\frac{1}{f}$ is also analytic. The chain rule also holds: if f is complex differentiable at p, and g is complex differentiable at $f(p)$, then $g \circ f$ is complex differentiable at p and $(g \circ f)'(p) = g'(f(p)) f'(p)$. Thus many aspects of complex differentiability are analogous to those for real differentiability. Some significant differences will become evident as we proceed.

We state without proof the following simple result, which is analogous to the real case. Many proofs are possible.

Lemma 2.2. *Suppose f is complex analytic on an open connected set Ω and $f'(z) = 0$ for all $z \in \Omega$. Then f is a constant.*

One often writes $f = u + iv$ in terms of the real and imaginary parts of f, and regards u and v as functions of (x, y). Consider, for example, the function $z \mapsto z^3$:

$$z^3 = (x+iy)^3 = x^3 - 3xy^2 + i(3x^2y - y^3) = u(x,y) + iv(x,y).$$

We noted in Example 2.8 of Chapter 3 that the level sets of u and v form orthogonal trajectories. We have the following general result:

Theorem 2.3 (Cauchy–Riemann equations). *Let $f(z) = u(x,y) + iv(x,y)$ be complex analytic on Ω. Then $u_x = v_y$ and $u_y = -v_x$ on Ω. The level sets of u, v form orthogonal trajectories at any point p where $f'(p) \neq 0$.*

PROOF. By definition of analyticity, $\lim_{h\to 0} \frac{f(z+h)-f(z)}{h}$ exists, where $h \in \mathbb{C}$. We take the special cases where $h \in \mathbb{R}$ and where h is purely imaginary. Thus both of the following limits exist, and they are equal:

$$\lim_{h\to 0} \frac{u(x+h,y) + iv(x+h,y) - (u(x,y) + iv(x,y))}{h} \tag{2.1}$$

$$\lim_{k\to 0} \frac{u(x,y+k) + iv(x,y+k) - (u(x,y) + iv(x,y))}{ik}. \tag{2.2}$$

(2.1) yields $u_x + iv_x$, and (2.2) yields $\frac{1}{i}u_y + \frac{1}{i}iv_y$. Therefore $u_x + iv_x = -iu_y + v_y$. Equating real and imaginary parts gives $u_x = v_y$ and $u_y = -v_x$.

Note that $f'(p) = 0$ if and only if both $\nabla u(p)$ and $\nabla v(p) = 0$. Hence, as long as $f'(p) \neq 0$, these gradients are non-zero and perpendicular. Thus the level sets are orthogonal. $\square$

Definition 2.4. Differentiable functions u, v of two variables satisfy the **Cauchy–Riemann** equations if $u_x = v_y$ and $u_y = -v_x$.

Definition 2.5. A real-valued twice differentiable function u defined on an open set in $\mathbb{R}^2$ is called **harmonic** if $u_{xx} + u_{yy} = 0$.

Corollary 2.6. *Let f be complex analytic with $f = u + iv$. If u, v are twice continuously differentiable (which is always true in this setting), then u and v are harmonic.*

PROOF. Using the Cauchy–Riemann equations and Theorem 2.2 of Chapter 3, we get

$$u_{xx} + u_{yy} = v_{yx} - v_{xy} = 0$$

$$v_{xx} + v_{yy} = -u_{yx} + u_{xy} = 0.$$

$\square$

REMARK. The property that f is complex analytic is very strong. It implies in particular that u and v are infinitely differentiable. We will not formally prove this result; it follows from the power series representation of an analytic function and basic analysis.

We mention without proof a converse of Theorem 2.1. See [A], [B], or [D2].

Theorem 2.7. *Suppose u, v are continuously differentiable real-valued functions on an open set Ω, and u, v satisfy the Cauchy–Riemann equations. Put $f = u + iv$. Then f is complex analytic on Ω.*

Let us return to two examples we have already seen:

EXAMPLE 2.2. Put $f(z) = z^2$. Then $f = u + iv$ where $u(x, y) = x^2 - y^2$ and $v(x, y) = 2xy$. The Cauchy–Riemann equations hold and the orthogonal trajectory result follows.

EXAMPLE 2.3. Put $f(z) = \log(z)$ on $0 < |z|$ and $-\pi < \theta < \pi$. The domain Ω of f is an open set. Since $\log(z) = \log|z| + i\theta$, we have

$$u(x, y) = \log|z| = \frac{1}{2}\log(x^2 + y^2)$$

$$v(x, y) = \tan^{-1}\left(\frac{y}{x}\right),$$

where $x \neq 0$. We obtain

$$u_x = \frac{x}{x^2 + y^2} \text{ and } u_y = \frac{y}{x^2 + y^2}$$

$$v_x = -\frac{y}{x^2 + y^2} \text{ and } v_y = \frac{x}{x^2 + y^2}.$$

By Theorem 2.2, f is complex analytic. The level sets of u are circles centered at 0 and the level sets of v are lines through 0 and hence intersect these circles orthogonally.

REMARK. The use of the inverse-tangent function in Example 2.3 is convenient, and nothing strange happens when $x = 0$. Nonetheless one sees that working with logs is subtle. The function f is called a *branch* of the logarithm. Had we defined Ω differently, we would have defined a different branch, and hence a different function.

EXERCISE 2.1. Put $u(x, y) = e^x \cos(y)$ and $v(x, y) = e^x \sin(y)$. Show that u, v satisfy the Cauchy–Riemann equations. What is $u + iv$?

EXERCISE 2.2. Put $f(z) = \frac{1}{z}$. Express f in the form $u(x,y) + iv(x,y)$ and verify the Cauchy–Riemann equations. Comment: It is generally better to work with z and $\bar{z}$ than with x and y. The next exercises indicate why.

EXERCISE 2.3. Introduce the formal differential operators

$$\frac{\partial}{\partial z} = \frac{1}{2}\left(\frac{\partial}{\partial x} - i\frac{\partial}{\partial y}\right)$$

$$\frac{\partial}{\partial \bar{z}} = \frac{1}{2}\left(\frac{\partial}{\partial x} + i\frac{\partial}{\partial y}\right).$$

Show that the Cauchy–Riemann equations are equivalent to $f_{\bar{z}} = \frac{\partial f}{\partial \bar{z}} = 0$.

EXERCISE 2.4. Show that $u_{xx} + u_{yy} = 4u_{z\bar{z}}$ and hence u is harmonic if and only if $u_{z\bar{z}} = 0$.

EXERCISE 2.5. If f is complex analytic on an open connected set, and $|f|$ is a constant, then f is a constant. Prove this statement by considering the Laplacian of $|f|^2 = u^2 + v^2$. Suggestion: Use z and $\bar{z}$ derivatives. Compare with using x and y.

EXERCISE 2.6. Put $p(x,y) = Ax^4 + Bx^3y + Cx^2y^2 + Dxy^3 + Ey^4$. Give a necessary and sufficient condition on these constants for p to be harmonic. Comment: The next two exercises explain the situation.

EXERCISE 2.7. Assume that $p(x,y)$ is a homogeneous polynomial of degree d:

$$p(x,y) = \sum_{j=0}^{d} c_j x^j y^{d-j}.$$

Under what condition is p harmonic? Suggestion: Think about the real and imaginary parts of z^d.

EXERCISE 2.8. For each positive integer d, show that the set of harmonic homogeneous polynomials of degree d is a vector space. Compute its dimension. Find a basis. Suggestion: Consider the previous exercise.

3. Complex integration

In this section we begin to develop the basic material on complex line integrals. We are rewarded in the next section by the Cauchy theory. Later in the chapter we use complex line integrals to evaluate real integrals that arise in applications.

The Cauchy–Riemann equations lead, via Green's theorem, to a proof of the Cauchy theorem. In complex analysis courses, one can prove this result directly without passing through Green's theorem, and therefore eliminating the stronger assumption that the partial derivatives of u and v are continuous. See [A,D2] for example.

The definition of a complex line integral is nothing new. Let γ be a piecewise-smooth curve in $\mathbb{C}$, and let $Pdx + Qdy$ be a smooth 1-form. We already know from Chapter 3 what $\int_{\gamma} Pdx + Qdy$ means and how to compute it. We introduce a new twist, by rewriting the integral in terms of dz and $d\bar{z}$. Here we put $dz = dx + idy$ and $d\bar{z} = dx - idy$. It follows that $dx = \frac{dz+d\bar{z}}{2}$ and $dy = \frac{dz-d\bar{z}}{2i}$. In nearly all of our work, the differential $d\bar{z}$ will not arise. The most common situation will be a differential 1-form of the type $f(z)dz$, where f is complex analytic.

We next state some simple inequalities about integrals that often get used in the subsequent examples.

Proposition 3.1. *Let f, g be integrable functions with $f(t) \leq g(t)$ on the interval $[a, b]$. Then*

$$\int_a^b f(t)dt \leq \int_a^b g(t)dt.$$

Proposition 3.2. *Let γ be a piecewise-smooth curve and let f be continuous on γ. Then*

$$\left| \int_\gamma f(z)dz \right| \leq \int_\gamma |f(z)dz|.$$

Proposition 3.3. *Let γ be a piecewise-smooth curve of length L. Suppose that $|f(z)| \leq M$ on γ. Then*

$$\left| \int_\gamma f(z)dz \right| \leq ML.$$

In the next section we use Green's theorem to prove that

$$\oint_\gamma f(z)dz = 0$$

when f is analytic on and inside a closed curve γ. When f fails to be analytic at even a single point, this conclusion need not hold. The decisive example is given by $f(z) = \frac{1}{z}$. Then f is complex-analytic on $\mathbb{C} - \{0\}$. Let γ be the unit circle, oriented positively. Put $z = e^{i\theta}$ to parametrize the unit circle. Then

$$\oint_\gamma \frac{dz}{z} = \int_0^{2\pi} \frac{ie^{i\theta}}{e^{i\theta}}d\theta = 2\pi i \neq 0.$$

In fact, if γ is any positively oriented simple closed curve winding around 0, then

$$\oint_\gamma \frac{dz}{z} = 2\pi i.$$

EXERCISE 3.1. Prove Propositions 3.1, 3.2, and 3.3.

EXERCISE 3.2. Evaluate the given line integrals:

- $\int_\gamma z \, dz$ where γ is a line segment from 0 to $1 + i$.
- $\int_\gamma \bar{z} dz$ where γ is a line segment from 0 to $1 + i$.
- $\int_\gamma z^n dz$ where γ is the unit circle traversed counterclockwise. Here n is an integer. The case $n = -1$ is special; it was computed above.

EXERCISE 3.3. Liouville's theorem states that if f is complex analytic on $\mathbb{C}$ and bounded, then f is a constant. One proof of this result uses the Cauchy integral formula. Prove the following key step in that proof. There is a constant C such that, for each $R > 0$, we have

$$\left| \int_{|\zeta|=R} \frac{f(\zeta)}{(\zeta - z)^2}d\zeta \right| \leq \frac{C}{R}.$$

4. The Cauchy theory

When f is analytic, and $f = u + iv$ is its expression into real and imaginary parts, we have seen that u and v are linked by the Cauchy–Riemann equations: $u_x = v_y$ and $u_y = -v_x$. Combined with Green's theorem, these equations yield Cauchy's theorem. (Note, however, the remark after the theorem.)

Theorem 4.1 (Cauchy's theorem). *Let f be complex analytic on an open set Ω. Let γ be a closed curve in Ω such that the interior of γ is also contained in Ω. Assume that $f = u + iv$, where u and v are continuously differentiable. (This hypothesis always holds when f is analytic.) Then*

$$\oint_\gamma f(z)dz = 0.$$

PROOF. Expressing f in terms of u, v we have

$$\int_\gamma f(z)dz = \int_\gamma (u+iv)(dx+idy) = \int_\gamma udx - vdy + i\int udy + vdx.$$

Now Green's theorem applies to give

$$\int_\gamma f(z)dz = \iint (-v_x - u_y)dxdy + i\iint (u_x - v_y)dxdy. \tag{3}$$

But each term in (3) vanishes by the Cauchy–Riemann equations. $\square$

REMARK. This proof of Cauchy's theorem tacitly assumes that u_x and u_y are continuous. This fact is proved in most complex analysis books. See [A], [B], or [D2] for example. Here we assume this continuity statement, and hence the hypotheses of Green's theorem hold.

Theorem 4.2 (Cauchy integral formula). *Suppose $f : \Omega \to \mathbb{C}$ is complex analytic on and inside a simple, positively oriented, closed curve γ. For each $p \in int(\gamma)$, we have*

$$f(p) = \frac{1}{2\pi i}\int_\gamma \frac{f(\zeta)}{\zeta - p}d\zeta.$$

PROOF. This proof needs to be thoroughly mastered by the student. Consider $\int_\gamma \frac{f(z)}{z-p}dz$. The function $z \to \frac{f(z)}{z-p}$ is complex analytic everywhere on Ω except at p. Hence Cauchy's theorem applies to it, as long as we do not wind around p. Let C_ϵ denote a circle of radius ϵ about p. Let γ_ϵ denote the curve shown in Figure 2; it does not wind around p. By Cauchy's theorem

$$0 = \int_{\gamma_\epsilon} \frac{f(z)}{z-p}dz.$$

The statement remains true as we let the vertical segments coalesce. But the line integrals over the vertical segments cancel in the limit. Therefore

$$0 = \int_{\gamma_\epsilon} \frac{f(z)}{z-p}dz = \int_\gamma \frac{f(z)}{z-p}dz - \int_{C_\epsilon} \frac{f(z)}{z-p}dz$$

and hence

$$\int_\gamma \frac{f(z)}{z-p}dz = \int_{C_\epsilon} \frac{f(z)}{z-p}dz.$$

Put $z = p + \epsilon e^{i\theta}$ for $0 \le \theta \le 2\pi$. The integral around the circle C_ϵ becomes

$$\int_0^{2\pi} \frac{f(p+\epsilon e^{i\theta})}{\epsilon e^{i\theta}}\epsilon i e^{i\theta}d\theta = \int_0^{2\pi} if(p+\epsilon e^{i\theta})d\theta,$$

for all sufficiently small ϵ. Let $\epsilon \to 0$. Then, since f is continuous at p,

$$\lim_{\epsilon \to 0}\int_0^{2\pi} if(p+\epsilon e^{i\theta})d\theta = 2\pi if(p). \tag{4}$$

The Cauchy integral formula follows.   □

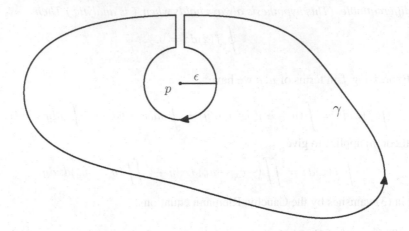

FIGURE 2. Proof of Cauchy integral formula

The theorem states something remarkable. We wish to know $f(p)$ for some p. We find it by integrating something over a curve surrounding p, without ever evaluating f at p. No such result is possible for general differentiable functions of real variables. (The real analogue of this result is the mean-value property of harmonic functions, which we do not discuss.) Since the Cauchy integral formula holds for each point inside γ, we often think of this point as a variable z and write

$$f(z) = \frac{1}{2\pi i} \int_\gamma \frac{f(\zeta)}{\zeta - z} d\zeta.$$

Let us also comment on (4). If we divide both sides by $2\pi i$, then (4) tells us that $f(p)$ is found by evaluating f on a small circle about p, averaging, and letting the radius tend to 0. See, for example, the discussion about square pulses in Section 8 of Chapter 7 and the discussion of approximate identities from Section 4 of Chapter 5 for occurrences of the same idea in engineering. Cauchy himself was trained as an engineer and his proof of the integral formula anticipated ideas such as the Dirac delta function.

Thus the proof of Cauchy's integral formula has two ideas. The integral around a rather general curve enclosing p equals an integral around a small circle about p. This part relies on f being complex analytic. We then evaluate this integral by letting the radius tend to 0, obtaining $f(p)$. This reasoning precisely parallels how one computes the electric flux through a surface surrounding a point charge. See Exercise 4.3 of Chapter 3 and Section 5 of Chapter 3.

EXAMPLE 4.1. Put $f(z) = \frac{1}{(z-b)}$. Assume $a \neq b$ and γ winds positively once around a, but does not enclose b. We obtain

$$\int_\gamma \frac{dz}{(z-a)(z-b)} = \int_\gamma \frac{f(z)}{z-a} dz = 2\pi i f(a) = 2\pi \frac{1}{a-b}.$$

EXAMPLE 4.2. Put $f(z) = e^z$. Assume γ winds once around -1. Then

$$\int_\gamma \frac{e^z}{z+1} dz = 2\pi i f(-1) = \frac{2\pi i}{e}.$$

If γ does not wind around -1, then the integral is 0 by Cauchy's theorem.

One of the most fundamental results in complex analysis is the existence of a convergent power series expansion.

Theorem 4.3. *Suppose $\Omega \subseteq \mathbb{C}$ is open and $f : \Omega \to \mathbb{C}$ is complex analytic. For each $p \in \Omega$, there is a power series representation*

$$f(z) = \sum_{n=0}^{\infty} c_n (z-p)^n.$$

This series converges on $\{|z - p| < R\}$, where R is the (minimum) distance from p to the boundary of Ω.

Before proving the theorem, we consider the crucial special case. In some sense, the Cauchy integral formula writes f as a superposition of geometric series. Hence we play a bit with the geometric series.

Consider $\frac{1}{1-z}$, which is complex analytic in $\mathbb{C} - \{1\}$. We know, for $|z| < 1$, that

$$\frac{1}{1-z} = \sum_{n=0}^{\infty} z^n.$$

What happens if we expand about an arbitrary point p?

In the region given by $|z - p| < |1 - p|$, we have

$$
\begin{aligned}
\frac{1}{1-z} &= \frac{1}{(1-p)-(z-p)} \\
&= \frac{1}{1-p} \frac{1}{1 - \frac{z-p}{1-p}} \\
&= \frac{1}{1-p} \sum_{n=0}^{\infty} \left(\frac{z-p}{1-p}\right)^n \\
&= \sum_{n=0}^{\infty} \frac{1}{(1-p)^{n+1}} (z-p)^n.
\end{aligned}
$$

The proof of the power series representation for a general f about a point p amounts to applying the Cauchy integral formula to the geometric series in the region where $|z - p| < |1 - p|$.

PROOF. Let γ be a circle about p contained in Ω. The Cauchy integral formula and then the geometric series yield

$$
\begin{aligned}
f(z) &= \frac{1}{2\pi i} \int_\gamma \frac{f(\zeta)}{\zeta - z} d\zeta \\
&= \frac{1}{2\pi i} \int_\gamma \frac{f(\zeta)}{(\zeta - p) - (z - p)} d\zeta \\
&= \frac{1}{2\pi i} \int_\gamma \frac{f(\zeta)}{\zeta - p} \frac{1}{1 - \frac{z-p}{\zeta-p}} d\zeta \\
&= \frac{1}{2\pi i} \int_\gamma \frac{f(\zeta)}{\zeta - p} \sum_{n=0}^{\infty} \left(\frac{z-p}{\zeta-p}\right)^n d\zeta.
\end{aligned}
$$

The last equality is valid if $\left|\frac{z-p}{\zeta-p}\right| < 1$. Interchanging the order of integration and summation gives

$$f(z) = \sum_{n=0}^{\infty} \int_\gamma \frac{1}{2\pi i} \frac{f(\zeta)}{(\zeta - p)^{n+1}} d\zeta \ (z-p)^n = \sum_{n=0}^{\infty} c_n (z-p)^n.$$

We refer to [A] or [D2] for the justification that we can interchange the order of integration and summation. Let δ denote the distance from p to the boundary of Ω. For $|z - p| < \delta$ we conclude:

$$f(z) = \sum_{n=0}^{\infty} c_n (z - p)^n.$$

Here the coefficient satisfies

$$c_n = \frac{1}{2\pi i} \int_{\gamma} \frac{f(\zeta)}{(\zeta - p)^{n+1}} d\zeta.$$

If $f^{(k)}$ denotes the k-th derivative of f, then $c_k = \frac{f^{(k)}(p)}{k!}$.   □

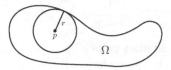

FIGURE 3. Proof of power series expansion

When f is complex analytic, its series representation implies that f is complex differentiable infinitely many times. We simply differentiate the series term-by-term inside the disk of convergence. The differentiation is justified because the series converges uniformly on smaller closed disks. The region of convergence for the series of derivatives is always the same as the region for the original series.

The proof gives an integral formula for the derivatives:

$$f^{(k)}(p) = \frac{k!}{2\pi i} \int_{\gamma} \frac{f(\zeta)}{(\zeta - p)^{k+1}} d\zeta. \tag{5}$$

The Cauchy integral formula is the special case $k = 0$.

REMARK. Suppose we are given an analytic function, expressed in a complicated manner, and we wish to estimate its Taylor coefficients. It might be difficult to compute derivatives. In some situations, it is easier to apply numerical methods to estimate (5).

The Cauchy integral formula and its consequences arise throughout this book, and the formula has applications to 21-st century work in circuits and control theory. A full appreciation of these applications requires the mathematical consequences discussed in the rest of this chapter. Furthermore, the ideas recur all over the place, as indicated by the following remark.

REMARK. Later in this book we discuss convolution in several different settings. One can regard the Cauchy integral formula as a convolution of f and $\frac{1}{z}$. The last step in the proof involves shrinking the curve to a point; this step is analogous to regarding evaluation of f at z as a Dirac delta function. In fact, Cauchy, who worked as an engineer early in his life, proved this theorem by 1825, before Maxwell was born, and long before Dirac.

EXERCISE 4.1. Find the radius of convergence of $\sum_{n=1}^{\infty} \frac{n^n z^n}{n!}$.

EXERCISE 4.2. Find an explicit formula for $z^2 \sum_{n=0}^{\infty} (1 + z^2)^{-n}$ and determine where it is valid.

5. Polynomials and power series

We first recall Cauchy's theorem and Cauchy's integral formula. We then sketch a proof of the fundamental theorem of algebra. Finally we briefly compare polynomials and power series.

Theorem 5.1 (Cauchy's theorem). *Let f be complex analytic on and inside a closed contour γ. Then*

$$\int_\gamma f(z)\, dz = 0.$$

In some sense, this theorem tells us that certain expressions define conservative vector fields. Both the real and imaginary parts of the differential 1-form $f(z)dz = (u+iv)(dx+idy)$ are *exact*, as defined in Chapter 3.

Theorem 5.2 (Cauchy integral formula). *Let f be complex analytic on and inside a positively oriented simple closed curve γ. For z inside γ we have*

$$f(z) = \frac{1}{2\pi i} \int_\gamma \frac{f(\zeta)}{\zeta - z} d\zeta.$$

Corollary 5.3 (Liouville's theorem). *A bounded analytic function, defined on all of $\mathbb{C}$, is constant.*

PROOF. Let f be bounded and complex analytic. To show f is a constant, it suffices by Lemma 2.2 to show that $f'(z) = 0$ for all z. By formula (5) when $k = 1$, for each simple closed curve γ with z in its interior,

$$f'(z) = \frac{1}{2\pi i} \int_\gamma \frac{f(\zeta)}{(\zeta - z)^2} d\zeta.$$

Let γ be a circle of radius R about z. Using the boundedness of f and estimating using Proposition 3.5 yield the inequality

$$|f'(z)| \leq \frac{C}{R}$$

for some constant C. (Recall Exercise 3.3.) Letting R tend to infinity in this inequality shows that $f'(z)$ must be 0. Since z was an arbitrary point, f' vanishes everywhere. □

We next state without proof a fundamental result in basic analysis. See, for example, [A] or [R] for a proof. We will combine this result with Liouville's theorem to give one of the standard proofs of the fundamental theorem of algebra.

Theorem 5.4. *Let K be a closed and bounded subset of $\mathbb{C}$. Suppose $f : K \to \mathbb{C}$ is continuous. Then f is bounded.*

Consider a polynomial $p(z)$. If p is never zero, then $\frac{1}{p}$ is continuous. Hence $\frac{1}{p}$ is bounded on any closed disk. Furthermore, $\frac{1}{p(z)}$ tends to 0 at infinity. See Example 8.1 of Chapter 2. We now can put these facts together to prove the fundamental theorem of algebra.

Theorem 5.5 (Fundamental theorem of algebra). *Let $p(z)$ be a non-constant polynomial in a complex variable z. Then there is a z_0 with $p(z_0) = 0$.*

PROOF. If $p(z)$ is never 0, then $\frac{1}{p}$ is complex analytic on all of $\mathbb{C}$. Since $\frac{1}{p(z)}$ tends to 0 at infinity, there is an $R > 0$ such that $\left|\frac{1}{p(z)}\right| \leq 1$ for $|z| > R$. The set $|z| \leq R$ is closed and bounded; since $\frac{1}{p}$ is continuous there, it is bounded. Thus $\frac{1}{p}$ is complex analytic on all of $\mathbb{C}$ and bounded. By Liouville's theorem, it is a constant, and hence p is also a constant. □

Corollary 5.6. *Each polynomial over $\mathbb{C}$ splits into linear factors.*

PROOF. If p is a constant, then we regard the conclusion to be true by convention. Otherwise, suppose p has degree at least 1 and $p(a) = 0$. By Exercise 5.6, the polynomial

$$z \mapsto p(z) = p(z) - p(a)$$

is divisible by $(z - a)$. The conclusion now follows by induction on the degree. $\square$

Exercises 5.5 and 5.6 amplify Corollary 5.6, and Exercise 5.7 extends Exercise 5.6 to complex analytic functions. The Cauchy integral formula has many additional corollaries, including formula (5) for the derivatives. One can use these formulas to generalize Liouville's theorem. A complex analytic function, defined on all of $\mathbb{C}$, that does not grow too fast at infinity must be a polynomial. (Exercise 5.4)

The power series representation is also a fundamental corollary of the Cauchy theory. Convergent power series generalize polynomials, but the analogue of the fundamental theorem of algebra fails. An analytic function given by a power series of infinite radius of convergence need not have a zero; consider e^z.

Let us summarize the relationship between the Cauchy theory and power series.

Theorem 5.7. *Assume $f : \Omega \to \mathbb{C}$ is complex analytic. For each $p \in \Omega$, the function f can be represented as a convergent power series:*

$$f(z) = \sum_{n=0}^{\infty} c_n (z - p)^n,$$

where the series converges for $\{z : |z - p| < \mathrm{dist}(p, \partial\Omega)\}$. In particular, if f is complex analytic on all of $\mathbb{C}$, then the series has infinite radius of convergence. Conversely, if for all $p \in \Omega$, there is an open ball about p on which

$$f(z) = \sum_{n=0}^{\infty} c_n (z - p)^n$$

and the series converges, then f is complex analytic on Ω.

A convergent power series defines a complex analytic function inside the region of convergence. This series is the *Taylor series* of the function. We sketch the ideas in analysis required to use Taylor series meaningfully. See [A], [D2], or [R] for more details.

Suppose $\sum_{n=0}^{\infty} a_n z^n$ converges for $|z| < R$. We can define a function f by

$$f(z) = \sum_{n=0}^{\infty} a_n z^n$$

there. The infinite series converges *uniformly* on any smaller closed disk given by $\{|z| \leq r < R\}$. Hence the series defines a complex differentiable function. Furthermore, taking its derivative term-by-term leads to a series with the same region of convergence. Therefore a complex analytic function is infinitely differentiable. Exercise 5.1 shows that (real) analyticity is a stronger property than infinite differentiability.

We saw in Chapter 1 the importance of exponentiating a linear operator, which we defined by using the power series. An operator-valued version of the Cauchy integral formula suggests how to define analytic functions of an arbitrary linear operator. Note that a rigorous justification of Definition 5.8 requires defining what we mean by an integral of a function whose values are linear transformations. For matrices, the idea is easy. By choosing a basis, we regard a linear map as a matrix; we can then integrate each component separately. If we replace L by a complex number z in Definition 5.8, then we are simply repeating the Cauchy integral formula. The formula then makes sense if L is a scalar times the identity operator, then more generally on each eigenspace, and then for a diagonalizable operator. Since arbitrary matrices are limits of diagonalizable matrices, we can define analytic functions of arbitrary matrices. The same idea applies even for operators on Hilbert spaces.

Definition 5.8 (Functions of operators). Let $L : \mathbb{C}^n \to \mathbb{C}^n$ be a linear map with spectrum $\sigma(L)$. Assume that f is complex analytic in an open set Ω containing $\sigma(L)$. Let γ be a positively oriented simple closed curve surrounding $\sigma(L)$. We define

$$f(L) = \frac{1}{2\pi i} \int_\gamma f(\zeta)(\zeta I - L)^{-1} d\zeta.$$

REMARK. If the analytic function f in Definition 5.8 is given by a convergent power series $\sum a_n z^n$, then $f(L)$ is given by the convergent series

$$\sum a_n L^n.$$

In particular e^L and $(I - L)^{-1}$ are defined using their series as expected.

EXERCISE 5.1. The purpose of this exercise is to show that there exist infinitely differentiable functions of a real variable whose Taylor series does not equal the function. For $x \leq 0$, put $f(x) = 0$. For $x > 0$, put $f(x) = e^{\frac{-1}{x}}$. Prove that f is infinitely differentiable at 0, with all derivatives equal to 0 there.

EXERCISE 5.2. Use Exercise 5.1 to prove that there is an infinitely differentiable function χ on the real line such that $\chi(x) = 0$ for $|x| > 2$ but $\chi(x) = 1$ for $|x| \leq 1$. Such functions are sometimes known as *bump* functions and arise in Fourier analysis and in physics.

EXERCISE 5.3. Prove that the radius of convergence of a power series is unchanged if we form a new series by taking the derivative term-by-term.

EXERCISE 5.4. Suppose that f is complex analytic on all of $\mathbb{C}$ and there is a constant C and a positive integer k such that $|f(z)| \leq C|z|^k$ for all z. Prove that f is a polynomial of degree at most k.

EXERCISE 5.5. Let $p(z)$ be a complex polynomial of degree d. Prove that p has at most d roots. Prove, assuming multiplicities are properly counted, that p has exactly d roots.

EXERCISE 5.6. Prove the following fact from high-school math. If a polynomial $p(z)$ vanishes at a, then it is divisible by $z - a$. In other words $\frac{p(z)}{(z-a)}$ is a polynomial. The crucial point is that $z^k - a^k$ factors.

EXERCISE 5.7. Let f be complex analytic on an open set Ω with $f(a) = 0$. Prove that there is an analytic function g on Ω with $f(z) = (z - a)g(z)$.

EXERCISE 5.8 (Difficult). Let L be a linear operator. Suppose f and g are complex analytic in an open ball including $\sigma(L)$. Using Definition 5.8, verify that $(fg)(L) = f(L)g(L)$.

6. Singularities

It often happens that f is complex analytic on an open set Ω except for finitely many points. These points are called **singularities**. We use the term *deleted neighborhood* of a point p for the set of points z with $0 < |z - p| < R$ for some R.

Definition 6.1. Suppose f is complex analytic in a deleted neighborhood of p. The point p is called a *singularity* or *singular point* of f. There are three possibilities.

(1) p is called **removable** if $\lim_{z \to p} \frac{f(z)}{z-p}$ exists.
(2) p is called a **pole of order** m if $(z - p)^m f(z)$ has a removable singularity at p.
(3) p is called an **essential singularity** if it is a singularity but is neither removable nor a pole.

REMARK. In the definition of a pole, m is a positive integer. It makes sense to regard a removable singularity as a pole of order 0. The author avoids this usage because often one considers also the order of a zero and it is awkward to say *zero of order zero*. Poles of order 1 are often called **simple poles**.

EXAMPLE 6.1. Put $f(z) = \frac{\sin z}{z}$. Then f has a removable singularity at 0. We can remove the singularity by putting $f(0) = 1 = \lim_{z \to 0} \frac{\sin z}{z}$. This extended function is complex analytic in all of $\mathbb{C}$.

Put $f(z) = \frac{z^2 - 9}{z - 3}$ if $z \neq 3$. Then $f(z) = z + 3$ everywhere except $z = 3$. We naturally remove the singularity by putting $f(3) = 6$.

EXAMPLE 6.2. The function $\frac{e^z}{z^5}$ has a pole of order 5 at 0. The function $f(z) = \frac{\cos(z) - 1}{z^4}$ has a pole of order 2 at 0. Despite the division by z^4, we easily see that $z^2 f(z)$ has a removable singularity at 0. In fact, $\lim_{z \to 0} z^2 f(z) = \frac{-1}{2}$.

EXAMPLE 6.3. $e^{\frac{-1}{z}}$ has an essential singularity at 0.

REMARK. The function $\log(z)$ is **not** analytic in any deleted neighborhood of 0. Hence, 0 is not a singularity for the logarithm. If we define $\log$ by excluding a branch cut (such as the interval $[-\infty, 0]$), then 0 is a boundary point of the domain of analyticity of $\log$. Thus 0 is not an isolated point in the region of definition.

When p is an isolated singularity for an analytic function f, we have a series representation called the *Laurent series* of f. We expand f in both positive and negative powers of $(z - p)$.

Theorem 6.2 (Laurent series). *Assume that f is complex analytic in an open set Ω, except for an isolated singularity at p. Then there is an annulus centered at p on which we can write*

$$f(z) = \sum_{n=-\infty}^{\infty} c_n (z - p)^n. \tag{6}$$

PROOF. The idea is to expand $\frac{1}{\zeta - z}$ in two ways. Write

$$\frac{1}{\zeta - z} = \frac{1}{\zeta - p - (z - p)}$$

as in the proof of the Cauchy integral formula. For $\frac{|z-p|}{|\zeta - p|} < 1$, we have

$$\frac{1}{\zeta - z} = \frac{1}{\zeta - p - (z - p)} = \sum_{n=0}^{\infty} \frac{(z - p)^n}{(\zeta - p)^{n+1}} \tag{7.1}$$

as before. We also have

$$\frac{1}{w - z} = \frac{1}{w - p - (z - p)} = \frac{1}{z - p} \frac{-1}{1 - \frac{w-p}{z-p}}.$$

For $\frac{|z-p|}{|w-p|} > 1$, we can write

$$\frac{1}{w - z} = -\sum_{n=0}^{\infty} (w - p)^n \frac{1}{(z - p)^{n+1}}. \tag{7.2}$$

Assume that the ball given by $\{z : |z - p| < R\}$ is contained in Ω. Assume $0 < \epsilon < r < R$. Consider two circles centered at p, denoted C_ϵ and C_r. We apply the Cauchy integral formula to a closed curve consisting of these two circles, with the inner circle negatively oriented. By the same trick as illustrated in Figure 2, we obtain $f(z)$ as the difference of two line integrals:

$$f(z) = \frac{1}{2\pi i} \int_{C_r} \frac{f(\zeta)}{\zeta - z} d\zeta - \frac{1}{2\pi i} \int_{C_\epsilon} \frac{f(w)}{w - z} dw. \tag{8}$$

Using (7.1) and (7.2), and proceeding as we did in establishing the power series expansion, we obtain the Laurent series expansion. Note that we require

$$\epsilon = |w - p| < |z - p| < |\zeta - p| = r$$

for both the series in (7.1) and (7.2) to converge.

One sometimes writes $a_n = c_n$ for $n \geq 0$ and $b_n = c_{-n}$ for $n \geq 1$. Then (6) becomes

$$f(z) = \sum_{n=0}^{\infty} a_n(z - p)^n + \sum_{n=1}^{\infty} b_n(z - p)^{-n}. \tag{9}$$

We obtain integral formulas for the coefficients. As in the proof of Theorem 4.3,

$$a_n = \frac{1}{2\pi i} \int_{C_r} \frac{f(\zeta)}{(\zeta - p)^{n+1}} d\zeta.$$

$$b_n = c_{-n} = \frac{1}{2\pi i} \int_{C_\epsilon} \frac{f(\zeta)}{(\zeta - p)^{-n+1}} d\zeta.$$

$\square$

Let f be complex analytic on Ω except at some point p. The Laurent expansion guarantees, in an annulus about p, that we can write $f(z) = \sum_{-\infty}^{\infty} c_n(z - p)^n$. It follows, for each appropriate curve γ, that

$$\oint_\gamma f(z)dz = 2\pi i c_{-1}.$$

Here *appropriate* means that γ is a simple, positively-oriented, closed curve in Ω enclosing p, and also that γ encloses no other singularities.

EXAMPLE 6.4. For γ a circle about 0 we have $\oint_\gamma \frac{e^z}{z^3} dz = \pi i$, since $e^z = 1 + z + \frac{z^2}{2} + \frac{z^3}{3!} + \cdots$.

The definition of a pole of order m at p is equivalent to: the Laurent series of f is

$$\frac{c_{-m}}{(z - p)^m} + \frac{c_{-m+1}}{(z - p)^{m-1}} + \cdots + \frac{c_{-1}}{z - p} + c_0 + \cdots,$$

and $c_{-m} \neq 0$. Here are three examples: $\frac{\cos(z)}{z^4}$ has a pole of order 4 at 0, $\frac{\cos(z)-1}{z^4}$ has a pole of order 2 at 0, and $\frac{\cos(z)-1+\frac{z^2}{2}}{z^4}$ has a removable singularity at 0.

The definition of essential singularity at p is equivalent to: the Laurent series of f has $c_n \neq 0$ for infinitely many negative n. For example,

$$e^{-\frac{1}{z}} = 1 - \frac{1}{z} + \frac{1}{2z^2} - \frac{1}{6z^3} + \cdots.$$

Even if f has an essential singularity at p, for appropriate γ we still have

$$\oint f(z)dz = 2\pi i c_{-1}.$$

Corollary 6.3 (Riemann's removable singularities theorem). *Suppose f is complex analytic in a deleted neighborhood of p and that f is bounded there. Then f has a removable singularity at p.*

PROOF. It suffices to show that each Laurent coefficient b_n vanishes. Assume $|f| \leq M$. But we have

$$|b_n| = \left| \frac{1}{2\pi i} \int_{C_\epsilon} \frac{f(\zeta)}{(\zeta - p)^{-n+1}} d\zeta \right|.$$

Parametrize the line integral using $\zeta = p + \epsilon e^{i\theta}$; hence $d\zeta = i\epsilon e^{i\theta} d\theta$. The ML-inequality yields $|b_n| \leq M\epsilon^n$. Letting ϵ tend to 0 proves that each b_n vanishes. □

EXERCISE 6.1. Assume that the log function is defined with the branch cut $-\pi < \theta < \pi$. What kind of singularity does $\frac{\log(z)}{z-1}$ have at 1? What happens if the branch cut is the positive real axis?

EXERCISE 6.2. What kind of singularity does $\sin(\frac{\pi}{z})$ have at 0? At what points does this function take the value 0?

EXERCISE 6.3. Put $f(z) = \frac{1}{z^2 - z^3}$. Find a Laurent series which is valid for $0 < |z| < 1$. Find a Laurent series which is valid for $1 < |z|$. Evaluate $\int_{|z|=a} f(z) dz$ for $a = \frac{1}{10}$ and for $a = 10$.

EXERCISE 6.4. Evaluate the integrals by inspection:

$$\int_{|z|=4} \frac{e^z}{z^7} dz$$

$$\int_{|z|=1} \frac{\sin z - z}{z^{11}} dz$$

$$\int_{|z|=17} \frac{\cos z}{z} dz.$$

EXERCISE 6.5. In each case, determine what kind of singularity f has at p:

(1) $f(z) = \frac{\cos(z)}{z}$ at $z = 0$.
(2) $f(z) = \frac{\cos(z) - 1 + \frac{z^2}{2}}{z^m}$ at $z = 0$. Answer for each positive integer m.
(3) $f(z) = e^z \sin(\frac{\pi}{z})$ at $z = 0$.
(4) $f(z) = \frac{g(z)}{z^2+1}$ at $z = i$ where g is analytic near i and $g(i) = 0$.
(5) $f(z) = \frac{g(z)}{z^2+1}$ at $z = i$ where g is analytic near i and $g(i) \neq 0$.

7. Residues

One of the reasons why complex variable theory is so useful in science is that complex variable methods enable us to evaluate many integrals rather easily. The previous section barely glimpses the scope of this technique. The starting point for discussion is the third Cauchy theorem in complex analysis, the Cauchy residue theorem.

Definition 7.1. Let f be analytic in a deleted neighborhood of p. The **residue** of f at p is the coefficient c_{-1} in the Laurent expansion (6) of f at p. We write $\mathrm{Res}(p)$ when f is given.

Theorem 7.2 (Cauchy Residue Theorem). *Suppose f is complex analytic on an open ball Ω, except at finitely many points. Let γ be a positively oriented simple closed curve in Ω. Then*

$$\oint_\gamma f(z) dz = 2\pi i \left(\sum \text{Residues at the singularities inside } \gamma \right).$$

When there are several singularities, we must be careful to compute the residue at each singularity. Many shortcuts exist for finding residues.

This theorem enables us to compute a large number of interesting integrals. Some of these integrals cannot be evaluated by the usual techniques of calculus. Others are tedious when using calculus. Many of these integrals arise in physics and engineering.

EXAMPLE 7.1. Integrals from 0 to 2π involving trig functions can often be computed by rewriting the integral as a line integral over the unit circle. Consider

$$\int_0^{2\pi} \frac{d\theta}{2 - \sin\theta} = \frac{2\pi}{\sqrt{3}}. \tag{10}$$

Put $z = e^{i\theta}$. Then $dz = ie^{i\theta}\,d\theta$ and hence $d\theta = \frac{dz}{iz}$ and $\sin\theta = \frac{z - z^{-1}}{2i}$. We have

$$\int_0^{2\pi} \frac{d\theta}{2 - \sin\theta} = \oint_{|z|=1} \frac{dz}{iz(2 - \frac{z - z^{-1}}{2i})} = \oint_{|z|=1} \frac{-2dz}{z^2 - 4iz - 1}.$$

Now we factor the denominator as $(z - \alpha)(z - \beta)$. These points are $i(2 \pm \sqrt{3})$; only one of these points is inside the unit circle. Hence we get

$$\oint_{|z|=1} \frac{-2dz}{(z - \alpha)(z - \beta)} = -4\pi i \frac{1}{\alpha - \beta} = \frac{2\pi}{\sqrt{3}}.$$

EXAMPLE 7.2. Integrals from $-\infty$ to ∞ of rational functions can often be evaluated as in the following examples:

$$J = \int_{-\infty}^{\infty} \frac{dx}{x^2 + 1} = \pi. \tag{11}$$

By symmetry, we note that

$$J = \lim_{R \to \infty} \int_{-R}^{R} \frac{dx}{x^2 + 1}.$$

The method is to introduce a simple closed curve γ_R as follows. We consider the interval from $-R$ to R on the real axis, followed by a semi-circle of radius R in the upper half-plane.

For R sufficiently large, we have

$$\int_{\gamma_R} \frac{dz}{z^2 + 1} = 2\pi i \operatorname{Res}(i) = \frac{2\pi i}{2i} = \pi.$$

Then we let R tend to infinity. After verifying that the integral along the semi-circle tends to 0, we conclude that the value of the original integral is π.

To verify the limit as R tends to infinity, we parametrize the integral over the semi-circle and see that its magnitude is at most a constant times $\frac{1}{R}$.

Here is a similar example, where the technique is the same, but evaluating the residue is slightly harder:

EXAMPLE 7.3.

$$\int_{-\infty}^{\infty} \frac{dx}{(x^2 + a^2)^2} = \frac{\pi}{2a^3}. \tag{12}$$

Using the same method as in Example 7.2, the answer is $2\pi i$ times the residue at ai. Note that,

$$\frac{1}{(z^2 + a^2)^2} = \frac{1}{(z + ai)^2(z - ai)^2} = \frac{f(z)}{(z - ai)^2}.$$

Here $f'(ai) = -2(z + ai)^{-3}|_{z=ai} = \frac{1}{4a^3 i}$. Hence the integral equals $2\frac{\pi i}{4a^3 i} = \frac{\pi}{2a^3}$.

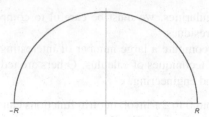

FIGURE 4. Contour for integration of rational functions

REMARK. In Example 7.3, we computed a residue by taking a derivative. Higher order poles require higher derivatives. For example, let f be complex analytic near p. Consider $\frac{f(z)}{(z-p)^m}$. The residue at p is $\frac{f^{(m-1)}(p)}{(m-1)!}$. See Exercise 7.2. This technique is useful when finding inverse Laplace transforms.

We can evaluate certain Fourier transforms using residues. We formally define the Fourier transform in Chapter 5. The next integral is a standard example.

EXAMPLE 7.4. Consider

$$\int_{-\infty}^{\infty} \frac{e^{iwx}}{x^2+1} dx = \pi e^{-w}, \text{ where } w > 0. \tag{13}$$

The integrand is complex analytic everywhere in the upper half-plane except as $z = i$. Again we use a contour from $-R$ to R together with a semi-circle, and let R tend to infinity. Using Lemma 8.1 (Jordan's lemma), we verify that the integral along the semi-circle tends to 0 as R tends to infinity. Hence the answer is $2\pi i$ times the residue there, namely

$$2\pi i \frac{e^{-w}}{2i} = \pi e^{-\omega}.$$

EXAMPLE 7.5. We can evaluate some integrals involving branch cuts. For $-1 < \alpha < 1$, we have

$$\int_0^{\infty} \frac{x^{\alpha}}{(x+1)^2} dx = \frac{\pi\alpha}{\sin(\pi\alpha)}. \tag{14}$$

We define z^{α} as the complex analytic function $z^{\alpha} = e^{\alpha \log(z)}$, where $\log(z) = \log|z| + i\theta$ and $0 < \theta < 2\pi$. Then $\frac{z^{\alpha}}{(z+1)^2}$ is complex analytic on $\mathbb{C} - \mathbb{R}_+$, except at -1. See Example 8.1 for the evaluation of (14).

We pause to review the ideas. Complex analysis, especially the residue theorem, enables us to evaluate many line integrals in our head. The basic principle combines the Laurent series expansion with the easy evaluation of $\int_{\gamma} \frac{dz}{(z-p)^k}$. Various clever techniques enable us to evaluate limits that arise.

Near a singularity at p, write f in a Laurent series that converges in the annulus $0 < |z - p| < r$. Let γ be a simple closed curve winding once around p that lies in this annulus. Then $\int_{\gamma} \frac{dz}{z-p} = 2\pi i$. But $\int_{\gamma} (z - p)^k dz = 0$ for $k \neq -1$. Therefore, if

$$f(z) = \sum_{n=-\infty}^{\infty} c_n(z-p)^n$$

for $0 < |z - p| < r$, then

$$\int_{\gamma} f(z) dz = 2\pi i c_{-1}. \tag{15}$$

EXAMPLE 7.6. We evaluate

$$J = \int_0^{2\pi} \frac{d\theta}{\frac{17}{4} + 2\cos\theta}.$$

The idea is to convert J to a line integral around the unit circle. If $z = e^{i\theta}$, then $dz = ie^{i\theta}d\theta$. Therefore $d\theta = \frac{dz}{iz}$ and $\cos\theta = \frac{z+z^{-1}}{2}$. Put $f(z) = \frac{1}{z+4}$. We have

$$\begin{aligned}
J &= \oint_{|z|=1} \frac{dz}{iz(\frac{17}{4} + z + \frac{1}{z})} \\
&= \frac{1}{i}\int_{|z|=1} \frac{dz}{z^2 + \frac{17}{4}z + 1} \\
&= \frac{1}{i}\int_{|z|=1} \frac{dz}{(z+4)(z+\frac{1}{4})} \\
&= \frac{1}{i}2\pi i f(-\frac{1}{4}) \\
&= \frac{8\pi}{15}.
\end{aligned}$$

We state as a corollary the easiest generalization of Example 7.2.

Corollary 7.3. *Let p, q be polynomials, with $q(x) \neq 0$ for all $x \in \mathbb{R}$. Assume $\deg(q) \geq \deg(p) + 2$. Then*

$$\int_{-\infty}^{\infty} \frac{p(x)}{q(x)}dx = 2\pi i \left(\sum \text{residues in the upper half-plane} \right).$$

EXAMPLE 7.7. Let γ_R be as in Figure 2.

$$\int_{-\infty}^{\infty} \frac{1}{(x^2+1)(x^2+4)}dx = \lim_{R\to\infty}\int_{\gamma_R} \frac{dz}{(z+i)(z-i)(z+2i)(z-2i)}dz$$

$$= 2\pi i \left(\frac{1}{2i \cdot 3i(-i)} + \frac{1}{3i \cdot i(-4i)} \right) = \frac{\pi}{2}.$$

In Example 7.7, the integrand $f(x)$ is an even function. Hence the integral from 0 to ∞ of f is half the integral from $-\infty$ to ∞. In the next section, we introduce a more difficult method to evaluate certain integrals from 0 to ∞.

EXERCISE 7.1. For appropriate values of the parameters, evaluate the following real integrals using the techniques of this section:

$$\int_{-\infty}^{\infty} \frac{dx}{(x^2+a^2)(x^2+b^2)}$$

$$\int_0^{2\pi} \sin^{2k}(x)dx$$

$$\int_0^{2\pi} \cos^{2k}(x)\sin^{2j}(x)dx$$

$$\int_0^{2\pi} \frac{dx}{a^2\cos^2(x) + b^2\sin^2(x)}.$$

EXERCISE 7.2. Assume that f is analytic near λ and let m be a positive integer. Show that the residue at λ of $\frac{f(z)}{(z-\lambda)^m}$ is $\frac{f^{m-1}(\lambda)}{(m-1)!}$. Use this result to find a formula for the residue of g at λ, assuming that g has a pole or order m there.

EXERCISE 7.3. For $z \neq 0$, put

$$f(z) = \frac{e^{\lambda z} - 1 - \lambda z}{z^3}.$$

What is the residue of f at 0?

8. More residue computations

We give more examples of integrals we can compute using complex variable techniques. The examples help clarify some subtle points and each should be carefully studied.

EXAMPLE 8.1. Fractional powers. Suppose $-1 < \alpha < 1$. Consider

$$\int_0^\infty \frac{x^\alpha}{(x+1)^2} = \lim_{R \to \infty, r \to 0} \int_r^R \frac{x^\alpha}{(x+1)^2} dx.$$

We define z^α as the complex analytic function $z^\alpha = e^{\alpha \log(z)}$ where $\log(z) = \log|z| + i\theta$, and $0 < \theta < 2\pi$. Then $\frac{z^\alpha}{(z+1)^2}$ is complex analytic on $\mathbb{C} - \mathbb{R}_+$, except at -1. On the complement of this branch cut, $\log(z)$ is complex analytic, hence z^α is also complex analytic there. By the Residue Theorem, we get

$$\oint_{C_{R,r}} \frac{z^\alpha}{(z+1)^2} dz = 2\pi i \mathrm{Res}(-1) =$$

$$\int_r^R \frac{x^\alpha}{(x+1)^2} dx + \int_{|z|=R} \frac{z^\alpha}{(z+1)^2} dz + \int_R^r e^{2\pi i \alpha} \frac{x^\alpha}{(x+1)^2} dx + \int_{|z|=r} \frac{z^\alpha}{(z+1)^2} dz$$

$$= I_1 + I_2 + I_3 + I_4.$$

We evaluate I_4 by parametrizing the circle:

$$I_4 = \int_0^{2\pi} \frac{r^{\alpha+1} i e^{i\theta}}{(re^{i\theta} + 1)^2} d\theta.$$

We are given $\alpha > -1$. When $r \to 0$, it follows that I_4 tends to 0.

We evaluate I_2 in a similar fashion.

$$I_2 = \int_0^{2\pi} \frac{(Re^{i\theta})^\alpha i Re^{i\theta} d\theta}{(Re^{i\theta} + 1)^2}.$$

We have $\alpha < 1$. Let $R \to \infty$; it follows that I_2 tends to 0. Thus, letting $r \to 0$ and $R \to \infty$ we see that I_2 and I_4 tend to 0. Also, $I_3 = e^{2\pi \alpha} J$. Thus, we have

$$J = \int_0^\infty \frac{x^\alpha}{(x+1)^2} dx = \frac{2\pi i}{1 - e^{2\pi i \alpha}} \mathrm{Res}(-1).$$

For the residue, note that $\frac{z^\alpha}{(z+1)^2}$ has a pole of order 2 at $z = -1$. The residue is $f'(-1)$, where $f(z) = z^\alpha$. Hence

$$J = \frac{2\pi i \alpha (-1)^{\alpha-1}}{1 - e^{2\pi i \alpha}}.$$

Since $(-1)^{\alpha-1} = e^{\pi i (\alpha-1)} = -e^{i\pi\alpha}$, we get the final answer of

$$\int_0^\infty \frac{x^\alpha}{(x+1)^2} = \frac{\pi\alpha}{\sin(\pi\alpha)}.$$

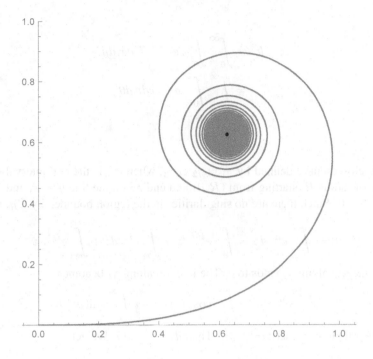

FIGURE 5. Roller coaster track

EXAMPLE 8.2. We consider equations for a roller coaster track. We will evaluate the integrals

$$\int_0^\infty \cos(x^2)dx$$

$$\int_0^\infty \sin(x^2)dx$$

at one time and indicate an application. Put

$$\gamma(t) = \int_0^t \cos(u^2) + i\sin(u^2)du = \int_0^t e^{iu^2}\,du.$$

Think of $\gamma(t)$ as the position of a moving particle in $\mathbb{C}$ at time t. Its derivative is the velocity vector

$$\gamma'(t) = v(t) = \cos(t^2) + i\sin(t^2)$$

and its acceleration vector is

$$\gamma''(t) = a(t) = -2t\sin(t^2) + i2t\cos(t^2).$$

The *curvature* of a smooth curve in $\mathbb{R}^2$ or $\mathbb{R}^3$ is $\frac{||v \times a||}{||v||^3}$. See Exercises 6.2 and 6.4 in Chapter 3. Here, $||v|| = 1$ and $v \times a = (0, 0, 2t)$. Note that v and a are orthogonal. Thus $||v \times a|| = ||a|| = 2t$. Hence the curvature of this curve at time t is $2t$. The curve spirals into a point as t tends to infinity. We find this point by evaluating the integral $J = \int_0^\infty e^{it^2}\,dt$. We claim that $J = \sqrt{\frac{\pi}{8}}$.

First we note that $k = \int_0^\infty e^{-x^2}\,dx = \frac{\sqrt{\pi}}{2}$. As usual, one finds k^2.

$$k^2 = \int_0^\infty \int_0^\infty e^{-x^2 - y^2}\, dx\, dy$$

$$= \int_0^\infty \int_0^{\frac{\pi}{2}} e^{-r^2} r\, dr\, d\theta$$

$$= \frac{\pi}{4}.$$

Since $k > 0$, we get $k = \frac{\sqrt{\pi}}{2}$.

Let C_R be the closed curve defined by $\gamma_1 \cup \gamma_2 \cup \gamma_3$, where γ_1 is the real interval $0 \le x \le R$, where γ_2 is a circular arc of radius R, starting from $(R, 0)$ and ending on the line $y = x$, and where γ_3 is the line segment from $R\frac{1+i}{\sqrt{2}}$ to 0. Since there are no singularities in the region bounded by C_R, we have

$$0 = \oint_{C_R} e^{iz^2}\, dz = \int_0^R e^{iz^2}\, dz + \int_{\gamma_2} e^{iz^2}\, dz + \int_{\gamma_3} e^{iz^2}\, dz.$$

Let $R \to \infty$. The integral along γ_2 tends to 0. The integral along γ_1 becomes

$$\int_0^\infty e^{iu^2}\, du = \int_0^\infty \cos(u^2)\, du + i \int_0^\infty \sin(u^2)\, du.$$

To evaluate the integral along γ_3, let $z = \frac{1+i}{\sqrt{2}} u$. Then $dz = \frac{1+i}{\sqrt{2}}\, du$. We get

$$\int_R^0 e^{iu^2}\, du = \frac{1+i}{\sqrt{2}} \int_R^0 e^{-u^2}\, du.$$

Letting R tend to infinity and using the computation of k above we get

$$\int_0^\infty e^{iu^2}\, du = (1+i)\sqrt{\frac{\pi}{8}}.$$

Therefore the point to which the roller track spirals is $(1+i)\sqrt{\frac{\pi}{8}}$.

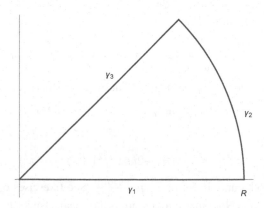

FIGURE 6. Contour for Example 7.2

The following subtle lemma gets used when verifying that certain contour integrals tend to zero.

Lemma 8.1 (Jordan's lemma). *Let C_R be the semi-circle in the upper half-plane of radius R and centered at* 0. *Then*

$$\lim_{R \to \infty} \int_{C_R} \frac{e^{iz}}{z} dz = \lim_{R \to \infty} \int_0^\pi e^{iR(\cos\theta + i\sin\theta)} i \, d\theta = 0.$$

PROOF. Since $|e^{iR\cos(\theta)}| = 1$, it suffices to show that $\lim_{R \to \infty} \int_0^\pi e^{-R\sin\theta} d\theta = 0$. The following step is valid because the sine function takes on the same values on the interval $(\frac{\pi}{2}, \pi)$ as it does on $(0, \frac{\pi}{2})$:

$$\int_0^\pi e^{-R\sin\theta} d\theta = 2 \int_0^{\frac{\pi}{2}} e^{-R\sin\theta} d\theta.$$

There is a positive constant c such that $\frac{\sin\theta}{\theta} \geq c > 0$ on $(0, \frac{\pi}{2})$. (This conclusion fails on the full interval.) Therefore $\sin\theta \geq c\theta$ on $(0, \frac{\pi}{2})$. Hence $-R\sin\theta \leq -Rc\theta$. Since the exponential function is increasing, $e^{-R\sin\theta} \leq e^{-Rc\theta}$. Integrating gives

$$\int_0^\pi e^{-R\sin\theta} d\theta = 2 \int_0^{\frac{\pi}{2}} e^{-R\sin\theta} d\theta \leq 2 \int_0^{\frac{\pi}{2}} e^{-Rc\theta} \, d\theta = 2 \frac{1 - e^{\frac{-Rc\pi}{2}}}{Rc},$$

which goes to 0 as $R \to \infty$. $\qquad\square$

REMARK. The proof of Jordan's lemma provides a quintessential example of analysis. We need to show that the absolute value of something is small as $R \to \infty$. To do so, we show that a **larger** quantity is small. What is the advantage of looking at something larger? It is algebraically easier to understand and hence we gain by considering it.

Here is the most well-known example which uses Jordan's lemma:

EXAMPLE 8.3. Put $f(x) = \frac{\sin(x)}{x} = \text{sinc}(x)$. Then $f(x)$ has a removable singularity at 0. The integral $\int_{-\infty}^{\infty} |f(x)| dx$ diverges, and hence the integral of f is conditionally convergent. We verify that the principal value of the integral is π:

$$\lim_{R \to \infty} \int_{-R}^{R} f(x) dx = PV \int_{-\infty}^{\infty} f(x) dx = \pi.$$

For a closed curve γ that does not enclose 0, we have $\oint_\gamma \frac{e^{iz}}{z} dz = 0$. Let γ be the union of four curves: the real interval $[-R, -\epsilon]$, the semi-circle of radius ϵ in the upper half-plane, the real interval $[\epsilon, R]$, and the semi-circle of radius R in the upper half-plane. This curve is known as an *indented contour*. See Figure 7.

By Jordan's lemma, $\int_{|z|=R} \frac{e^{iz}}{z} dz \to 0$ as $R \to \infty$. When $\epsilon \to 0$, we see that $\int_{|z|=\epsilon} \frac{e^{iz}}{z} dz \to -\pi i$. (The minus sign appears because the circle is traversed with negative orientation.) Since $\int_\gamma \frac{e^{iz}}{z} dz = 0$, it follows that

$$PV \int_{-\infty}^{\infty} \frac{e^{iz}}{z} dz = -\lim_{\epsilon \to 0} \int_{|z|=\epsilon} \frac{e^{iz}}{z} dz = i\pi.$$

Therefore

$$\int_{-\infty}^{\infty} \frac{\sin(x)}{x} dx = \text{Im} \int_{-\infty}^{\infty} \frac{e^{ix}}{x} dx = \text{Im}(i\pi) = \pi.$$

This integral arises in various applications, but it is especially valuable for us because it illustrates conditional convergence. We discuss this integral again in the next chapter. We have seen that

$$PV \int_{-\infty}^{\infty} \frac{\sin(x)}{x} dx = \pi,$$

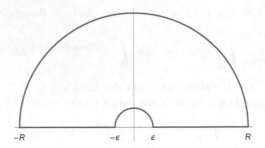

$$-R \qquad\qquad -\epsilon \quad \epsilon \qquad\qquad R$$

FIGURE 7. Indented contour for Example 8.3

but

$$\int_{-\infty}^{\infty} \left| \frac{\sin(x)}{x} \right| dx$$

diverges.

The following remark, due to Riemann, deserves some thought. An analogous statement holds for conditionally convergent integrals. Furthermore, the integrals and infinite series that arise in signal processing are often conditionally convergent. Several of the Fourier series discussed in Chapter 5 illustrate this behavior.

REMARK. Let $\{a_n\}$ be a sequence of real numbers. If $\sum a_n$ converges, but $\sum |a_n|$ diverges, then we can rearrange the terms to make the rearranged series converge to any number. See Exercises 8.6 and 8.7.

We end this chapter with a theorem which evaluates certain real integrals over the positive real axis. First consider

$$J = \int_0^{\infty} \frac{dx}{q(x)}$$

where q is a polynomial of degree at least two, and $q(x) > 0$ for $x \geq 0$. If q were *even*, then the value of J would be half the integral over the whole real line and could be found by Corollary 7.3. In general we can find J by introducing logarithms.

Consider the branch of the logarithm defined by $\log(z) = \log(|z|) + i\theta$, where $0 < \theta < 2\pi$. We will integrate on the positive real axis twice, using $\theta = 0$ once and then using $\theta = 2\pi$ the second time, as suggested by the contour $\gamma_{\epsilon,R}$ from Figure 8. We consider the line integral

$$J_{\epsilon,R} = \int_{\gamma_{\epsilon,R}} \frac{\log(z)dz}{q(z)}.$$

For R sufficiently large and ϵ sufficiently small, the value of $J_{\epsilon,R}$ is $2\pi i$ times the sum of the residues of $\left(\frac{\log(z)}{q(z)} \right)$ in the entire complex plane. The integrals on the circular arcs tend to 0 as ϵ tends to 0 and R tends to ∞. What happens on the positive real axis? Note that $\log(z) = \log(|z|)$ on the top part, but $\log(z) = \log(|z|) + 2\pi i$ on the bottom part. The bottom part is traversed in the opposite direction from the top part; hence everything cancels except for the term

$$\int_R^{\epsilon} \frac{2\pi i \, dz}{q(z)}.$$

After canceling out the common factor of $2\pi i$, and including the residues in all of $\mathbb{C}$ in the sum, we find

$$I = -\sum \operatorname{Res}\left(\frac{\log(z)}{q(z)}\right).$$

We state a simple generalization of this calculation as the next theorem.

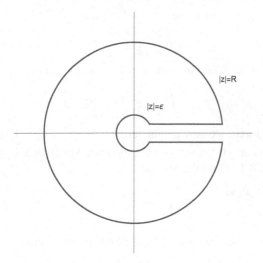

FIGURE 8. A useful contour

Theorem 8.2. *Let $\frac{f}{q}$ be a rational function, where the degree of q is at least two more than the degree of f, and suppose that $q(x) \neq 0$ for $x \geq 0$. Then the integral in (16) converges and*

$$\int_0^\infty \frac{f(x)}{q(x)}dx = -\sum \operatorname{Res}(\frac{\log(z)f(z)}{q(z)}). \qquad (16)$$

The sum in (16) is taken over all roots of q in $\mathbb{C}$.

EXAMPLE 8.4. First we compute

$$I = \int_0^\infty \frac{dx}{x^2+1}$$

by this method. There are two singularities. Using (16) we obtain

$$I = -\left(\frac{\log(i)}{2i} + \frac{\log(-i)}{-2i}\right) = -(\frac{\pi}{4} - \frac{3\pi}{4}) = \frac{\pi}{2}.$$

Example 8.4 can be done by Corollary 7.3, but the next example cannot.

EXAMPLE 8.5. Next we consider $p(x) = x^2 + x + 1$. We wish to find

$$I = \int_0^\infty \frac{dx}{x^2+x+1}.$$

Notice that $z^3 - 1 = (z-1)(z^2+z+1)$, and hence the poles are at the complex cube roots of unity ω and $\overline{\omega}$. Note that $2\omega + 1 = i\sqrt{3}$ and $2\overline{\omega} + 1 = -i\sqrt{3}$. By (16) we obtain

$$I = -\left(\frac{\log(\omega)}{2\omega+1} + \frac{\log(\overline{\omega})}{2\overline{\omega}+1}\right) = -\left(\frac{2\pi}{3\sqrt{3}} - \frac{4\pi}{3\sqrt{3}}\right) = \frac{2\pi}{3\sqrt{3}}.$$

REMARK. The same technique can be used to evaluate integrals involving powers of the logarithm. Include an extra logarithm and proceed as above.

EXERCISE 8.1. Assume $a > 0$ and n is a positive integer. Choose the branch of logarithm used in Theorem 8.2. Find the residue of $\frac{(\log(z))^n}{(z+a)^2}$ at $z = -a$.

EXERCISE 8.2. Assume $a > 0$. Evaluate
$$\int_0^\infty \frac{\log(x)}{(x+a)^2} dx.$$
Use the same contour as in Figure 8, but this time consider the integral
$$\int_{\gamma_{\epsilon,R}} \frac{(\log(z))^2}{(z+a)^2} dz.$$
On the bottom part of the real axis, we now get the term $(\log(z) + 2\pi i)^2$. Expanding the square gives three terms, and only the first term cancels the term on the top part. Verify that $I = \frac{\log(a)}{a}$. Note: The pole at $-a$ is of order two. Use Exercise 8.1 to find the residue.

EXERCISE 8.3. Difficult. Find
$$\int_0^\infty \frac{(\log(x))^n}{(x+1)^2} dx.$$

EXERCISE 8.4. Verify the following using the methods of this section.
$$\int_{-\infty}^\infty \frac{x\sin(x)}{x^2+1} dx = \frac{\pi}{e}$$
$$\int_{-\infty}^\infty \frac{x^2 \sin^2(x)}{(x^2+1)^2} dx = \frac{\pi}{4}(1 + e^{-2}).$$

EXERCISE 8.5. For complex numbers a_j (possibly repeated), put
$$f(z) = \frac{1}{\prod_{j=1}^d (z - a_j)}.$$
Find the sum of all the residues of f. Explain fully.

EXERCISE 8.6. Consider the series $\sum_{n=1}^\infty a_n = \sum_{n=1}^\infty \frac{(-1)^{n+1}}{n}$. Find its sum. Then consider the series
$$a_1 + a_3 + a_2 + a_5 + a_7 + a_4 + a_9 + a_{11} + a_6 + \cdots$$
obtained by reordering (take two positive terms, then one negative term, then two positive terms, and so on) the same terms. Show that the sum is different.

EXERCISE 8.7. Prove Riemann's remark.

EXERCISE 8.8. For $-1 < m < 1$, use the contour in Figure 7 to evaluate
$$\int_0^\infty \frac{dx}{x^m(x^2+1)}.$$

EXERCISE 8.9. For $-1 < m < 1$ and $0 < b < a$, use the contour in Figure 8 to verify that
$$\int_0^\infty \frac{x^m dx}{(x+a)(x+b)} = \frac{a^m - b^m}{a - b} \frac{\pi}{\sin(\pi m)}.$$
What happens when b tends to a?

CHAPTER 5

Transform methods

Many scientific or mathematical problems are difficult as posed, but simplify when expressed in different notation or coordinates. We discussed this idea for linear transformations in Chapter 1; the notion of similar matrices provides an especially illuminating example. In Chapter 5 we will see several new situations where changing one's perspective leads to powerful methods. We consider the Laplace transform, Fourier series, the Fourier transform, and generating functions. For example, the Laplace transform converts differentiation into multiplication. One passes from the *time domain* into the *frequency domain* to simplify the solution of many ordinary differential equations.

1. Laplace transforms

The Laplace transform is often used in engineering when solving ordinary differential equations. We start with a function $t \to f(t)$, typically defined for $t \geq 0$. We think of t as time and call the domain of f the **time domain**. The Laplace transform of f, written $\mathcal{L}f$, is a function of a complex variable s. When we are working with this function, we say we are in the **frequency domain**. The reference [KM] states that "fluency in *both* time- and frequency-domain methods is necessary".

Let us recall the precise meaning of improper integrals. We define

$$\int_0^\infty g(t)dt = \lim_{R\to\infty} \int_0^R g(t)dt. \tag{1}$$

When the limit in (1) exists, we say that the improper integral *converges*. One also considers improper integrals of the form $\int_a^b f(t)dt$ where f has a singularity at one or both of the endpoints. If, for example, f is not defined at a, we write

$$\int_a^b f(t)dt = \lim_{\epsilon\to 0} \int_{a+\epsilon}^b f(t)dt,$$

where the limit is taken using positive values of epsilon.

Definition 1.1 (Laplace transform). Let $f : [0,\infty) \to \mathbb{R}$, and assume that $\int_0^\infty e^{-st}f(t)dt$ converges. We write

$$F(s) = (\mathcal{L}f)(s) = \int_0^\infty e^{-st}f(t)dt, \tag{2}$$

and call F or $F(s)$ the **Laplace transform** of f. We allow s to be complex in (2).

EXAMPLE 1.1. Here are three explicit examples of Laplace transforms. Assume $t \geq 0$.
- For $a \in \mathbb{R}$, put $f(t) = e^{at}$. Then $(\mathcal{L}f)(s) = F(s) = \frac{1}{s-a}$. We require $\mathrm{Re}(s) > a$.
- If $f(t) = \sin(at)$, then $(\mathcal{L}f)(s) = F(s) = \frac{a}{s^2+a^2}$.
- If $f(t) = \frac{1}{\sqrt{t}}$ for $t > 0$, then $(\mathcal{L}f)(s) = F(s) = \sqrt{\frac{\pi}{s}}$.

The function $t \mapsto \frac{1}{\sqrt{t}}$ is defined only for $t > 0$, and the integral (2) is therefore improper also at 0. In this case, the improper integral converges at both 0 and ∞, and hence the Laplace transform is defined.

Consider the example $f(t) = e^{at}$. The integral defining $F(s)$ makes sense only for $\mathrm{Re}(s) > a$. Yet the formula for $F(s)$ makes sense for $s \neq a$. This phenomenon is a good example of analytic continuation. We briefly discuss this concept. Given a piecewise continuous function on $[0, \infty)$, the integral in (2) converges for s in some subset of $\mathbb{C}$. This **region of convergence** is often abbreviated **ROC**. When f grows faster than exponentially at infinity, the **ROC** is empty. The **ROC** will be non-empty when f is piecewise continuous and does not grow too fast at zero and infinity, the endpoints of the interval of integration. See Exercise 1.8. The function $F(s)$ is often complex analytic in a region larger than one might expect. Analytic continuation is significant in number theory, where the zeta and Gamma functions are defined by formulas and then extended to larger sets by continuation.

EXAMPLE 1.2. Put $f(t) = e^t$ for $t \geq 0$. The integral (2) defining the Laplace transform converges when $\int_0^\infty e^{t(1-s)}dt$ exists. Thus we require $\mathrm{Re}(s) > 1$. See Exercise 8.3 of Chapter 2. Hence the **ROC** is the half-plane to the right of the line $\mathrm{Re}(s) = 1$ in $\mathbb{C}$. Here $F(s) = \frac{1}{s-1}$ is analytic everywhere except at the pole where $s = 1$.

REMARK. For many signals $f(t)$ arising in engineering, the Laplace transform $F(s)$ is a rational function of s. The location of the poles determines properties of the signal. One can invert such Laplace transforms by using the partial fractions decomposition of $F(s)$. Example 1.3 illustrates the idea, and Example 1.5 handles the general case.

EXAMPLE 1.3. Given that $F(s) = \frac{s+1}{s^2(s-1)}$, we use partial fractions to write

$$F(s) = \frac{s+1}{s^2(s-1)} = \frac{2}{s-1} + \frac{-1}{s^2} + \frac{-2}{s}.$$

By linearity (Theorem 1.2) and the examples below we obtain

$$f(t) = 2e^t - t - 2$$

for $t \geq 0$. Theorem 1.3 below uses residues to provide a general method for inverting Laplace transforms.

We summarize some facts about $\mathcal{L}$. First we have the easy but useful result:

Theorem 1.2. *Let $\mathcal{L}$ denote the Laplace transform operation.*
(1) *$\mathcal{L}$ is linear: $\mathcal{L}(f + g) = \mathcal{L}(f) + \mathcal{L}(g)$ and $\mathcal{L}(cf) = c\mathcal{L}(f)$.*
(2) *If f is differentiable, then $\mathcal{L}(f')(s) = s\mathcal{L}(f)(s) - f(0)$.*
(3) *If f is twice differentiable, then*

$$\mathcal{L}(f'')(s) = s\mathcal{L}(f')(s) - f'(0) = s^2\mathcal{L}(f)(s) - sf(0) - f'(0).$$

(4) *$(\mathcal{L}f)(s + c) = \mathcal{L}(e^{-ct}f(t))$.*

PROOF. The reader can easily check statements (1) and (4). We use integration by parts to prove statements (2) and (3). To prove (2), note that

$$\mathcal{L}(f')(s) = \int_0^\infty e^{-st}f'(t)dt$$
$$= e^{-st}f(t)\big|_0^\infty - \int_0^\infty (-s)e^{-st}f(t)dt$$
$$= -f(0) + s\int_0^\infty e^{-st}f(t)dt$$
$$= s(\mathcal{L}f)(s) - f(0).$$

To prove (3), apply (2) to f'. □

The next result is considerably more difficult, and we will not prove it here. At the end of Section 5 we indicate how to derive this theorem from the Fourier inversion formula.

Theorem 1.3. *The Laplace transform $\mathcal{L}$ is injective: if $\mathcal{L}(f) = \mathcal{L}(g)$, then f and g agree except on a set of measure 0. Furthermore, for sufficiently large a, and with the integral interpreted correctly,*

$$f(t) = \frac{1}{2\pi i} \int_{a-i\infty}^{a+i\infty} e^{st} F(s)\,ds.$$

This theorem often enables us to invert the Laplace transform using residues. One chooses a sufficiently large such that the curve in Figure 1 encloses all the singularities and then applies Cauchy's residue theorem. One of course requires hypotheses on f guaranteeing that the integral over the semi-circular arc tends to 0 as $r \to \infty$.

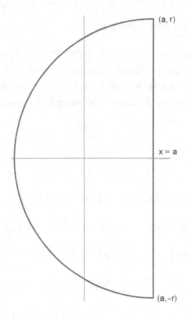

FIGURE 1. Inverting the Laplace transform

EXAMPLE 1.4. We use Theorem 1.3 to find the inverse Laplace transform of $\frac{1}{s-\lambda}$, where $\lambda > 0$. Choose the contour pictured in Figure 1 to enclose λ. The Cauchy integral formula or residue theorem yields

$$\frac{1}{2\pi i} \int_{\gamma_r} e^{st} F(s)\,ds = \frac{1}{2\pi i} \int_{\gamma_r} \frac{e^{st}}{s-\lambda}\,ds = e^{\lambda t}.$$

By the methods of the previous chapter, the integral along the circular part of the curve tends to 0 as r tends to infinity, and we obtain the result. The condition $t \geq 0$ is presumed.

The next example fully illustrates the power of the residue method. Using Corollary 1.6, which is elementary, we can prove Theorem 1.3 in the special case when $F(s)$ is rational.

EXAMPLE 1.5. We can easily find the inverse Laplace transform of an arbitrary rational function $F(s)$ using Theorem 1.3. The first step is to find the poles. Then, using partial fractions, write

$$F(s) = \sum \frac{c_j}{(s-\lambda_j)^{n_j}}.$$

By linearity it suffices to determine the inverse transform of

$$\frac{1}{(s-\lambda)^p}$$

for each positive integer p. Theorem 1.3 and the power series for the exponential function combine to give the answer. (One could also use Exercise 7.2. from Chapter 4.)

$$f(t) = \operatorname{Res}\frac{e^{st}}{(s-\lambda)^p} = \operatorname{Res} e^{\lambda t}\frac{e^{(s-\lambda)t}}{(s-\lambda)^p} = e^{\lambda t}\frac{t^{p-1}}{(p-1)!}.$$

Now go back to Example 1.3 and see how easy it is.

We next introduce the Gamma function, which (by Proposition 1.5 below) generalizes the factorial function. As a consequence, the Gamma function appears, sometimes unexpectedly, throughout pure and applied mathematics. Its definition as an integral leads to one method for proving Stirling's formula, an asymptotic approximation for $n!$ when n is large. The simplest version of Stirling's formula says that

$$\lim_{n\to\infty}\frac{n!}{n^n e^{-n}\sqrt{2\pi n}} = 1.$$

See [T] for a proof.

Definition 1.4 (Gamma function). For $p > -1$, we put $\Gamma(p+1) = \int_0^\infty e^{-t}t^p dt$.

Note when $-1 < p < 0$ that the integral is improper but convergent at 0.

Proposition 1.5. *If $p \in \mathbb{N}$, then $\Gamma(p+1) = p!$. Also, $\Gamma(\frac{1}{2}) = \sqrt{\pi}$.*

PROOF. Note that $\Gamma(1) = \int_0^\infty e^{-t}dt = 1$. Integration by parts yields $\Gamma(p+1) = p\Gamma(p)$. Hence, by induction, $\Gamma(p+1) = p!$.

Next, we show that $\Gamma(\frac{1}{2}) = \int_0^\infty e^{-t}t^{\frac{-1}{2}} dt = \sqrt{\pi}$. We already know that

$$\sqrt{\pi} = \int_{-\infty}^\infty e^{-x^2}dx = 2\int_0^\infty e^{-x^2}dx. \tag{3}$$

Make the change of variables $t = x^2$ in the far right-hand side of (3). Then $x = \sqrt{t}$ and $dx = \frac{1}{2\sqrt{t}}dt$. We obtain $\sqrt{\pi} = 2\int_0^\infty e^{-t}\frac{1}{2\sqrt{t}}dt = \Gamma(\frac{1}{2})$. □

Corollary 1.6. *Suppose $f(t) = t^p$ for $p > -1$. Then the Laplace transform $F(s)$ satisfies*

$$F(s) = \frac{\Gamma(p+1)}{s^{p+1}}.$$

PROOF. Replace t by st in the definition of the Gamma function. See Exercises 1.2 and 1.13. □

1.1. Convolution and the Laplace transform. Convolution will appear quite often in the rest of this book. We next introduce this topic and discuss its relationship with the Laplace transform.

We begin with grade school arithmetic. Suppose we want to find $493 * 124$ in base ten. We consider the functions $4x^2 + 9x + 3$ and $x^2 + 2x + 4$. Their product is given by $h(x) = 4x^4 + 17x^3 + 37x^2 + 42x + 12$. Setting $x = 10$ gives

$$493 * 124 = h(10) = 40000 + 17000 + 3700 + 420 + 12 = 61132.$$

The coefficients can be obtained either by expanding the product of the polynomials, or by convolution, as we next note.

Consider more generally the product of two power series:

$$\left(\sum_{j=0}^{\infty} a_j z^j \right) \left(\sum_{k=0}^{\infty} b_k z^k \right) = \sum_{n=0}^{\infty} c_n z^n,$$

where $c_n = \sum_0^n a_{n-k} b_k$. We call the sequence $\{c_n\}$ the **convolution** of the sequences $\{a_n\}$ and $\{b_n\}$. We can interpret the convolution as a weighted average.

We have a similar definition for the convolution of functions defined on intervals of the real line.

Definition 1.7. Suppose f and g are integrable on $[0, \infty)$. The function $f * g$, defined by

$$(f * g)(x) = \int_0^x f(x - t) g(t) dt,$$

is called the **convolution** of f and g.

In Definition 1.7 we integrated over the interval $[0, x]$ to make the formula look like the formula for convolution of sequences. We could, however, integrate over $[0, \infty)$ or even $(-\infty, \infty)$ without changing anything. The reason is that we may assume $f(x) = g(x) = 0$ when $x < 0$. With this assumption we have

$$(f * g)(x) = \int_0^x f(x - t) g(t) dt = \int_0^\infty f(x - t) g(t) dt = \int_{-\infty}^\infty f(x - t) g(t) dt.$$

See Exercises 1.11 and 1.12 and the Remark after Example 1.6 for related information.

The assumption that f and g vanish for $x < 0$ is natural in many applications. Linear time-invariant systems, written **LTI** systems, are given by convolution. In this case, if the input signal is given by g, then the output signal is given by $h * g$, where h is called the **impulse response**. This **LTI** system is **causal** when $h(x) = 0$ for $x < 0$. In this setting, one usually writes the variable as t (for time) rather than as x. The Laplace transform of h is known as the **transfer function** of the **LTI** system. See [KM] for additional discussion. See also Exercise 4.13 later in this chapter. Section 8 of Chapter 7 considers linear time-invariant systems in detail.

The following result is the analogue for Laplace transforms of the product of infinite series.

Theorem 1.8. *For integrable f, g we have $F(s)G(s) = \mathcal{L}(f * g)$. Thus $\mathcal{L}^{-1}(FG) = f * g$.*

PROOF. Assuming the integrals converge, we have

$$F(s)G(s) = \int_0^\infty e^{-su} f(u) du \int_0^\infty e^{-sv} g(v) dv = \int_0^\infty \int_0^\infty e^{-s(u+v)} f(u) g(v) du\, dv$$

$$= \int_0^\infty e^{-st} \left(\int_0^t f(t - v) g(v) dv \right) dt = \mathcal{L}(f * g).$$

In the last step we changed variables by putting $u = t - v$. Then $du = dt$ and $0 \le v \le t$. □

1.2. Solving ODE and integral equations. We next show how to use the Laplace transform for solving ODE and integral equations.

EXAMPLE 1.6. Solve the ODE $y' - 2y = e^{2t}$ given $y(0)$.

1) We solve using **variation of parameters**. We first solve the homogeneous equation $y' - 2y = 0$, obtaining $y(t) = ce^{2t}$. Next we vary the parameter c; thus we assume $y(t) = c(t)e^{2t}$ for some function c. Then $y'(t) = (c'(t) + 2c(t))e^{2t}$. Therefore

$$e^{2t} = y' - 2y = c'(t)e^{2t}.$$

Thus $c'(t) = 1$ and hence $c(t) = t + y(0)$. We conclude that $y(t) = (t + y(0))e^{2t}$. In this derivation, there is no assumption that $t \geq 0$.

2) We solve using the **Laplace transform**. We start by transforming both sides of the equation:

$$\mathcal{L}(y' - 2y) = \mathcal{L}(e^{2t}) = \frac{1}{s - 2}.$$

By Theorem 1.2 we also have

$$\mathcal{L}(y' - 2y) = \mathcal{L}(y') - \mathcal{L}(2y) = s\mathcal{L}(y) - y(0) - 2\mathcal{L}(y) = (s - 2)\mathcal{L}(y) - y(0).$$

Equating the two expressions for the Laplace transform of $y' - 2y$ gives

$$(s - 2)\mathcal{L}(y) = y(0) + \frac{1}{s - 2}.$$

We solve this algebraic equation for $\mathcal{L}(y)$ to get

$$\mathcal{L}(y) = \frac{y(0)}{s - 2} + \frac{1}{(s - 2)^2}.$$

Assuming that $t \geq 0$, we invert the Laplace transform to obtain $y = y(0)e^{2t} + te^{2t} = (y(0) + t)e^{2t}$.

REMARK. The Laplace transform method assumes that $t \geq 0$. This issue comes up several times in the sequel. Often one introduces the Heaviside function, written $H(t)$ or $u(t)$. This function is 1 for positive t and 0 for negative t. Given a function f defined for all t, the function $t \mapsto u(t)f(t)$ is then defined for $t > 0$. The author is a bit skeptical of this approach, having seen errors arising from careless use of it. See, for example, Exercises 1.6 and 1.7. The Heaviside function is named for Oliver Heaviside, not because the function is heavy on the right-side of the axis. See [H] for interesting information about Heaviside.

EXAMPLE 1.7. Solve $y(x) = x^3 + \int_0^x \sin(x - t)y(t)dt$. Again we solve the equation in two ways. First we convert to an ODE and solve it. Second we use the Laplace transform directly.

1) Differentiating twice gives an ODE:

$$y'(x) = 3x^2 + 0 + \int_0^x \cos(x - t)y(t)dt,$$

$$y''(x) = 6x + y(x) - \int_0^x \sin(x - t)y(t)dt = 6x + x^3.$$

Therefore $y'(x) = \frac{1}{4}x^4 + 3x^2 + C_1$ and $y(x) = \frac{1}{20}x^5 + x^3 + C_1 x + C_2$. The initial condition $y(0) = 0$ implies that $C_2 = 0$. The condition $y'(0) = 0$ implies that $C_1 = 0$. Hence $y(x) = \frac{1}{20}x^5 + x^3$.

2) Write $y(x) = x^3 + \int_0^x \sin(x-t)y(t)dt$. Hence $y = x^3 + h(x)$, where $h(x)$ is the convolution of the sine function and y. Apply the Laplace transform on both sides of the equation and use Theorem 1.8 to get

$$(\mathcal{L}y)(s) = \mathcal{L}(x^3) + \mathcal{L}(h)$$

$$= \frac{6}{s^4} + \mathcal{L}(\sin x)\mathcal{L}(y)$$

$$= \frac{6}{s^4} + \frac{1}{s^2+1}(\mathcal{L}y)(s).$$

Therefore

$$(\mathcal{L}y)(s)\left(1 - \frac{1}{s^2+1}\right) = \frac{6}{s^4}.$$

Hence $(\mathcal{L}y)(s) = \frac{6}{s^4} + \frac{6}{s^6}$ and thus $y = \frac{1}{20}x^5 + x^3$. As before, this method presumes that $x \geq 0$.

1.3. Additional properties of Laplace transforms. Next we use the Laplace transform to investigate the sinc function and to evaluate several interesting integrals.

The **sinc** function is defined by $\mathrm{sinc}(x) = \frac{\sin x}{x}$ for $x \neq 0$ and $\mathrm{sinc}(0) = 1$. Figure 2 shows the graph of $\mathrm{sinc}(x)$ for $-5 \leq x \leq 5$, illustrating the local behavior near 0. Figure 3 shows the graph of $\mathrm{sinc}(x)$ for $-100 \leq x \leq 100$, illustrating the global behavior but distorting the behavior near 0. Note that sinc is an even function. Note also that sinc is complex analytic everywhere in $\mathbb{C}$, because the removable singularity at 0 has been removed! This function also satisfies the following infinite product formulas:

$$\mathrm{sinc}(x) = \frac{\sin(x)}{x} = \prod_{n=1}^{\infty}\left(1 - \frac{x^2}{\pi^2 n^2}\right).$$

$$\mathrm{sinc}(x) = \prod_{j=1}^{\infty}\cos\left(\frac{x}{2^j}\right)$$

The sinc function arises in signal processing because it is the Fourier transform of an indicator function. See Example 4.1 later in this chapter.

The integral $\int_{-\infty}^{\infty}\frac{\sin x}{x}dx$ is conditionally convergent. In Example 8.3 of Chapter 4 we showed that its principal value is π. We give a different derivation here. The method applies in many situations, as indicated in the subsequent remark and in Example 1.8. Let J denote the principal value of the integral, namely

$$J = \lim_{R\to\infty}\int_{-R}^{R}\frac{\sin x}{x}dx.$$

Then we have

$$J = 2\int_0^{\infty}\frac{\sin(x)}{x}dx = 2\int_0^{\infty}\sin(x)\int_0^{\infty}e^{-xt}dt\,dx.$$

Now interchange the order of integration and use Exercise 1.3 to get

$$J = 2\int_0^{\infty}\int_0^{\infty}e^{-xt}\sin(x)dx\,dt = 2\int_0^{\infty}\frac{1}{1+t^2}dt = \pi.$$

REMARK. Put $\langle u, v\rangle = \int_0^{\infty}u(x)v(x)dx$. The technique used in the previous computation says

$$\langle \mathcal{L}u, v\rangle = \langle u, \mathcal{L}v\rangle.$$

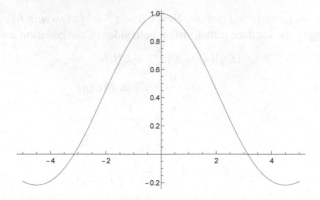

FIGURE 2. The sinc function near 0

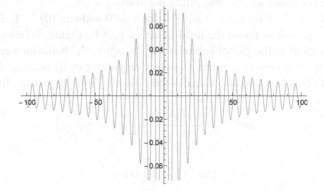

FIGURE 3. The sinc function globally

In the special case when $v = f$ and $u = 1$ we obtain

$$\int_0^\infty \frac{f(t)}{t} dt = \int_0^\infty F(s) ds.$$

The trick in integrating sinc is thus to write $\frac{1}{x}$ as the Laplace transform of something, namely 1, and then move the Laplace transform to the sine function. We next combine this technique with Theorem 8.2 of Chapter 4 to compute another interesting integral.

EXAMPLE 1.8. For $a, b > 0$, we claim

$$\int_0^\infty \frac{\cos(at) - \cos(bt)}{t} dt = \ln(b) - \ln(a) = \ln\left(\frac{b}{a}\right).$$

Note first that the Laplace transform of $\cos(at)$ is $\frac{s}{s^2+a^2}$. Of course $\frac{1}{t}$ is the Laplace transform of 1. By the remark, the value of the integral is therefore the same as that of

$$\int_0^\infty \left(\frac{s}{s^2 + a^2} - \frac{s}{s^2 + b^2}\right) ds = \int_0^\infty \frac{s(b^2 - a^2)}{(s^2 + a^2)(s^2 + b^2)} ds.$$

We evaluate this integral using Theorem 8.2 of Chapter 4. By that result, the value is

$$- \sum \text{residues} \left(\log(s) \frac{s(b^2 - a^2)}{(s^2 + a^2)(s^2 + b^2)} \right). \tag{4}$$

Here $0 < \theta < 2\pi$ defines the branch of the logarithm. There are four singularities, at $\pm ai$ and $\pm bi$. We obtain the following values:

$$\text{res}_{ai} = \log(ai) \frac{(ai)(b^2 - a^2)}{2ai(b^2 - a^2)} = \frac{1}{2} \ln(a) + \frac{i\pi}{2}$$

$$\text{res}_{-ai} = \log(-ai) \frac{(-ai)(b^2 - a^2)}{-2ai(b^2 - a^2)} = \frac{1}{2} \ln(a) + \frac{3i\pi}{2}$$

$$\text{res}_{bi} = \log(bi) \frac{(bi)(b^2 - a^2)}{2bi(a^2 - b^2)} = -\frac{1}{2} \ln(b) - \frac{i\pi}{2}$$

$$\text{res}_{-bi} = \log(-bi) \frac{(-bi)(b^2 - a^2)}{-2bi(a^2 - b^2)} = -\frac{1}{2} \ln(b) - \frac{3i\pi}{2}.$$

Adding the results and using the minus sign in (4) gives the answer.

EXAMPLE 1.9. We find the inverse Laplace transform of $\frac{1}{s(s+1)}$ in two ways. Using partial fractions, we first write

$$\frac{1}{s(s+1)} = \frac{1}{s} - \frac{1}{s+1}.$$

Example 1.1 and linearity yield the inverse Laplace transform $1 - e^{-t}$. Alternatively, the method in the remark before Example 1.8 gives

$$\mathcal{L}^{-1} \left(\frac{F(s)}{s} \right)(t) = \int_0^t f(x) \, dx.$$

Put $f(x) = e^{-x}$; then $F(s) = \frac{1}{s+1}$ and we obtain

$$\mathcal{L}^{-1} \left(\frac{1}{s(s+1)} \right) = \mathcal{L}^{-1} \left(\frac{F(s)}{s} \right)(t) = \int_0^t f(x) \, dx = \int_0^t e^{-x} \, dx = 1 - e^{-t}.$$

REMARK. The reader may wonder what sorts of functions of s can be Laplace transforms. Suppose that f has exponential growth. In other words, an inequality of the form

$$|f(t)| \le M e^{ct}$$

holds for some real c and positive M. Then, for $\text{Re}(s) > c$, the integral defining $F(s)$ satisfies the simple inequality

$$|F(s)| \le \left| \int_0^\infty e^{-st} f(t) \, dt \right| \le M \int_0^\infty |e^{-(s-c)t}| \, dt = \frac{M}{\text{Re}(s) - c}.$$

As a consequence, assuming $\text{Re}(s) > c$, we have $\lim_{s \to \infty} = 0$.

Corollary 1.9. *A rational function of s is a Laplace transform if and only if the degree of the numerator is smaller than the degree of the denominator. In particular, the only polynomial that is a Laplace transform is identically zero.*

EXERCISE 1.1. Show that the critical points of the sinc function are those x for which $\text{sinc}(x) = \cos(x)$.

EXERCISE 1.2. Assuming that s is real, use the change of variables $t = \frac{u}{s}$ to verify for $p > -1$ that

$$\mathcal{L}(t^p) = \int_0^\infty e^{-st} t^p dt = \frac{\Gamma(p+1)}{s^{p+1}}.$$

See Exercise 1.13 for a method to verify this formula when $\operatorname{Re}(s) > 0$. Also note the result when $p = \frac{-1}{2}$. What is the **ROC**?

EXERCISE 1.3. Verify that the Laplace transform of $\sin(at)$ is $\frac{a}{s^2+a^2}$ and the Laplace transform of $\cos(at)$ is $\frac{s}{s^2+a^2}$. Suggestion: Use complex exponentials.

EXERCISE 1.4. Find the inverse Laplace transform of the following functions in the frequency domain.

$$\frac{a}{s^2 - a^2}$$

$$\frac{1}{s(s^2 - a^2)}$$

$$\frac{e^{-as}}{s^2}$$

$$\frac{1}{(s-a)(s-b)(s-c)}.$$

EXERCISE 1.5. Verify (4) of Theorem 1.2.

EXERCISE 1.6. Solve the differential equation: $f'(t) = f(t)$ for $t < 0$, and $f'(t) = 1$ for $t > 0$, and f continuous at 0. Check your answer.

EXERCISE 1.7. Solve the integral equation

$$f(x) = 1 + x + \int_0^x f(t)dt.$$

Do NOT assume $f(x) = 0$ for $x < 0$. If you use Laplace transforms, then you might make this assumption without realizing it.

EXERCISE 1.8. Assume that f is piecewise continuous on the open interval $(0, \infty)$. Assume that f satisfies both of the following growth estimates. Show that the Laplace transform of f exists for s with sufficiently large real part.

- For some α with $\alpha < 1$, the function $t^\alpha |f(t)|$ is bounded near 0.
- For some β, the function $e^{-\beta t}|f(t)|$ is bounded near ∞.

EXERCISE 1.9. Let p_j be positive integers and λ_j be real numbers. Suppose a signal is given by

$$f(t) = \sum_{j=0}^n c_j t^{p_j} e^{\lambda_j t}.$$

What can you say about the Laplace transform $F(s)$?

EXERCISE 1.10. Consider the Lambert W-function from Exercise 9.6 in Chapter 3. Use the Gamma function to show that

$$\int_0^\infty W(x) x^{\frac{-3}{2}} dx = \sqrt{8\pi}.$$

EXERCISE 1.11. Define $u(t)$ by $u(t) = 1$ for $t > 0$ and $u(t) = 0$ otherwise. Compute $u * u$.

EXERCISE 1.12. Define $u(t)$ as in Exercise 1.11. For $a > b > 0$ find

$$e^{-at}u(t) * e^{-bt}u(t).$$

What happens when $a \to b$? What happens when $a, b \to 0$?

EXERCISE 1.13. Exercise 1.2 found a formula for the Laplace transform of t^p, assuming that s is real. Prove that the formula remains valid for s with $\mathrm{Re}(s) > 0$. Suggestion: Start with $\int_0^R e^{-st} t^p \, dt$ and make the same change of variables, obtaining a line integral along a segment in $\mathbb{C}$. Connect this segment to the interval $[0, R]$ via a circular arc. Combine Cauchy's theorem with an estimate along the circular arc to show that the limit along the real axis equals the limit along the other line. Compare with the proof of Jordan's lemma (Lemma 8.1 from Chapter 4).

EXERCISE 1.14. Put $f(t) = t$ for $0 \le t \le 1$ and $f(t) = 2 - t$ for $1 \le t \le 2$ and $f(t) = 0$ otherwise. Show that the Laplace transform of f is given by

$$F(s) = \frac{e^{-2s}(e^s - 1)^2}{s^2}.$$

Comment: The function f is called the *tent* function or a *triangular window*. It arises in signal processing, where the tent represents an idealized signal.

EXERCISE 1.15. Suppose that $F(s)$ is analytic in the right half-plane and has no poles on the imaginary axis. Show that $\lim_{t \to \infty} f(t) = 0$. Can you make a stronger statement?

EXERCISE 1.16. Suppose $f(t)$ is real for $t \ge 0$. What statement can you make about the possible poles of $F(s)$? Suggestion: compute $F(\bar{s})$.

EXERCISE 1.17. Suppose $F(s) = \frac{e^{-as}}{s}$. Show that $f(t) = 0$ for $t < a$ and $f(t) = 1$ for $t > a$.

EXERCISE 1.18. Suppose $F(s) = \frac{e^{-2s}(e^s - 1)^2}{s^2}$, as in Exercise 1.14. Show that the only singularity of F is removable.

EXERCISE 1.19. Use Stirling's formula to evaluate

$$\lim_{n \to \infty} \frac{\sqrt{n}\binom{2n}{n}}{2^{2n}}.$$

2. Generating functions and the Z-transform

Let $\{a_n\}$ be a sequence of real or complex numbers. We gather the information provided by all the terms into one piece of information as follows. Consider the formal series

$$f(z) = \sum_{n=0}^{\infty} a_n z^n \tag{5.1}$$

$$F(z) = \sum_{n=0}^{\infty} a_n \left(\frac{1}{z}\right)^n. \tag{5.2}$$

When the series in (5.1) converges, f is called the **ordinary generating function** of the sequence a_n. Also, F is called the **Z-transform** of the sequence. Notice that $F(z) = f(\frac{1}{z})$. Thus using generating functions and using the Z-transforms are essentially the same.

The relationship between sequences and their generating functions is not very different from the relationship between the decimal expansion of a number in the interval $(0, 1)$ and the number itself. For example, consider the repeating decimal

$$.493493493.... = \frac{493}{999}.$$

We write the decimal expansion as

$$\sum_{n=1}^{\infty} a_n \frac{1}{10^n},$$

where $a_1 = 4$, $a_2 = 9$, and $a_3 = 3$. Also, for each n, we have $a_{n+3} = a_n$. The information provided by the sequence of decimal digits is the same as the information provided by the single rational number $\frac{493}{999}$. In fact, the decimal digits of any rational number in $[0, 1]$ satisfy a linear recurrence relation. Generating functions amount to replacing $\frac{1}{10}$ with a variable z. This example suggests that we can use generating functions to solve recurrence relations.

If $f(z) = \sum_{n=0}^{\infty} a_n z^n$ and the series converges for $|z| < R$, then f is an analytic function for $|z| < R$. Hence $a_n = \frac{f^{(n)}(0)}{n!}$. Thus, when the series converges, we can recover the sequence from its generating function, and we have two formulas for a_n:

- $a_n = \frac{f^{(n)}(0)}{n!}$ (Taylor coefficient)
- $a_n = \frac{1}{2\pi i} \int_{|z|=\epsilon} \frac{f(z)}{z^{n+1}} dz$ (Cauchy integral formula).

We have a similar integral formula for recovering the sequence $\{a_n\}$ from its Z-transform. Suppose $h(z) = \sum_{n=0}^{\infty} a_n (\frac{1}{z})^n$. Then $h(z) = f(\frac{1}{z})$. By the change of variables $w = \frac{1}{z}$, we get $z = \frac{1}{w}$. Hence $dz = -\frac{1}{w^2} dw$ and

$$a_n = \frac{1}{2\pi i} \int_{|w|=\frac{1}{\epsilon}} h(w) w^{n-1} \, dw.$$

REMARK. If F is the generating function for a sequence $\{a_k\}$, and G is the generating function for the sequence $\{b_k\}$, then FG is the generating function for the convolution sequence given by

$$c_n = \sum_{k=0}^{n} a_{n-k} b_k.$$

See the paragraph preceding Definition 1.7.

A sequence and its generating function provide the same information. Passing from one to the other is similar to the changes in perspective we have seen when diagonalizing matrices, applying the Laplace transform, and so on.

EXAMPLE 2.1. We find the generating function $f(z)$ and the Z-transform $F(z)$ for several sequences.

(1) Put $a_n = 1$ for all n. Then

$$f(z) = \sum_{n=0}^{\infty} a_n z^n = \frac{1}{1-z} \text{ if } |z| < 1.$$

$$F(z) = f\left(\frac{1}{z}\right) = \sum_{n=0}^{\infty} a_n \left(\frac{1}{z}\right)^n = \frac{z}{z-1} \text{ if } |z| > 1.$$

(2) Put $a_n = 0$ for all $n \geq 1$ and $a_0 = 1$. Then $f(z) = \sum_{n=0}^{\infty} a_n z^n = 1$.

(3) Put $a_n = \lambda^n$. Then

$$f(z) = \sum_{n=0}^{\infty} a_n z^n = \frac{1}{1 - \lambda z} \text{ if } |\lambda z| < 1.$$

$$F(z) = \frac{z}{z - \lambda} \text{ if } |\frac{\lambda}{z}| < 1.$$

We illustrate the use of generating functions in several examples of varying difficulty. First we solve a constant coefficient linear recurrence relation.

EXAMPLE 2.2. Find a formula for a_n where $a_{n+1} - 2a_n = n$ with $a(0) = 1$.

Method 1. We solve this recurrence in a manner similar to how we solve constant coefficient ODE. We solve the homogeneous equation $a_{n+1} - 2a_n = 0$ by assuming a solution of the form λ^n. We get $\lambda = 2$ and $a_n = 2^n$.

The general solution is of the form $K2^n + f(n)$, where $f(n)$ is a particular solution and K is a constant. We try $f(n) = bn + c$ and find $b = c = -1$. Thus the general solution is $c2^n - n - 1$ for some c. Using $a_0 = 1$ gives $c = 2$. Therefore $a_n = 2^{n+1} - n - 1$.

Method 2. We solve this recurrence using generating functions. Put $f(z) = \sum a_n z^n$. We use the recurrence to find a formula for f. We have

$$f(z) = \sum_{n=0}^{\infty} a_n z^n = a_0 + \sum_{n=0}^{\infty} a_{n+1} z^{n+1}.$$

Now we use $a_{n+1} = 2a_n + n$ to obtain

$$f(z) = a_0 + \sum_{n-0}^{\infty} (2a_n + n)z^{n+1} = a_0 + 2zf(z) + \sum_{n=0}^{\infty} nz^{n+1}.$$

Therefore

$$f(z)\,(1 - 2z) = a_0 + \sum_{n=0}^{\infty} nz^{n+1} = a_0 + \frac{z^2}{(1 - z)^2}. \tag{6}$$

Simplifying (6) and putting $a_0 = 1$ gives

$$f(z) = \frac{a_0}{1 - 2z} + \frac{z^2}{(1 - z)^2(1 - 2z)} = \frac{1 - 2z + 2z^2}{(1 - 2z)(1 - z)^2}. \tag{7}$$

We must next expand (7) in a series. To do so, use partial fractions to write

$$f(z) = \frac{2}{1 - 2z} - \frac{1}{(1 - z)^2}. \tag{8}$$

The coefficient of z^n in the expansion of the right-hand side of (8) is $2(2^n) - (n + 1)$. Thus, as before, $a_n = 2^{n+1} - n - 1$.

Using generating functions for solving such recurrences involves combining the method of partial fractions with our knowledge of the geometric series. It is often useful to obtain additional series by differentiating the geometric series. For example, when $|z| < 1$,

$$\sum_{n=0}^{\infty} (n + 2)(n + 1)z^n = \sum_{n=0}^{\infty} n(n - 1)z^{n-2} = \frac{2}{(1 - z)^3}. \tag{9}$$

To obtain (9), simply differentiate the geometric series twice. The following remark recalls some facts about power series and justifies the technique in (9).

REMARK. Consider a series $\sum a_n z^n$ whose radius of convergence is R. Thus $\sum a_n z^n$ converges uniformly and absolutely on $|z| < R$, and diverges for $|z| > R$. There is a formula, called the root test, for R in terms of the a_n:

$$\frac{1}{R} = \limsup_{n \to \infty} |a_n|^{\frac{1}{n}}.$$

This test always works. The ratio test does not always apply, because the limit of the ratios of successive terms might not exist. When this limit does exist, we have

$$\frac{1}{R} = \lim_{n \to \infty} |\frac{a_{n+1}}{a_n}|.$$

When $\sum a_n z^n$ has radius of convergence R, this series defines an analytic function f in the set $|z| < R$. If we differentiate the series term-by-term, the new series will have the same radius of convergence, and

$$f'(z) = \sum n a_n z^{n-1}$$

holds if $|z| < R$.

EXAMPLE 2.3. There is a you-tube video alleging that

$$\sum_{n=1}^{\infty} n = 1 + 2 + 3 + \cdots = \frac{-1}{12}.$$

Furthermore, the physicists who produced the video claim that this "result" arises in string theory. We state here what is actually true. First consider the formula

$$\sum_{n=0}^{\infty} e^{-ns} = \frac{1}{1 - e^{-s}},$$

which is obtained by the geometric series in e^{-s} and is valid for $|e^{-s}| < 1$. Hence the formula

$$\sum_{n=0}^{\infty} n e^{-ns} = \frac{e^s}{(1 - e^s)^2}, \tag{$*$}$$

obtained by differentiation, is also valid there. The function $\frac{e^s}{(1-e^s)^2}$ has a pole of order 2 at the origin. Its Laurent series is given by

$$\frac{e^s}{(1 - e^s)^2} = \frac{1}{s^2} + \frac{-1}{12} + \frac{s^2}{240} - \frac{s^4}{6048} + \cdots$$

Thus

$$\lim_{s \to 0} \left(\frac{e^s}{(1 - e^s)^2} - \frac{1}{s^2} \right) = \frac{-1}{12}.$$

If we formally set $s = 0$ in the left-hand side of $(*)$, then we get $\sum n$. Since the series diverges at $s = 0$, this step is invalid. But the number $\frac{-1}{12}$ is the limit, if we first subtract off the divergent part! See also Exercise 2.15.

2.1. Fibonacci numbers. We next use generating functions to analyze one of the most storied sequences: the Fibonacci numbers.

The Fibonacci numbers F_n are defined by $F_0 = F_1 = 1$ and $F_{n+2} = F_{n+1} + F_n$. The reader has surely seen the first few values:

$$1, 1, 2, 3, 5, 8, 13, 21, 34, 55, 89, 144, \ldots$$

These numbers arise in both art and science. They even play a role in photo-shopping!

One of the most famous facts about the Fibonacci numbers is that

$$\lim_{n \to \infty} \frac{F_{n+1}}{F_n} = \phi = \frac{1 + \sqrt{5}}{2}.$$

Many proofs of this limit are known. We prove it after first using generating functions to find an explicit formula for F_n. We then sketch an alternate approach using linear algebra. See also Exercise 2.22.

Theorem 2.1 (Binet's formula). $F_n = \frac{1}{\sqrt{5}}(\phi^{n+1} - \psi^{n+1})$, where $\phi = \frac{1+\sqrt{5}}{2}$ and $\psi = \frac{1-\sqrt{5}}{2}$.

PROOF. Let $P(z) = \sum F_n z^n$ be the generating function. Since $F_{n+2} = F_{n+1} + F_n$, we have

$$\sum F_{n+2} z^n = \sum F_{n+1} z^n + \sum F_n z^n.$$

We write this information in the form

$$\frac{1}{z^2} \sum F_{n+2} z^{n+2} = \frac{1}{z} \sum F_{n+1} z^{n+1} + \sum F_n z^n.$$

In this way we can get $P(z)$ into the notation. Using the values of F_0 and F_1, we obtain

$$\frac{1}{z^2}(P(z) - 1 - z) = \frac{1}{z}(P(z) - 1) + P(z).$$

Solving for $P(z)$ gives an explicit formula for the generating function:

$$P(z) = \frac{1}{1 - z - z^2} = \sum F_n z^n.$$

This formula yields Binet's formula for the coefficients by combining partial fractions with the geometric series. For appropriate a, b, c we have

$$\frac{1}{1 - z - z^2} = \frac{ac}{1 - az} - \frac{bc}{1 - bz}, \tag{10}$$

and hence the coefficient of z^n in the series expansion is given by $c(a^{n+1} - b^{n+1})$. Since (10) holds for $z = 0$ we get $c = \frac{1}{a-b}$. Since $ab = -1$ and $a + b = 1$ we obtain $a = \frac{1+\sqrt{5}}{2} = \phi$ and $b = \frac{1-\sqrt{5}}{2}$. Therefore $c = \frac{1}{\sqrt{5}}$. Rewriting in terms of ϕ and ψ yields Binet's formula. $\qquad\square$

Corollary 2.2. *The radius of convergence of the series $\sum F_n z^n$ is $\frac{1}{\phi}$.*

PROOF. The radius of convergence is the distance of the nearest pole to 0. The poles are the roots of $z^2 + z - 1 = 0$, and hence are at $-\phi$ and $-\psi$. Notice that $\psi < 0$, but that $-\psi = \frac{1}{\phi} < \phi$. Therefore the radius of convergence is $\frac{1}{\phi}$, which equals $\frac{-1+\sqrt{5}}{2}$. $\qquad\square$

Corollary 2.3. *The limit of the ratios of successive Fibonacci numbers is the golden ratio ϕ:*

$$\lim_{n \to \infty} \frac{F_{n+1}}{F_n} = \phi.$$

PROOF. Since $\left|\frac{\psi}{\phi}\right| < 1$, we obtain

$$\frac{F_{n+1}}{F_n} = \frac{\phi^{n+1}}{\phi^n} \frac{1 - (\frac{\psi}{\phi})^{n+1}}{1 - (\frac{\psi}{\phi})^n} \to \phi.$$

$\square$

We continue to analyze Fibonacci numbers, using methods from other parts of this book. By definition, the Fibonacci sequence solves the difference equation $F_{n+2} = F_{n+1} + F_n$ with $F_0 = F_1 = 1$. As we did with ODE, we can guess a solution λ^n and obtain the characteristic equation

$$\lambda^{n+2} = \lambda^{n+1} + \lambda^n.$$

The (non-zero) roots are of course $\frac{1 \pm \sqrt{5}}{2}$, namely ϕ and ψ. The general solution then satisfies

$$F_n = c_1 \left(\frac{1 + \sqrt{5}}{2}\right)^n + c_2 \left(\frac{1 - \sqrt{5}}{2}\right)^n,$$

where the constants c_1 and c_2 are chosen to agree with the initial conditions. A small amount of algebra yields Binet's formula. See Exercise 2.20.

Again, in analogy with ODE, we can regard the second-order recurrence equation as a system of first-order equations. To do so, consider the matrix L defined by $L = \begin{pmatrix} 1 & 1 \\ 1 & 0 \end{pmatrix}$. The matrix L is chosen in order to have

$$\begin{pmatrix} F_{n+2} \\ F_{n+1} \end{pmatrix} = L \begin{pmatrix} F_{n+1} \\ F_n \end{pmatrix}.$$

Note that its eigenvalues are ϕ and ψ. By matrix multiplication we therefore have

$$\begin{pmatrix} F_{n+1} \\ F_n \end{pmatrix} = L^n \begin{pmatrix} F_1 \\ F_0 \end{pmatrix} = L^n \begin{pmatrix} 1 \\ 1 \end{pmatrix}.$$

By our work on linear algebra in Chapter 1, we can write $L = PDP^{-1}$, where D is diagonal with ϕ and ψ on the diagonal. Then $L^n = PD^n P^{-1}$, and (after finding P explicitly) we obtain another derivation of Binet's formula. Exercise 2.21 asks for the details.

2.2. Bernoulli numbers. We wish to discuss a more difficult use of generating functions. Our goal is Theorem 2.5, which finds infinitely many sums all at once.

First we note a simple lemma for higher derivatives of a product.

Lemma 2.4 (Leibniz rule). *Let D denote differentiation. If f and g are p times differentiable, then*

$$D^p(fg) = \sum_{k=0}^{p} \binom{p}{k} D^k f \, D^{p-k} g.$$

PROOF. The proof is a simple induction on p, which we leave to the reader in Exercise 2.9. $\square$

Consider the formulas:

$$1 + 2 + 3 + \cdots + n = \frac{n(n+1)}{2}$$

$$1^2 + 2^2 + 3^2 + \cdots + n^2 = \frac{n^3}{3} + \frac{n^2}{2} + \frac{n}{6}$$

$$1^p + 2^p + 3^p + \cdots + n^p = \frac{n^{p+1}}{p+1} + \frac{n^p}{2} + \text{lower order terms.}$$

One way to establish such formulas is to assume that $\sum_{j=1}^{n} j^p$ is a polynomial of degree $p+1$ in n and then to compute enough values to solve for the undetermined coefficients. Given the correct formula, one can establish it by induction. It is remarkable that a general method exists that finds all these formulas simultaneously.

We approach this problem by beginning with the finite geometric series and taking derivatives. We have:

$$\sum_{j=0}^{n-1} e^{jx} = \frac{1 - e^{nx}}{1 - e^x}.$$

Now differentiate p times and evaluate at 0. We get

$$(\frac{d}{dx})^p \left(\frac{1 - e^{nx}}{1 - e^x} \right)(0) = \sum_{j=0}^{n-1} j^p.$$

To evaluate the left-hand side, we introduce Bernoulli numbers. These numbers have many applications in number theory and combinatorics. Also they were the subject of one of the first computer algorithms. See the Wikipedia page, which seems reliable on this topic.

We define B_n, the n-th **Bernoulli number**, by

$$\sum_{n=0}^{\infty} \frac{B_n}{n!} z^n = \frac{z}{e^z - 1}. \tag{11}$$

Thus $\frac{z}{e^z-1}$ is the generating function of the sequence $\frac{B_n}{n!}$. Note that $B_0 = 1$, that $B_1 = \frac{1}{2}$, and $B_2 = \frac{1}{6}$.

Theorem 2.5. *Let N and p be positive integers. Then*

$$\sum_{j=0}^{N-1} j^p = \sum_{k=0}^{p} \binom{p+1}{k+1} B_{p-k} \frac{N^{k+1}}{p+1}.$$

PROOF. By the finite geometric series,

$$\sum_{j=0}^{N-1} e^{jz} = \frac{e^{Nz} - 1}{e^z - 1} = \frac{e^{Nz} - 1}{z} \frac{z}{e^z - 1} = f(z)g(z). \tag{12}$$

We define f and g to be the factors in (12). We divide and multiply by z to ensure that f and g have valid series expansions near 0. Taking p derivatives, evaluating at 0, and using the Leibniz rule, we get

$$\sum_{j=0}^{N-1} j^p = (\frac{d}{dz})^p (f(z)g(z))(0) = \sum_{k=0}^{p} \binom{p}{k} f^{(k)}(0) g^{(p-k)}(0).$$

We have $f^{(k)}(0) = \frac{N^{k+1}}{k+1}$, and $g^{(p-k)}(0) = B_{p-k}$, by formula (11) defining the Bernoulli numbers. We therefore conclude:

$$\sum_{j=0}^{N-1} j^p = \sum_{k=0}^{p} \binom{p}{k} B_{p-k} \frac{N^{k+1}}{k+1} = \sum_{k=0}^{p} \binom{p+1}{k+1} B_{p-k} \frac{N^{k+1}}{p+1}. \tag{13}$$

$\square$

REMARK. The Bernoulli numbers also arise in connection with the Riemann zeta function. Assuming that $\mathrm{Re}(s) > 1$, the formula

$$\zeta(s) = \sum_{n=1}^{\infty} \frac{1}{n^s}$$

defines the zeta function. The method of **analytic continuation** leads to a function (still written ζ) that is complex analytic except at 1, where it has a pole. The formula $\zeta(-s) = \frac{-B_{s+1}}{s+1}$ holds when s is a positive integer. The strange value of $\frac{-1}{12}$ for $\sum n$ in Example 2.3 arises as $-\frac{B_2}{2}$, by formally putting $s = 1$ in the series definition. This definition requires $\mathrm{Re}(s) > 1$. The method of assigning values to divergent series in this manner occurs in both number theory and mathematical physics; the method is sometimes known as **zeta function regularization**.

2.3. Using generating functions. We end this section by comparing the method of generating functions in solving constant coefficient linear recurrences with the method of the Laplace transform in solving constant coefficient linear ODE.

We think of the Laplace transform $\mathcal{L}$ as an injective linear map from a space of functions V in the time domain to a space of functions W in the frequency domain:

$$\mathcal{L}f(s) = \int_0^{\infty} e^{-st} f(t) dt.$$

Similarly we can let V be the space of sequences and W the space of formal power series. We define an injective linear map $\mathbf{T} : V \to W$ by $\mathbf{T}(\{a_n\}) = \sum_{n=0}^{\infty} a_n z^n$.

EXAMPLE 2.4. Consider the ODE $y'' - 5y' + 6y = 0$ and the recurrence $a_{n+2} - 5a_{n+1} + 6a_n = 0$. We assume the initial conditions $y(0) = 1$ and $y'(0) = 1$ for y and the initial conditions $a_0 = 1$ and $a_1 = 1$ for the sequence. We solve each problem in two ways:

Let $T(y) = y'' - 5y' + 6y$. We wish to find the null space of T. It is 2-dimensional. We find a basis for $\mathcal{N}(T)$ by guessing $y = e^{\lambda x}$ and using the characteristic equation

$$(\lambda^2 - 5\lambda + 6)e^{\lambda x} = 0.$$

Here $\lambda = 2$ or $\lambda = 3$. The general solution is $c_1 e^{2x} + c_2 e^{3x}$. Assuming the initial conditions, we find

$$y(x) = 2e^{2x} - e^{3x}.$$

The second approach uses the Laplace transform. We have

$$\mathcal{L}(y'') - 5\mathcal{L}(y') + 6\mathcal{L}(y) = 0.$$

Using the initial conditions and Theorem 1.2, we obtain

$$(s^2 - 5s + 6)\mathcal{L}(y)(s) = s - 4$$

and hence

$$\mathcal{L}(y)(s) = \frac{s - 4}{(s - 2)(s - 3)} = \frac{-1}{s - 3} + \frac{2}{s - 2}$$

Inverting the Laplace transform gives $y = 2e^{2t} - e^{3t}$. As usual, we are assuming $t \geq 0$.

Let us next solve the constant coefficient recurrence

$$a_{n+2} - 5a_{n+1} + 6a_n = 0.$$

In the first method, we guess $a_n = \lambda^n$ for $\lambda \neq 0$. We obtain the equation $\lambda^n(\lambda^2 - 5\lambda + 6) = 0$ whose roots are $\lambda = 2$ and $\lambda = 3$. Thus $a_n = c_1 2^n + c_2 3^n$. Using the initial conditions $a_0 = 1$ and $a_1 = 1$, we obtain the equations $1 = c_1 + c_2$ and $1 = 2c_1 + 3c_2$. Finally we get $a_n = 2^{n+1} - 3^n$.

In the second method, we use generating functions. Consider the map T from sequences to formal power series. We apply it to the recurrence:

$$\sum_{n=0}^{\infty} a_{n+2} z^n - 5 \sum_{n=0}^{\infty} a_{n+1} z^n + 6 \sum_{n=0}^{\infty} a_n z^n = 0.$$

Let $A(z) = \sum_{n=0}^{\infty} a_n z^n$. Then the recurrence yields

$$\frac{1}{z^2}(A(z) - a_0 - a_1 z) - \frac{5}{z}(A(z) - a_0) + 6A(z) = 0.$$

We solve this equation for $A(z)$:

$$A(z)(6z^2 - 5z + 1) = 1 - 4z$$

$$A(z) = \frac{1 - 4z}{6z^2 - 5z + 1} = \frac{2}{1 - 2z} - \frac{1}{1 - 3z}.$$

By expanding each geometric series, we find $a_n = 2^{n+1} - 3^n$.

REMARK. In Example 2.4, the Laplace transform of y is given by $\frac{s-4}{(s-2)(s-3)}$ and the generating function A is given by $A(z) = \frac{1-4z}{6z^2-5z+1}$. In we replace z by $\frac{1}{s}$, we get $A(\frac{1}{s}) = \mathcal{L}(y)(s)$. Thus using the Z-transform is formally identical to using the Laplace transform.

We have placed these methods side-by-side to illustrate the basic idea mentioned in the introduction to the chapter. We often simplify a problem by expressing it in different notation. For linear ODE and recurrences, we apply injective linear maps to obtain equivalent algebraic problems. Furthermore, the role of the Laplace transform in solving ODE is precisely analogous to the role of generating functions in solving recurrences.

We next prove a general result about the generating function of a polynomial sequence, and exhibit it for the simple example n^2. The much harder Example 2.6 gives the generating function for the rational sequence $\frac{1}{2n+1}$.

Theorem 2.6. *Let $p(n)$ be a polynomial of degree d. Put $a_n = p(n)$. Then the generating function f for $\{a_n\}$ is a polynomial of degree $d + 1$ in the variable $\frac{1}{1-z}$ with no constant term.*

PROOF. The result is linear algebra! The idea is to choose the right basis for the space of polynomials. By definition,

$$f(z) = \sum_{n=0}^{\infty} p(n) z^n.$$

We wish to find f explicitly. This series converges for $|z| < 1$. By the proof of Theorem 8.1 from Chapter 1, there are constants c_j such that

$$p(x) = \sum_{j=0}^{d} c_j \binom{x+j}{j}.$$

Substituting this expression for p and then interchanging the order of summation yields

$$f(z) = \sum_{n=0}^{\infty} \sum_{j=0}^{d} c_j \binom{n+j}{j} z^n = \sum_{j=0}^{d} c_j \sum_{n=0}^{\infty} \binom{n+j}{j} z^n. \tag{14}$$

But $\sum_{n=0}^{\infty} \binom{n+j}{j} z^n$ is $\frac{1}{j!}$ times the j-th derivative of the geometric series, and hence equals $\frac{1}{(1-z)^{j+1}}$. Thus (14) defines a polynomial in $\frac{1}{1-z}$, without a constant term. $\qquad \square$

EXAMPLE 2.5. Put $a_n = n^2$. Then $a_n = (n+2)(n+1) - 3(n+1) + 1$ and therefore

$$\sum_0^\infty n^2 z^n = \sum (n+2)(n+1)z^n - 3\sum (n+1)z^n + \sum z^n = \frac{2}{(1-z)^3} - \frac{3}{(1-z)^2} + \frac{1}{1-z}.$$

The polynomial $n \mapsto n^2$ gets transformed into the polynomial $t \mapsto 2t^3 - 3t^2 + t$, with t evaluated at $t = \frac{1}{1-z}$.

EXAMPLE 2.6. We find the generating function for the sequence $\frac{1}{2n+1}$. For $0 < x < 1$ we claim that

$$\sum_{n=0}^\infty \frac{x^n}{2n+1} = \frac{1}{2\sqrt{x}}\left(\ln(1+\sqrt{x}) + \ln(1-\sqrt{x})\right) = \frac{1}{2\sqrt{x}}\ln\left(\frac{1+\sqrt{x}}{1-\sqrt{x}}\right).$$

Again we rely on the geometric series. This time we integrate, but things are a bit messy. First we have formulas when $|x| < 1$:

$$\ln(1-x) = -\int_0^x \frac{1}{1-t}\,dt = -\int_0^x \sum_0^\infty t^n\,dt = -\sum_0^\infty \frac{x^{n+1}}{n+1}.$$

$$\ln(1+x) = \sum_0^\infty \frac{(-1)^n x^{n+1}}{n+1}.$$

Subtract the equations and use $\ln(a) - \ln(b) = \ln(\frac{a}{b})$ to get

$$\ln\left(\frac{1+x}{1-x}\right) = 2\sum_{n=0}^\infty \frac{x^{2n+1}}{2n+1} = 2x\sum_{n=0}^\infty \frac{(x^2)^n}{2n+1}.$$

Finally, replace x by $\sqrt{x}$ and divide by $2\sqrt{x}$ to obtain the claimed formula.

Exercise 2.19 asks you to derive the formula in another way.

REMARK. Theorem 2.6 does not have a simple generalization to rational sequences. It is difficult to find formulas for the generating functions of rational sequences. If $a_n = \frac{1}{n^s}$, for example, then the generating function is called the *polylogarithm* function. When $s = 2$ this function is called the *dilogarithm* or *Spence's function*. It appears in particle physics. See the Wolfram World webpage for the short list of values for which the dilogarithm can be computed explicitly.

EXERCISE 2.1. Solve the recurrence $a_{n+1} - 5a_n = 1$ with $a_0 = 1$.

EXERCISE 2.2. Solve the recurrence $a_{n+2} - 5a_{n+1} + 6a_n = 1$ with $a_0 = 2$ and $a_1 = 1$.

EXERCISE 2.3. Solve the recurrence $a_{n+1} - a_n = \frac{-1}{n(n+1)}$ given $a_1 = 1$.

EXERCISE 2.4. Find the series expansion of $\frac{1}{(1-z)^d}$, when d is a positive integer.

EXERCISE 2.5. Put $p(x) = a_0 + a_1 x + a_2 x^2 + a_3 x^3$. Find $\sum_{n=0}^\infty p(n)z^n$.

EXERCISE 2.6. (A non-linear recurrence.) Put $a_{n+1} = \frac{a_n}{2} + \frac{2}{a_n}$. If $a_1 > 0$, find $\lim_{n\to\infty} a_n$.

EXERCISE 2.7. Prove the combinatorial identity

$$\frac{\binom{p}{k}}{k+1} = \frac{\binom{p+1}{k+1}}{p+1}.$$

used in (13). Suggestion: Count something two ways.

EXERCISE 2.8. Verify by any method that the two terms of highest order in $\sum_{j=1}^n j^p$ are $\frac{n^{p+1}}{p+1} + \frac{n^p}{2}$. Suggestion: Consider approximations to the integral $\int_0^1 x^p dx$.

EXERCISE 2.9. Prove the Leibniz rule in Lemma 2.4.

EXERCISE 2.10. Use the method of undetermined coefficients to find $\sum_{j=1}^n j^5$. Comment: Carrying out all the details is a bit tedious and makes one appreciate Theorem 2.5.

EXERCISE 2.11. Prove the following necessary and sufficient condition on a sequence $\{a_n\}$ for its generating function $\sum_{n=0}^\infty a_n z^n$ to be a rational function. See pages 79-82 in [D2] for a detailed discussion. The generating function is rational if and only if the infinite matrix

$$\begin{pmatrix} a_0 & a_1 & a_2 & \cdots \\ a_1 & a_2 & a_3 & \cdots \\ a_2 & a_3 & a_4 & \cdots \\ \cdots & & & \end{pmatrix}$$

has finite rank.

EXERCISE 2.12. Consider a power series whose coefficients repeat the pattern $1, 1, 0, -1, -1$. Thus the series starts

$$1 + z - z^3 - z^4 + z^5 + z^6 - z^8 - z^9 + \cdots$$

Write the sum explicitly as a rational function in lowest terms.

EXERCISE 2.13. Consider the series, with first term $4z$, whose coefficients repeat the pattern $4, 9, 3$. Thus (for $|z| < 1$) put

$$f(z) = 4z + 9z^2 + 3z^3 + 4z^4 + 9z^5 + 3z^6 + \cdots$$

Find $f(\frac{1}{10})$. (The answer requires no computation!) Can you also find $f(z)$ with no computation?

EXERCISE 2.14. Let $\{a_n\}$ be a sequence of real or complex numbers. We say that this sequence is **Abel summable** to L if

$$\lim_{r \to 1} \sum_{n=0}^\infty a_n r^n = L.$$

Here the limit is taken as r increases to 1. Show that the sequence $1, 0, 1, 0, 1, 0, \ldots$ is Abel summable to $\frac{1}{2}$.

Using summation by parts, Lemma 3.2 in the next section, one can prove: if $\sum a_n$ converges to L, then $\{a_n\}$ is Abel summable to L. Show that $\{\frac{(-1)^n}{2n+1}\}$ is Abel summable to $\frac{\pi}{4}$. Since the series is convergent, we obtain the impressive result

$$\sum_{n=0}^\infty \frac{(-1)^n}{2n+1} = \frac{\pi}{4}.$$

EXERCISE 2.15. Consider the sequence $1, 0, -1, 1, 0, -1, 1, 0, -1, \ldots$. Show that it is Abel summable and find the Abel limit.

EXERCISE 2.16. Read the remark after Theorem 2.5. Presuming that the formula $\zeta(-s) = \frac{-B_{s+1}}{s+1}$ is valid when ζ is defined by the series, what values does one obtain for $1+1+1+\cdots$ and for $1+4+9+16+\cdots$?

EXERCISE 2.17. After looking at Example 2.3, express $\sum_{n=0}^\infty n^2 e^{-ns}$ and $\sum_{n=0}^\infty n^3 e^{-ns}$ as Laurent series. Following the reasoning in Example 2.3 (in other words, disregarding the singular part) establish the dubious results that $\sum_{n=1}^\infty n^2 = 0$ and that $\sum_{n=1}^\infty n^3 = \frac{1}{120}$.

EXERCISE 2.18. Find the generating function for the sequence with $a_n = \frac{1}{n}$ for $n \geq 1$.

EXERCISE 2.19 (Difficult but rewarding). Determine the generating function in Example 2.6 in the following manner. First write

$$f(x) = \sum_{n=0}^{\infty} \frac{x^n}{2n+1} = \frac{1}{2} \sum_{n=0}^{\infty} \frac{x^n}{n+\frac{1}{2}} = \frac{1}{2\sqrt{x}} \sum_{n=0}^{\infty} \frac{x^{n+\frac{1}{2}}}{n+\frac{1}{2}} = \frac{1}{2\sqrt{x}} g(x).$$

Differentiating the series for g shows that $g'(x) = \frac{1}{\sqrt{x}(1-x)}$. Use elementary calculus to find g explicitly.

EXERCISE 2.20. Complete the second proof of Binet's formula suggested after Corollary 2.3.

EXERCISE 2.21. Diagonalize the matrix $L = \begin{pmatrix} 1 & 1 \\ 1 & 0 \end{pmatrix}$. Find L^n in terms of both Fibonacci numbers and in terms of ϕ and ψ. By taking determinants, verify that

$$F_{n+1}F_{n-1} - (F_n)^2 = (-1)^{n+1}.$$

REMARK. One can derive many other identities involving Fibonacci numbers using powers of L.

EXERCISE 2.22. Consider a general second-order linear recurrence

$$x_{n+2} = bx_{n+1} + cx_n$$

with x_0 and x_1 known.

• Find the matrix L such that $L \begin{pmatrix} x_{n+2} \\ x_{n+1} \end{pmatrix} = \begin{pmatrix} x_{n+1} \\ x_n \end{pmatrix}$.
• Find the eigenvalues of L.
• Find the condition on b, c that guarantees these eigenvalues are distinct.
• Given this condition, find a formula for x_n.
• When $b = c = x_0 = x_1 = 1$, verify that your formula agrees with Binet's formula.

3. Fourier series

In the next chapter we will systematically study Hilbert spaces and orthonormal expansion. We follow the historical order here by first introducing Fourier series; a Fourier series is a specific orthonormal expansion in terms of complex exponentials. We study in detail the Fourier series of the sawtooth function. We also note that the root mean square value of a current is simply the norm $||\mathcal{I}||$, thereby beginning to connect the abstract ideas to engineering.

REMARK. In recent years much work in signal processing and related areas uses **wavelet analysis** rather than Fourier series or Fourier transforms. One replaces complex exponentials (sines and cosines) by other sets of basis functions, called *wavelets*. We refer to [Dau], [M], and [N] for quite different perspectives on this important topic.

Consider the interval $[0, 2\pi]$. For f, g continuous on this interval we define their inner product by

$$\langle f, g \rangle = \frac{1}{2\pi} \int_0^{2\pi} f(t)\overline{g(t)}dt.$$

The complex exponentials ϕ_n given by $\phi_n(t) = e^{int}$ for $n \in \mathbb{Z}$ form an orthonormal system:

$$\langle \phi_m, \phi_n \rangle = 0 \text{ if } m \neq n.$$

$$\langle \phi_n, \phi_n \rangle = 1 \text{ for all } n.$$

The *Kronecker delta* δ_{jk} is a convenient notation when working with orthonormality. The symbol δ_{jk} equals 1 when $j = k$ and equals 0 otherwise. Thus orthonormality can be expressed by saying that

$$\langle \phi_m, \phi_n \rangle = \delta_{mn}.$$

We check the orthonormality of the ϕ_n:

$$\langle \phi_n, \phi_m \rangle = \langle e^{int}, e^{imt} \rangle = \frac{1}{2\pi} \int_0^{2\pi} e^{int} e^{-imt} dt.$$

If $n \neq m$, the integral is 0. If $n = m$, the integral is 1. The system of complex exponential functions is also *complete*, a concept we postpone until Chapter 6. Completeness implies that 0 is the only function orthogonal to each ϕ_n.

Definition 3.1. Let g be integrable on $[0, 2\pi]$. Its n-th **Fourier coefficient** is the complex number given by $\hat{g}(n) = \langle g, \phi_n \rangle$. The **Fourier series** of g is the formal sum

$$\sum_{n=-\infty}^{\infty} \hat{g}(n) e^{int}.$$

A **Fourier series** is a formal sum $\sum_{n=-\infty}^{\infty} c_n e^{inrt}$. Here the coefficients c_n are complex, the constant r is real, and t is a real variable.

The constant r is chosen to be $\frac{2\pi}{P}$ when we are working on the interval $[0, P]$. Here P is the period. Usually we will assume $P = 2\pi$.

Convergence of Fourier series is a deep and difficult topic. We illustrate some subtle issues by considering the convergence of the Fourier series

$$\sum_{n=1}^{\infty} \frac{\sin(nx)}{n}.$$

Figures 4 and 5 show the graphs of the first three and the first five terms in this series. Already one can guess that the limit function is a sawtooth.

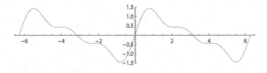

FIGURE 4. Fourier series: three terms

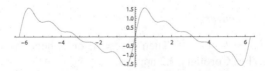

FIGURE 5. Fourier series: five terms

The following formula holds for $0 < x < 2\pi$:

$$\sum_{n=1}^{\infty} \frac{\sin(nx)}{n} = \frac{\pi - x}{2}. \tag{15}$$

The series is conditionally convergent and we must be careful. By extending this function periodically we obtain the sawtooth function. Notice that the series sums to 0 when x is an integer multiple of 2π. Hence (15) fails at these points.

To prove convergence of the series, we begin with **summation by parts**, an algebraic formula analogous to integration by parts. The partial sums B_N in Lemma 3.2 are analogous to integrals, and the differences $a_{n+1} - a_n$ are analogous to derivatives.

Lemma 3.2 (Summation by parts). *Let $\{a_j\}$ and $\{b_j\}$ be sequences of complex numbers. Define the partial sum by $B_N = \sum_{j=1}^{N} b_j$. Then, for each positive integer N,*

$$\sum_{n=1}^{N} a_n b_n = a_N B_N - \sum_{n=1}^{N-1} \left((a_{n+1} - a_n) B_n \right).$$

PROOF. The proof is by induction on N and is left to the reader.   □

We give two examples illustrating the lemma before we use it meaningfully:

$$a_1 b_1 + a_2 b_2 = a_2(b_1 + b_2) - (a_2 - a_1)b_1$$

$$a_1 b_1 + a_2 b_2 + a_3 b_3 = a_3(b_1 + b_2 + b_3) - ((a_2 - a_1)b_1 + (a_3 - a_2)(b_1 + b_2)).$$

Corollary 3.3. *Assume $0 < a_{j+1} < a_j$ for all j and $\lim a_n = 0$. Assume also that there is a constant C such that $|B_N| = |\sum_{j=1}^{N} b_j| \le C$ for all N. Then $\sum_{j=1}^{\infty} a_j b_j$ converges.*

PROOF. Using summation by parts,

$$\sum_{j=1}^{N} a_k b_k = a_N B_N - \sum_{n=1}^{N-1} (a_{n+1} - a_n) B_n. \tag{16}$$

We show that the right-hand side of (16) converges. Since $a_N \to 0$ and $|B_N| \le C$, it follows that $a_N B_N \to 0$. The hypotheses bound the other term in (16):

$$| \sum_{n=1}^{N-1} (a_{n+1} - a_n) B_n | \le \sum_{i=1}^{N-1} |a_{n+1} - a_n|\, |B_n| \le C \sum_{i=1}^{N-1} (a_n - a_{n+1}). \tag{17}$$

The right-hand side of (17) telescopes and hence converges. Therefore the left-hand side of (16) also converges as N tends to infinity.   □

Corollary 3.4. $\sum \frac{(-1)^{n+1}}{n}$ *converges.*

PROOF. Put $a_n = \frac{1}{n}$, $b_n = (-1)^{n+1}$. Then $B_N = 0$ or 1, hence $|B_N| \le 1$ for all N. Obviously $a_{n+1} < a_n$ and $\lim a_n = 0$. Thus Corollary 3.3 applies.   □

Theorem 3.5. *For each $x \in \mathbb{R}$, the series $\sum_{n=1}^{\infty} \frac{\sin nx}{n}$ converges.*

PROOF. We use summation by parts and Corollary 3.3. We first show that there is a constant C (independent of N) such that $|\sum_{n=1}^{N} \sin(nx)| \leq C$. By definition of sine, we have

$$\sin(nx) = \frac{e^{inx} - e^{-inx}}{2i}.$$

Using the triangle inequality, we get

$$|\sum_{n=1}^{N} \sin(nx)| \leq \frac{1}{2}\left(|\sum_{n=1}^{N} e^{inx}| + |\sum_{n=1}^{N} e^{-inx}|\right).$$

We bound these two terms by using the finite geometric series. If $e^{ix} \neq 1$, then

$$\sum_{n=1}^{N} e^{inx} = e^{ix} \sum_{n=0}^{N-1} e^{inx} = e^{ix}\frac{1 - e^{iNx}}{1 - e^{ix}}.$$

Therefore, assuming that $e^{ix} \neq 1$, the finite geometric series gives

$$|\sum_{n=1}^{N} e^{inx}| = \frac{|1 - e^{iNx}|}{|1 - e^{ix}|} \leq \frac{2}{|1 - e^{ix}|}.$$

Replacing x by $-x$ gives a similar estimate on the other term. Therefore, if $e^{ix} \neq 1$, then

$$|\sum_{n=1}^{N} \sin(nx)| \leq \frac{2}{|1 - e^{ix}|} = C(x).$$

This constant is independent of N. By Corollary 3.3, the series converges. If $e^{ix} = 1$, then $x = 2\pi k$ and $\sin(nx) = 0$. For such x, clearly $\sum \frac{\sin(nx)}{n} = 0$. Thus the series converges for all x. $\square$

Next we find the Fourier series for $\frac{\pi - x}{2}$; we warn the reader that we have not yet established conditions under which the Fourier series of a function converges to that function.

Suppose f is defined on $[0, 2\pi]$. Consider

$$\sum_{n=-\infty}^{\infty} \langle f, e^{inx}\rangle e^{inx}, \tag{18}$$

where $\langle f, e^{inx}\rangle = \frac{1}{2\pi}\int_0^{2\pi} f(t)e^{-int}dt$. The series in (18) is called the Fourier series of f.

When f is the function $\frac{\pi - x}{2}$ we obtain

$$\langle \frac{\pi - x}{2}, e^{inx}\rangle = \frac{1}{2\pi}\frac{1}{2}\int_0^{2\pi} (\pi - x)e^{-inx}dx = J_n.$$

Here $J_n = 0$ when $n = 0$. If $n \neq 0$, then

$$J_n = \frac{1}{2\pi}\frac{1}{2}\int_0^{2\pi} (\pi - x)e^{-inx}dx = \frac{1}{2in}.$$

The sawtooth function results from extending f periodically to all of $\mathbb{R}$. We see that its Fourier series is

$$\sum_{n\neq 0} \frac{1}{2in}e^{inx} = \sum_{n=1}^{\infty} \frac{1}{2i}\frac{e^{inx}}{n} + \sum_{n=-1}^{-\infty} \frac{1}{-2i}\frac{e^{inx}}{n} = \sum_{n=1}^{\infty} \frac{\sin(nx)}{n}. \tag{18.1}$$

We repeat; at this stage we do not know that the Fourier series represents the function. See [D3] and especially [SS] for general results guaranteeing that the Fourier series of a function converges to the function. The formula (18.1) does converge to f on the open interval $(0, 2\pi)$ but not at the endpoints.

REMARK. Consider a Fourier series $\sum_{n=-\infty}^{\infty} c_n e^{inx}$. If this series converges to an integrable function f, defined on the interval $[0, 2\pi]$, then we have

$$\|f\|^2 = \frac{1}{2\pi} \int_0^{2\pi} |f(x)|^2 dx = \sum_n |c_n|^2. \tag{19}$$

This result is known as Parseval's formula. The result follows easily from the Hilbert space theory described in Chapter 6, because the series represents the orthonormal expansion of f in terms of complex exponentials.

EXAMPLE 3.1 (Root mean square). The norm $\|f\|$ is called the **root mean square** value of f and is often written **rms**. This concept plays a significant role in electrical engineering. For a constant current $\mathcal{I}$, the **power** P dissipated by a resistance R satisfies $P = \mathcal{I}^2 R$. If the current varies over time, then we can compute the **average power** P_a by the formula $P_a = \|\mathcal{I}\|^2 R$, where $\|\mathcal{I}\|$ is the root mean square value of the current. We can also express the average power in terms of the root mean square value of the voltage as $\frac{\|V\|^2}{R}$. Eliminating the resistance R gives the average power as $P_a = \|\mathcal{I}\|\,\|V\|$. This formula is valid for any periodic function.

Next we discuss the relationship between Fourier series and Laplace transforms. First consider the Laplace transform of the exponential e^{int}. When $\mathrm{Re}(s) > 0$, we have

$$\int_0^\infty e^{-st} e^{int} dt = \lim_{R\to\infty} \int_0^R e^{t(in-s)} dt = \lim_{R\to\infty} \frac{e^{R(in-s)} - 1}{in - s} = \frac{1}{s - in}.$$

Assume next that f is a trig polynomial:

$$f(t) = \sum_{-d}^{d} a_n e^{int}.$$

We regard f as a finite Fourier series; it defines a periodic function of t. Its Laplace transform is given by

$$F(s) = \int_0^\infty e^{-st} f(t) dt = \sum_{-d}^{d} a_n \int_0^\infty e^{int-st} dt = \sum_{-d}^{d} \frac{a_n}{s - in}.$$

This formula for $F(s)$ shows that F has poles at the points $s = in$ and the coefficients c_n are the residues at these poles. All the poles are on the imaginary axis. The analogous statement holds when f is given by a Fourier series

$$f(t) = \sum_{-\infty}^{\infty} a_n e^{int},$$

assuming enough convergence that the Laplace transform of the infinite sum is the sum of the Laplace transforms of the terms. In the frequency domain we have poles only on the imaginary axis, and the residues at these poles are the Fourier coefficients of the function in the time domain.

EXERCISE 3.1. The average value of $\sin$ on the interval $[0, 2\pi]$ is obviously 0. Hence, for alternating current, we use instead the root mean square value to compute power. Show that the root mean square value of $\sin(x)$ equals $\frac{1}{\sqrt{2}}$. (Remember the constant $\frac{1}{2\pi}$ outside the integral.) Prove this result without doing any integrals. Hint: The answer is the same for cos. Then use $\sin^2 + \cos^2 = 1$.

EXERCISE 3.2. Put $f(x) = \sum_{-N}^{N} c_n e^{inx}$. Verify that $||f||^2 = \sum_{-N}^{N} |c_n|^2$.

EXERCISE 3.3. Verify that the series converges for all real x.

$$g(x) = \sum_{n=1}^{\infty} \frac{\sin(nx)}{n^2}.$$

Use Mathematica or something similar to graph the partial sums S_N for $N = 8$, $N = 20$, and $N = 100$.

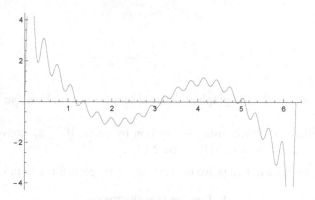

FIGURE 6. The graph of a partial sum for g

EXERCISE 3.4. Find the Fourier series of x^2 on the interval $[-\pi, \pi]$. The series converges at each x; what information do you obtain when $x = 0$? Do the same computations for x^4.

EXERCISE 3.5. Find the Fourier series of $|\sin(x)|$ on $[0, \pi]$. Use a change of variable to find the series for $|\cos(x)|$ on $[\frac{-\pi}{2}, \frac{\pi}{2}]$.

EXERCISE 3.6. Suppose $f(t)$ defines a square wave. Thus $f(t) = 1$ on $(0, \pi)$ and $f(t) = -1$ on $(-\pi, 0)$, and we extend f to be periodic of period 2π. Show that the Laplace transform $F(s)$ is given by

$$F(s) = \frac{1}{s}\left(\frac{2}{1 + e^{-\pi s}} - 1\right).$$

What kind of singularity does this function have at $s = 0$?

EXERCISE 3.7. Use summation by parts to verify that the series converges for all real x.

$$\sum_{n=2}^{\infty} \frac{\sin(nx)}{\ln(x)}.$$

EXERCISE 3.8 (Difficult). Consider the function

$$f(x) = \sum_{n=0}^{\infty} \frac{\cos(nx)}{\ln(n+2)}.$$

Using summation by parts twice, and using the concavity of the logarithm function, prove that this Fourier series converges to a non-negative function f. Figure 7 shows the graph of the partial sum S_N with $N = 100$.

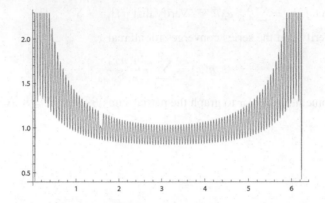

FIGURE 7. The graph of a partial sum for f from Exercise 3.8

EXERCISE 3.9 (Difficult). Prove, using summation by parts: if $\sum a_n$ converges, then $\{a_n\}$ is Abel summable to the same value. Refer also to Exercise 2.12.

EXERCISE 3.10. Use induction to prove the summation by parts formula in Lemma 3.2.

4. Fourier transforms on $\mathbb{R}$

Fourier series apply when functions are periodic. Analyzing signals in the real-world requires a continuous analogue of Fourier series, known as the Fourier transform. To define it, we will replace infinite sums with integrals. Often these integrals will be conditionally convergent and require considerable care. The Fourier transform is used throughout science, and we barely scratch the surface of both the theory and the applications. We also restrict to the one-dimensional situation. See [T] and [F2] for more information

Let $f : \mathbb{R} \to \mathbb{C}$ be a continuous (or piecewise continuous) function. We say that f is in $L^1(\mathbb{R})$ if f is absolutely integrable:

$$||f||_{L^1} = \int_{-\infty}^{\infty} |f(x)|dx < \infty.$$

In this section we will not worry much about the existence of the integrals involved.

Definition 4.1 (Fourier transform). For $f \in L^1(\mathbb{R})$, and for $w \in \mathbb{R}$, we put

$$\hat{f}(w) = \frac{1}{\sqrt{2\pi}} \int_{-\infty}^{\infty} f(x)e^{-ixw}dx. \tag{20}$$

The function $\hat{f}$ is called the Fourier transform of f.

Some authors omit the constant $\frac{1}{\sqrt{2\pi}}$ in front of the integral in (20); others include a factor of 2π in the exponential. This factor in the exponential amounts to replacing frequency, measured in cycles per seconds, by angular frequency, measured in radians per second. For both typographical and conceptual reasons, we sometimes write $\mathcal{F}(f)$ instead of $\hat{f}$. We write both w and ξ for the variable on the Fourier transform side of things. When working with Fourier transforms, we often think of the variable x as position and the variable ξ as momentum. Note the contrast with the interpretation of the Laplace transform variables. The variable x can also denote time, in which case the w variable is measured in reciprocal seconds.

We note some basic properties of the Fourier transform in Theorem 4.2. There we write $C(\mathbb{R})$ for the continuous functions from $\mathbb{R}$ to $\mathbb{C}$. Many of the more difficult results in the theory are proved by first working in a space, called the *Schwartz space*, of smooth functions with rapid decay at infinity.

Theorem 4.2. *The Fourier transform $\mathcal{F}$ is linear from $L^1(\mathbb{R})$ to $C(\mathbb{R})$.*

PROOF. The defining properties of linearity $\mathcal{F}(f+g) = \mathcal{F}(f) + \mathcal{F}(g)$ and $\mathcal{F}(cf) = c\mathcal{F}(f)$ follow immediately from (20). Suppose $f \in L^1(\mathbb{R})$. To show that $\hat{f}$ is continuous, we need to show that

$$|\hat{f}(w+h) - \hat{f}(w)| \tag{21}$$

tends to 0 as h does. A simple estimate shows that

$$|\hat{f}(w+h) - \hat{f}(w)| \leq \int_{-\infty}^{\infty} |f(x)| \, |e^{-ihx} - 1| \, dx \leq 2||f||_{L^1}. \tag{22}$$

By a basic theorem in measure theory, these inequalities guarantee that (21) tends to 0 as h does. Therefore $\hat{f}$ is continuous. The proof implies the stronger statement that it is uniformly continuous. $\square$

REMARK. The previous proof is subtle because interchanging two limits requires care. We want to take the limit as h tends to 0 of an integral depending on h. Naively putting $h = 0$ is invalid, because the limit of an integral is not necessarily the integral of the limit. See Exercise 4.9. The far right-hand estimate in (22), by the dominated convergence theorem, allows us to do so in this proof.

Consider the two simple operations $x \mapsto -x$ on $\mathbb{R}$ and $w \mapsto \overline{w}$ on $\mathbb{C}$. The first is *time reversal* and the second is complex conjugation. Doing either of them twice results in the identity transformation. These two simple operations get intertwined when the Fourier transform is involved. First we have the next easy result; the inversion formula (Theorem 4.4 below) also involves time reversal.

Proposition 4.3. *If $f \in L^1$, then $\overline{\hat{f}(w)} = \hat{\overline{f}}(-w)$.*

PROOF. Left to the reader in Exercise 4.3. $\square$

REMARK. When we regard the variable x as time, we can consider the function $x \mapsto f(x-p) = f_p(x)$ as the function f **delayed** by p. A change of variables in the integral (See Exercise 4.2), implies that

$$\mathcal{F}(f_p)(\xi) = e^{-i\xi p}\mathcal{F}(f)(\xi).$$

For Laplace transforms, the time and frequency domains have different flavors. For Fourier transforms, things are symmetric in position and momentum. In particular, for nice functions f, the Fourier inversion formula holds. For its statement we regard a function as **nice** if it is infinitely differentiable and decays rapidly at infinity. For $c > 0$, the Gaussian e^{-cx^2} is nice. Any polynomial times a Gaussian is also nice.

Before proving the inversion formula, we recall the notion of a bump function from Exercise 5.2 of Chapter 4. Such a function is infinitely differentiable at every point, yet vanishes outside of a bounded interval. Bump functions provide additional examples of smooth functions with rapid decay at infinity.

Theorem 4.4 (Fourier inversion formula). *Let $f : \mathbb{R} \to \mathbb{R}$ be a smooth function with rapid decay at infinity. Then $\hat{f}$ is also smooth of rapid decay. Furthermore*

$$f(x) = \frac{1}{\sqrt{2\pi}} \int_{-\infty}^{\infty} \hat{f}(w)e^{ixw}dw. \tag{23}$$

PROOF. We sketch the proof of (23). Define $\mathcal{F}^*$ by changing the sign of the exponential in (20) as in (23). Then (23) says that $\mathcal{F}^*\mathcal{F} = I$. Hence we need to prove that

$$f(x) = \frac{1}{2\pi} \int_{-\infty}^{\infty} \int_{-\infty}^{\infty} f(t)e^{-i\xi(t-x)}dt \, d\xi. \tag{23.1}$$

We introduce the Gaussian convergence factor $e^{\frac{-\epsilon^2 \xi^2}{2}}$ to enable us to interchange the order of integration. For $\epsilon > 0$ we therefore consider

$$\frac{1}{2\pi} \int_{-\infty}^{\infty} \int_{-\infty}^{\infty} e^{\frac{-\epsilon^2 \xi^2}{2}} f(t) e^{-i\xi(t-x)} d\xi \, dt. \qquad (23.2)$$

When integrating first with respect to ξ, we have the Fourier transform of a Gaussian (with $\sigma = \frac{1}{\epsilon}$) evaluated at $t - x$. Using the computation in Example 4.2., (23.2) becomes

$$\frac{1}{\sqrt{2\pi}} \frac{1}{\epsilon} \int_{-\infty}^{\infty} e^{\frac{-(t-x)^2}{2\epsilon^2}} f(t) dt. \qquad (23.3)$$

Now make the change of variables $s = \frac{t-x}{\epsilon}$ to obtain

$$\frac{1}{2\pi} \int_{-\infty}^{\infty} \int_{-\infty}^{\infty} e^{\frac{-\epsilon^2 \xi^2}{2}} f(t) e^{-i\xi(t-x)} d\xi \, dt = \frac{1}{\sqrt{2\pi}} \int_{-\infty}^{\infty} e^{\frac{-s^2}{2}} f(x + \epsilon s) ds. \qquad (23.4)$$

Let ϵ tend to 0 in (23.4). Assuming that the limit of an integral is the integral of the limit, and recalling that $\int_{-\infty}^{\infty} e^{\frac{-s^2}{2}} ds = \sqrt{2\pi}$, we obtain (23.1) and the inversion formula (23). References [T], [F2], [D3] include more details and explanation.   □

REMARK. Formula (23) remains valid when f is piecewise differentiable and $\int_{-\infty}^{\infty} |f(x)| dx$ is finite.

The famous Riemann–Lebesgue lemma applies to both Fourier series and Fourier transforms. For the circle, it states that the Fourier coefficients a_n of an integrable function tend to zero as n tends to infinity. For the real line, it states that the Fourier transform of an arbitrary L^1 function vanishes at ∞. Applications of Fourier analysis study how the decay rate of Fourier coefficients or Fourier transforms is related to regularity properties of the functions. The faster the Fourier transform tends to zero at infinity, the smoother the function must be. The proof of the lemma (in either setting) requires some integration theory, and hence we only provide two sketches, in the Fourier transform setting. The proof for Fourier series is quite similar.

Theorem 4.5 (Riemann–Lebesgue lemma). *If $f \in L^1(\mathbb{R})$, then $\lim_{|\xi| \to \infty} \hat{f}(\xi) = 0$.*

PROOF. We sketch two approaches. Suppose first that f is an indicator function of a bounded interval. Thus $f(x) = 1$ on $[a, b]$ and $f(x) = 0$ elsewhere. Compare with Example 4.1 below. By elementary computation,

$$\hat{f}(\xi) = \frac{1}{\sqrt{2\pi}} \int_a^b e^{-ix\xi} dx = \frac{1}{\sqrt{2\pi}} \frac{1}{(-i\xi)} (e^{-ib\xi} - e^{-ia\xi}).$$

Therefore there is a constant C such that

$$|\hat{f}(\xi)| \le \frac{C}{|\xi|}.$$

If next f is a step function, which by definition is a finite linear combination of indicator functions, the same inequality (for a different value of C) follows easily. By approximating f by a step function, one can prove that the conclusion holds for f Riemann integrable or even for $f \in L^1$.

The second approach is to observe that the result is easy if f is continuously differentiable and vanishes outside a bounded set; an integration by parts shows that

$$i\xi \hat{f}(\xi) = \widehat{f'}(\xi).$$

Since f' is continuous and vanishes outside a bounded set, $\widehat{f'}$ is a bounded function. Hence, for such functions, there is again a constant C such that

$$|\hat{f}(\xi)| \leq \frac{C}{|\xi|}.$$

Again we can approximate f by such functions and establish the result. $\qquad\square$

REMARK. The proof of (23) relies on a profound idea in mathematics and physics. The Gaussian factor is being used as an **approximate identity**. An approximate identity is a one-parameter family of well-behaved functions G_ϵ such that (for continuous f) $\lim_\epsilon (G_\epsilon * f)(x) = f(x)$. We do not specify where ϵ is tending, because in various situations the parameter tends to different values. Often it tends to infinity. For the Poisson kernel, mentioned in Section 6 of Chapter 7, the parameter tends to 1. In all these cases, evaluation of a function at x becomes a limit of convolutions. Readers who know about the Dirac delta "function" should note that approximate identities are one way to make this notion precise. See Exercise 4.10 for the simplest example of an approximate identity used in electrical engineering. We define and discuss approximate identities in more detail in subsection 4.1. See also subsection 4.3.

The right-hand side of formula (23) is almost the Fourier transform of $\hat{f}$, except for a change in sign. It follows for nice functions that applying the Fourier transform twice to a function reverses the direction in x:

$$(\mathcal{F}^2 g)(x) = g(-x).$$

Applying the Fourier transformation four times therefore yields the identity operator.

This discussion shows that f and $\hat{f}$ convey equivalent information, but from different perspectives. In engineering the pair of functions $f, \hat{f}$ is known as a **Fourier transform pair**. As with the Laplace transform, the Fourier transform converts differentiation into multiplication.

Proposition 4.6. *Suppose G and G' are in L^1. Then $(i\xi)\hat{G}(\xi) = \widehat{G'}(\xi)$. Equivalently we can write, where M denotes multiplication by $i\xi$,*

$$G' = (\mathcal{F}^{-1} M \mathcal{F})(G). \tag{24}$$

PROOF. See Exercise 4.3. $\qquad\square$

The higher dimensional analogue of the Fourier transform is quite similar to the one-dimensional situation. These remarks of this section suggest that the Fourier transform is particularly useful when solving certain linear differential equations, both ordinary and partial. We refer to [T] and [F2] for these applications.

The next example shows that the Fourier transform of an *indicator function* is a sinc function.

EXAMPLE 4.1. Put $f(x) = 1$ if $-1 \leq x \leq 1$ and $f(x) = 0$ otherwise. Then

$$\hat{f}(w) = \frac{1}{\sqrt{2\pi}} \frac{i}{w} (e^{-iw} - e^{iw}) = \sqrt{\frac{2}{\pi}} \frac{\sin(w)}{w}.$$

We study the Fourier transform of a Gaussian. First we need a lemma.

Lemma 4.7. *The value of $\int_{-\infty}^{\infty} e^{-(x+iy)^2} dx$ is independent of y and equals $\int_{-\infty}^{\infty} e^{-x^2} dx = \sqrt{2\pi}$.*

PROOF. Integrate the function e^{-z^2} around a rectangle as follows. The vertices of the rectangle are at $R, R+iy, -R+iy$, and $-R$. The function is complex analytic everywhere and hence the integral is 0. The integrals along the vertical parts tend to 0 as R tends to infinity, because, for c_y independent of R,

$$\left| \int_0^y e^{-(R+it)^2} dt \right| \leq \int_0^y e^{-R^2+t^2} dt = c_y e^{-R^2}.$$

Hence $\int_{-\infty}^{\infty} e^{-(x+i0)^2} dx = \int_{-\infty}^{\infty} e^{-(x+iy)^2} dx$. $\qquad\qquad\qquad\qquad\qquad\qquad\qquad\qquad$ $\square$

EXAMPLE 4.2. Put $f(x) = e^{-\frac{1}{2}\frac{x^2}{\sigma^2}}$. We compute the Fourier transform:

$$\hat{f}(w) = \frac{1}{\sqrt{2\pi}} \int_{-\infty}^{\infty} e^{-\frac{1}{2}\frac{x^2}{\sigma^2}} e^{-ixw} dx = \frac{1}{\sqrt{2\pi}} \int_{-\infty}^{\infty} e^{-\frac{1}{2}(\frac{x^2}{\sigma^2}+2ixw)} dx$$

$$= \frac{1}{\sqrt{2\pi}} \int_{-\infty}^{\infty} e^{-\frac{1}{2}(\frac{x}{\sigma}+iw\sigma)^2} e^{-\frac{1}{2}w^2\sigma^2} dx = \frac{1}{\sqrt{2\pi}} \left(\int_{-\infty}^{\infty} e^{-\frac{x^2}{\sigma^2}} dx \right) e^{-\frac{1}{2}w^2\sigma^2} = \sigma e^{-\frac{1}{2}w^2\sigma^2}.$$

We regard the function $G(x,\sigma) = \frac{1}{\sqrt{2\pi\sigma^2}}e^{-\frac{1}{2}\frac{x^2}{\sigma^2}}$ as the probability density function of a Gaussian with mean 0 and variance σ^2. Its Fourier transform is $\frac{\sigma}{\sqrt{2\pi}}e^{-\frac{1}{2}w^2\sigma^2}$, which is the density function of a Gaussian $G(w, \frac{1}{\sigma})$ with mean 0 and variance $\frac{1}{\sigma^2}$. In other words, the Fourier transform of a Gaussian is a Gaussian with inverted variance. Figure 8 depicts Gaussians where σ equals $\frac{1}{4}$, 1, and 4.

Example 4.2 is closely connected with probability and quantum mechanics. See Sections 4 and 5 of Chapter 7. When $\sigma = 1$, the function $G(x, 1)$ determines the *standard normal* probability distribution. Its Fourier transform is itself. When σ tends to 0, the density is closely clustered about its mean. When σ tends to infinity, things become increasingly spread. The *Heisenberg uncertainty principle* from quantum mechanics puts a quantitative lower bound on the difference between measuring the position first and the momentum second versus measuring these in the other order. Exercise 4.14 determines additional eigenfunctions of the Fourier transform and suggests further connections with probability theory and quantum mechanics.

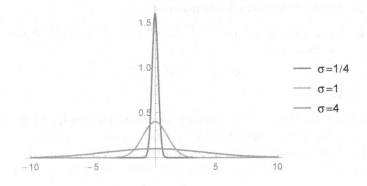

FIGURE 8. Gaussians with different variances

EXAMPLE 4.3. In Example 7.4 of Chapter 4 we used residues to show that the Fourier transform of $\frac{1}{x^2+1}$ is $\sqrt{\frac{\pi}{2}}e^{-|w|}$. Many other examples are computed using residues.

4.1. Convolution and approximate identities. The Fourier transform exhibits a close relationship with convolution. Approximate identities, which provide a rigorous view of the Dirac delta function, are closely connected with convolution and with applications later in this book.

We develop some of these ideas next. For functions $f, g \in L^1$, we put

$$(f * g)(x) = \int_{-\infty}^{\infty} f(x - t)g(t)dt.$$

REMARK. For $f, g \in L^1$, one has an easy version of Young's inequality:

$$\|f * g\|_{L^1} \leq \|f\|_{L^1} \|g\|_{L^1}.$$

For readers who know what L^p spaces are, one generalization is the following:

$$\|f * g\|_{L^r} \leq \|f\|_{L^p} \|g\|_{L^q}$$

whenever $1 \leq p, q, r \leq \infty$ and $\frac{1}{p} + \frac{1}{q} = 1 + \frac{1}{r}$. Taking $p = q = r = 1$ gives Young's inequality. Taking $q = 1$ and $p = r = \infty$ gives

$$\|f * g\|_{L^\infty} \leq \|f\|_{L^\infty} \|g\|_{L^1}.$$

Therefore, if f is bounded and g is absolutely integrable, then $f * g$ is bounded. This result is fundamental to BIBO stability; see Section 8 of Chapter 7. See [F1] for a proof of the generalized Young's inequality and related results.

See Exercise 4.13 for the basic properties of convolution. In particular, complex exponentials are eigenfunctions of the map $f \mapsto f * h$, a crucial fact in applied mathematics. The following analogue of Theorem 1.8 also holds.

Theorem 4.8. *Assume* f, g *are in* L^1. *Then*

$$\hat{f}\hat{g} = \frac{1}{\sqrt{2\pi}}(f * g)\hat{.}$$

PROOF. By definition and then putting $x = t - y$ in the inner integral we get

$$\hat{f}(\xi)\hat{g}(\xi) = \frac{1}{2\pi} \int_{-\infty}^{\infty} \int_{-\infty}^{\infty} f(x)g(y)e^{-i(x+y)\xi}dx\,dy = \frac{1}{2\pi} \int_{-\infty}^{\infty} \int_{-\infty}^{\infty} f(t-y)g(y)e^{-it\xi}dt\,dy.$$

Now interchange the order of integration and the result follows. $\qquad\square$

Corollary 4.9. *There is no* L^1 *function* h *such that* $h * f = f$ *for all* $f \in L^1$.

PROOF. If such an h exists, then Theorem 4.8 yields $\sqrt{2\pi}\hat{h}\hat{f} = \hat{f}$ for every f. Hence $\hat{h} = \frac{1}{\sqrt{2\pi}}$, a non-zero constant. Then h cannot be in L^1 because of the Riemann–Lebesgue lemma, Theorem 4.5. $\qquad\square$

Thus there is no identity operator for convolution. Instead we use approximate identities, as described in Definition 4.10. In pure mathematics, one never worries about evaluating a function at x. One just writes $f(x)$. In physics and applied mathematics, one samples f near x many times and takes an average. The next definition and theorem show how convolution integrals do the job. We can evaluate a function at x using the limit formula in Theorem 4.11. Although the limit as η tends to 0 of the functions K_η does not exist as an ordinary function, one often writes $\delta(x)$ for this limit, and calls it the Dirac delta function. Distribution theory provides a rigorous way of understanding this object. See for example [T]. We can be completely rigorous by working only with approximate identities.

Definition 4.10 (Approximate identity). An **approximate identity** is a family of functions $K_\eta : \mathbb{R} \to \mathbb{R}$ satisfying the following properties.

- For each $\eta > 0$, the function K_η is integrable and $\int_{-\infty}^{\infty} K_\eta(x)dx = 1$.
- There is a constant C such that $\int_{-\infty}^{\infty} |K_\eta(x)|dx \leq C$.
- For every $a > 0$, we have

$$\lim_{\eta \to 0} \int_{|x| \geq a} |K_\eta(x)|\,dx = 0.$$

The next result justifies the terminology. We also note that one can define approximate identities in a similar way on the unit circle or in higher dimensions. By placing stronger conditions on the functions K_η, one also can make the limit in Theorem 4.11 exist almost everywhere in x.

Theorem 4.11. *Let K_η be an approximate identity. Assume $f : \mathbb{R} \to \mathbb{C}$ is continuous at x. Then*

$$\lim_{\eta \to 0} (K_\eta * f)(x) = f(x).$$

If also each K_η is infinitely differentiable, then $K_\eta * f$ is infinitely differentiable, and hence f is the pointwise limit of infinitely differentiable functions.

PROOF. See Exercise 4.11.   □

REMARK. The proof of Theorem 4.11 is not difficult; furthermore, essentially the same proof arises throughout mathematics. Exercise 4.12 asks for proofs of two additional examples:

- Assume $\lim_n a_n = L$. Then $\lim_N \frac{1}{N} \sum_{j=1}^{N} a_j = L$.
- Put $F(x) = \int_0^x f(t)dt$. Assume f is continuous. Then F is differentiable and $F'(x) = f(x)$.

In Chapter 6 we will discuss the Fourier transform on $L^2(\mathbb{R})$, the square integrable functions. By definition we have

$$||f||_{L^2}^2 = \int_{-\infty}^{\infty} |f(x)|^2 dx.$$

When we regard f as a signal, this squared norm is the **total signal energy**. Plancherel's theorem implies that $||f||_{L^2}^2 = ||\hat{f}||_{L^2}^2$, providing a second way to compute the signal energy. See Exercise 6.5 of Chapter 6 for an interesting example. The expression $|\hat{f}(\omega)|^2$ is called the **energy spectrum**. Reference [KM] shows graphs of the energy spectra for (the first few notes of) Beethoven's fifth symphony and the Beatles song "A Hard Days Night". In both cases one also sees the Riemann–Lebesgue lemma in action; the energy spectra are negligible beyond a certain value of ω.

4.2. Connecting Fourier and Laplace transforms. We close our chapter on transform methods with comments that tie together these transforms, Fourier series, and the inversion formula.

Assume that f is piecewise differentiable and that $\int_{-\infty}^{\infty} |f(x)|dx$ converges. Then the Fourier inversion formula (23) holds. In order to apply this result to find an inversion formula for the Laplace transform, we consider a function f such that $f(x) = 0$ for $x < 0$. We apply (23) to the function $x \mapsto e^{-cx} f(x)$ and use the definition of the Fourier transform. We obtain

$$e^{-cx} f(x) = \frac{1}{2\pi} \int_{-\infty}^{\infty} \int_{0}^{\infty} f(t) e^{-ct} e^{-i\xi(t-x)} dt d\xi.$$

We rewrite this formula as

$$f(x) = \frac{1}{2\pi} \int_{-\infty}^{\infty} e^{x(i\xi+c)} \int_{0}^{\infty} f(t) e^{-t(c+i\xi)} dt d\xi.$$

To make things look like the Laplace transform, put $s = c + i\xi$ and change variables. Thus $d\xi = \frac{ds}{i}$. The integral along the real line becomes an integral along a line parallel to the imaginary axis. Using $F(s)$ to denote the Laplace transform, we obtain

$$f(x) = \frac{1}{2\pi i} \int_{c-i\infty}^{c+i\infty} e^{xs} \int_{0}^{\infty} e^{-ts} f(t) dt \, ds = \frac{1}{2\pi i} \int_{c-i\infty}^{c+i\infty} e^{xs} F(s) ds. \qquad (25)$$

This result is Theorem 1.3 for inverting the Laplace transform. It applies when c is chosen to make $\int_0^\infty e^{-ct} |f(t)| dt$ finite.

It is sometimes convenient to regard the Fourier transform as a continuous analogue of Fourier series. Start with the interval $[0, l]$. Consider distinct real numbers a, b. When $l(a - b)$ is an integer multiple of 2π, the functions e^{iax} and e^{ibx} are orthogonal on this interval. When $a \neq b$, we have

$$\int_0^l e^{iax} e^{-ibx} dx = \int_0^l e^{ix(a-b)} dx = \frac{e^{il(a-b)} - 1}{i(a - b)} = 0.$$

When $a = b$, the integral equals l. For Fourier series, the orthogonal functions are parametrized by the integers. We can regard the Fourier transform integral (20) as a limit when $l \to \infty$, and where we now use $e^{ix\xi}$ for all real ξ. The crucial point is that

$$\lim_{l \to \infty} \frac{1}{2l} \int_{-l}^l e^{iax} e^{-ibx} dx = 0$$

when $a \neq b$ and equals 1 when $a = b$. See Exercise 4.16.

Using this idea we give a formal derivation of the Fourier inversion formula analogous to the formula in Theorem 1.3 for inverting the Laplace transform, formally derived above.

For a function A to be determined, put $f(x) = \int_{-\infty}^\infty A(u) e^{ixu} du$. We want A to be $\sqrt{\frac{1}{2\pi}}$ times $\hat{f}$. Multiplying by e^{-ixy} we obtain

$$e^{-ixy} f(x) = \int_{-\infty}^\infty A(u) e^{ix(u-y)} du.$$

Integrating both sides with respect to x we obtain

$$\sqrt{2\pi} \hat{f}(y) = \int_{-\infty}^\infty e^{-ixy} f(x) dx = \int_{-\infty}^\infty \int_{-\infty}^\infty A(u) e^{ix(u-y)} du\, dx. \tag{26}$$

We regard the inner integral in (26) as the limit of integrals on $[-l, l]$ as l tends to infinity. We formally interchange the order of integration and get

$$\int_{-\infty}^\infty e^{-ixy} f(x) dx = \lim_{l \to \infty} \int_{-\infty}^\infty \frac{e^{il(u-y)} - e^{-il(u-y)}}{i(u - y)} A(u) du.$$

Using the definition of $\sin(lt)$ and then changing variables with $v = (u - y)l$ yields

$$\int_{-\infty}^\infty e^{-ixy} f(x) dx = \lim_{l \to \infty} \int_{-\infty}^\infty 2\frac{\sin(v)}{v} A\left(y + \frac{v}{l}\right) dv. \tag{27}$$

Recall that $\int_{-\infty}^\infty \frac{\sin(v)}{v} dv = \pi$. Formally plugging $l = \infty$ into (27) shows that

$$\int_{-\infty}^\infty e^{-ixy} f(x) dx = 2\pi A(y),$$

which is the Fourier inversion formula. The proof we gave of the Fourier inversion formula is similar, but there we introduced a Gaussian factor to create an approximate identity and to allow the interchange of order in the double integral. In the more formal derivation here, the use of the sinc function $\frac{\sin(x)}{x}$ is harder to justify, because this function is not absolutely integrable. The second property in Definition 4.10 fails.

4.3. A bit more on the Dirac delta function. We have defined the Dirac delta function as a functional: $\delta(f) = f(0)$. Thus δ is a function, but its domain is a space of functions rather than the real numbers. Nonetheless, in applied math, physics, and engineering, one often writes expressions such as $\delta(x - a)$ to abbreviate the functional for which

$$\delta(x - a)\, (f) = \int_{-\infty}^{\infty} \delta(x - a)\, f(x)\, dx = f(a). \tag{28}$$

To understand (28), one can simply change variables in the (formal) integral. If we put $x - a = t$, then $dt = dx$ and the integral becomes

$$\int_{-\infty}^{\infty} \delta(x - a)\, f(x)\, dx = \int_{-\infty}^{\infty} \delta(t)\, f(t + a)\, dt = f(a).$$

We would like to replace the function $x \mapsto x - a$ by more general functions, thereby evaluating expressions such as

$$\int_{-\infty}^{\infty} \delta(g(x))\, f(x)\, dx.$$

This procedure of formally changing variables is easily justified by using approximate identities. We obtain the following result. Further generalizations are possible as well. See Exercise 4.19.

Theorem 4.12. *Let* $g : \mathbb{R} \to \mathbb{R}$ *be a continuously differentiable bijection with* $g' \neq 0$. *Then*

$$\int_{-\infty}^{\infty} \delta(g(x))\, f(x)\, dx = \int_{-\infty}^{\infty} \delta(t)\, f(g^{-1}(t))\, \frac{dt}{|g'(g^{-1}(t))|}.$$

In other words,

$$(\delta \circ g)(f) = \frac{f(g^{-1}(0))}{|g'(g^{-1}(0))|}. \tag{29}$$

PROOF. (Sketch) We choose an approximate identity, and apply the change of variables formula in the integral. Then we take limits. Following common practice, we do not bother introducing the approximation to the delta function. Hence we formally change variables in the integral. Put $t = g(x)$ and thus $x = g^{-1}(t)$ and $dt = g'(x)dx$. Therefore we have

$$dx = \frac{dt}{g'(x)} = \frac{dt}{g'(g^{-1}(t))}.$$

Now changing variables in the integral gives

$$\int_{-\infty}^{\infty} \delta(g(x))\, f(x)\, dx = \int_{-\infty}^{\infty} \delta(t)\, f(g^{-1}(t))\, \frac{dt}{|g'(g^{-1}(t))|}. \tag{30}$$

But (30) involves the usual delta function evaluated at t, and hence the right-hand side of (30) is evaluated by setting $t = 0$ in the integrand, yielding (29). $\qquad\square$

EXAMPLE 4.4. Assume $a \neq 0$. Put $g(x) = ax + b$. Then $g^{-1}(t) = \frac{t-b}{a}$. The theorem gives

$$\int_{-\infty}^{\infty} \delta(ax + b)\, f(x)\, dx = \frac{1}{|a|}\, f\left(\frac{-b}{a}\right).$$

The special cases when $a = 1$ or when $b = 0$ arise quite often.

EXERCISE 4.1. Compute the Fourier transform of $e^{-|x|}$.

EXERCISE 4.2. Verify the conclusion about delay in the Remark after Proposition 4.3.

EXERCISE 4.3. Verify Proposition 4.3.

EXERCISE 4.4. Suppose f is an even function and $\hat{f}$ exists. What can you say about $\hat{f}$? What if f is an odd function?

EXERCISE 4.5. Put $f(x) = x^{a-1}e^{-x}$ for $x > 0$ and $f(x) = 0$ otherwise. Determine the condition on a such that $f \in L^1$. In this case, find $\hat{f}$.

EXERCISE 4.6. Put $f(x) = e^{-x}$ for $x > 0$ and 0 for $x \leq 0$. Find $\hat{f}$.

EXERCISE 4.7. Let M denote multiplication by $i\xi$. Let D denote differentiation. Show that

$$D = \mathcal{F}^{-1}M\mathcal{F}. \tag{31}$$

Use (31) to give a definition of functions of differentiation.

EXERCISE 4.8. Put $f(x) = 1 - |x|$ for $|x| \leq 1$ and $f(x) = 0$ otherwise. Compute $\hat{f}$.

EXERCISE 4.9. Put $f_n(x) = 0$ when $|x| \leq \frac{1}{n}$ or when $|x| \geq \frac{3}{n}$. Put $f(\frac{\pm 2}{n}) = n$ and make f linear otherwise. Graph f_n and find $\hat{f_n}$. Show that

$$\int_{-\infty}^{\infty} \lim_{n \to \infty} f_n(x)dx \neq \lim_{n \to \infty} \int_{-\infty}^{\infty} f_n(x)dx.$$

EXERCISE 4.10 (Square pulse). Define $p_\eta(t)$ by $p_\eta(t) = 0$ for $|t| > \frac{\eta}{2}$ and $p_\eta(t) = \frac{1}{\eta}$ otherwise. Assume that f is continuous at t_0. Show that $\lim_{\eta \to 0}(p_\eta * f)(t_0) = f(t_0)$. In other words, the pulse behaves like an approximate identity. Suggestion: Use the ϵ - δ definition of a limit.

EXERCISE 4.11. Prove Theorem 4.11.

EXERCISE 4.12. Prove both statements in the Remark following Theorem 4.11. See if you can explain why the proofs of these statements and of Theorem 4.11 are so much alike. Compare also with the proof of the Cauchy integral formula from Chapter 4.

EXERCISE 4.13. Verify the following properties of convolution, crucial for **LTI** systems.
- Linearity: $h * (f + g) = h * f + h * g$ and $h * (cf) = c(h * f)$.
- Commutivity: $h * f = f * h$
- Associativity $(h * f) * g = h * (f * g)$
- The exponentials $e^{i\omega t}$ are eigenfunctions of the linear map $f \to h * f$.
- What are the eigenvalues corresponding to these eigenfunctions?

EXERCISE 4.14. Let p_η be the square pulse from Exercise 4.10. Let $g(t) = t^m$ for a positive integer m. Compute $p_\eta * g$. Check your answer by finding the limit as η tends to 0.

EXERCISE 4.15 (Hermite polynomials). Define polynomials $H_n(x)$ by

$$e^{2xt-t^2} = \sum_{n=0}^{\infty} H_n(x)\frac{t^n}{n!}. \tag{32}$$

- Find $H_n(x)$ for $n = 0, 1, 2, 3$.
- Let $L = x - \frac{d}{dx}$. Let M denote multiplication by $e^{\frac{-x^2}{2}}$. Find ML^nM^{-1}.
- Show that the Fourier transform of $e^{\frac{-x^2}{2}}H_n(x)$ is $(-i)^n e^{\frac{-\xi^2}{2}}H_n(\xi)$. The functions $e^{\frac{-x^2}{2}}H_n(x)$ are thus eigenfunctions of the Fourier transform. Suggestion: Multiply both sides of (32) by $e^{\frac{-x^2}{2}}$.
- Show that the coefficients of each H_n are integers.

REMARK. The Hermite polynomials arise in conjunction with the harmonic oscillator. Figure 9 shows the graphs of functions ψ_n, where

$$\psi_n(\xi) = c_n H_n(\xi) e^{\frac{-\xi^2}{2}}.$$

These functions are eigenfunctions of the Schrödinger equation from quantum mechanics. Section 5 of Chapter 7 gives an introduction to quantum mechanics. See [Gr] or [RS] for additional information. See also Exercise 5.9 of Chapter 7 for more on the Hermite polynomials.

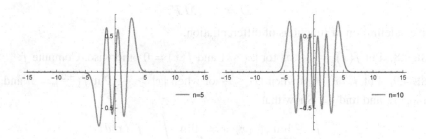

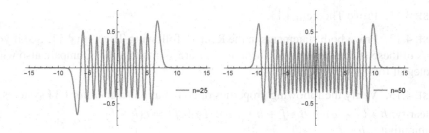

FIGURE 9. Stationary states

EXERCISE 4.16. For $n \in \mathbb{N}$ and $a \neq 0$, evaluate $\int_{-\infty}^{\infty} \delta(ax + b)\, x^n\, dx$.

EXERCISE 4.17. Verify the following limit. For $a \neq b$,

$$\lim_{l \to \infty} \frac{1}{2l} \int_{-l}^{l} e^{iax} e^{-ibx} dx = 0.$$

For $a = b$, the limit is 1.

EXERCISE 4.18. (Difficult) Use residues to find the Fourier transform of $\frac{1}{\cosh(x)}$.

EXERCISE 4.19. Explain and derive the formula $\delta \circ g = \sum_j \frac{\delta(x - r_j)}{|g'(r_j)|}$. Here the r_j are the roots of g.

CHAPTER 6

Hilbert spaces

An inner product space is a complex vector space with an inner product whose properties abstract those of the inner product $\sum_{j=1}^{n} z_j \overline{w}_j$ in $\mathbb{C}^n$. Given the inner product, we can define a norm, and use it to measure the lengths of the vectors in the space. It therefore makes sense to consider convergent sequences. The space is complete if and only if all Cauchy sequences converge. A Hilbert space is a complete inner product space. In infinite dimensions, convergence issues are much more delicate than many readers will be used to. In fact, it took mathematicians almost a century to go from Fourier's initial ideas to the modern framework of Hilbert spaces. Hence this chapter will be considerably harder than the previous ones. The ideas remain fundamental to modern science.

One of the most useful ideas in this chapter is orthonormal expansion. We will see that a Fourier series is an orthonormal expansion in terms of complex exponentials. Furthermore, a wide class of second-order differential equations, known as Sturm–Liouville equations, also lead to systems of orthogonal functions and orthonormal expansions. We discuss examples and applications of Hilbert space theory, such as least squares regression, the basics of quantum mechanics, and Legendre polynomials, in Chapter 7.

1. Inner products and norms

Let H be a complex vector space. An inner product is a function $\langle \, , \, \rangle$ that takes two elements from H and computes a complex number. Here are the defining properties:

Definition 1.1. An inner product on a complex vector space H is a function $\langle \, , \, \rangle : H \times H \to \mathbb{C}$ satisfying the following properties:

- $\langle u + v, w \rangle = \langle u, w \rangle + \langle v, w \rangle$.
- $\langle cu, w \rangle = c \langle u, w \rangle$.
- $\langle u, w \rangle = \overline{\langle w, u \rangle}$ (Hermitian symmetry)
- If $u \neq 0$, then $\langle u, u \rangle > 0$. (positive definiteness)

When H is an inner product space, we put $||u||^2 = \langle u, u \rangle$. Then $||u||$ is called the *norm* of u and satisfies the following properties:

- $||cu|| = |c| \, ||u||$.
- $||u|| \geq 0$ for all u, and equals 0 if and only if $u = 0$.
- $||u + v|| \leq ||u|| + ||v||$. (The triangle inequality)

Definition 1.2. A real or complex vector space is called a *normed linear space* or *normed vector space* if there is a function $|| \ ||$ satisfying the above three properties.

The norm need not arise from an inner product. In this book, however, we will be primarily concerned with inner products spaces.

In any normed vector space, we define the distance between u and v to be $||u - v||$. The triangle inequality justifies this terminology and enables us to discuss limits.

Theorem 1.3. *The following hold in an inner product space.*

- $|\langle u, v \rangle| \leq ||u|| \; ||v||$. *(Cauchy–Schwarz inequality)*
- $||u + v|| \leq ||u|| + ||v||$. *(triangle inequality)*
- $||u + v||^2 + ||u - v||^2 = 2||u||^2 + 2||v||^2$. *(the parallelogram law)*

PROOF. We leave the parallelogram law to the reader. The triangle inequality follows from the Cauchy–Schwarz inequality because

$$||u + v||^2 = ||u||^2 + 2\mathrm{Re}(\langle u, v \rangle) + ||v||^2 \leq ||u||^2 + 2||u|| \; ||v|| + ||v||^2 = (||u|| + ||v||)^2$$

holds if and only if $\mathrm{Re}(\langle u, v \rangle) \leq ||u|| \; ||v||$. Since $\mathrm{Re}(z) \leq |z|$ for $z \in \mathbb{C}$, we see that the Cauchy–Schwarz inequality implies the triangle inequality. To prove Cauchy–Schwarz, note first that it holds if $v = 0$. Assume $v \neq 0$. We know that $0 \leq ||u + tv||^2$ for all complex t. Put $t = -\frac{\langle u, v \rangle}{||v||^2}$ to get

$$0 \leq ||u||^2 - 2\frac{|\langle u, v \rangle|^2}{||v||^2} + \frac{|\langle u, v \rangle|^2 ||v||^2}{||v||^2} = ||u||^2 - \frac{|\langle u, v \rangle|^2}{||v||^2}, \tag{1}$$

which gives the Cauchy–Schwarz inequality. $\qquad\square$

The proof of the Cauchy–Schwarz inequality suggests orthogonal projection. Put $\pi_v(u) = \frac{\langle u, v \rangle}{||v||^2} v$. Then $\pi_v(u)$ is the orthogonal projection of u onto v. See Figure 1. If we regard $u + tv$ as a complex line, then the inequality states the obvious: the point on the line closest to the origin has non-negative length. We apply this inequality many times in the rest of the book. The delightful book [S] has many additional uses of the Cauchy–Schwarz inequality.

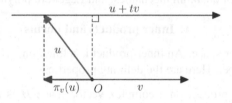

FIGURE 1. Orthogonal projection

The norm enables us to measure the length of vectors and thereby discuss limits. To do so we recall several fundamental facts from basic analysis. Let $\{x_n\}$ be a sequence of real numbers. Recall the definition of a Cauchy sequence:

Definition 1.4. A sequence $\{x_n\}$ of real numbers is a **Cauchy sequence** if, for all $\epsilon > 0$, there is an N such that $n, m \geq N$ implies that $|x_n - x_m| < \epsilon$.

We compare the meaning of Cauchy sequence with the meaning of limit: $\{x_n\}$ converges to L if and only if the terms are eventually all close to L, and $\{x_n\}$ is a Cauchy sequence if and only if the terms are eventually all close to each other. It was the genius of Cauchy to formulate the second condition, enabling us to prove that a sequence has a limit even when we do not know what the limit is.

Theorem 1.5. *Convergence of sequences:*
  (1) *A bounded monotone sequence of real numbers has a limit.*
  (2) *A bounded sequence of real numbers always has a convergent subsequence.*
  (3) *A sequence of real numbers has a limit if and only if it is a Cauchy sequence.*

Item (3) of Theorem 1.5 and the definition of the norm imply the following result. Its conclusion need not hold for infinite-dimensional inner product spaces.

Proposition 1.6. *A sequence in $\mathbb{C}^n$ converges if and only if it is a Cauchy sequence.*

Definition 1.7. Let V be an inner product space (or more generally a normed vector space). A sequence $\{z_n\}$ in V is called a **Cauchy sequence** if, for each $\epsilon > 0$, there is an N such that $m, n \geq N$ implies $||z_m - z_n|| < \epsilon$. A sequence $\{z_n\}$ *converges* to w if, for each $\epsilon > 0$, there is an N such that $n \geq N$ implies $||z_n - w|| < \epsilon$.

We also need to define convergent infinite series.

Definition 1.8. Let V be a normed vector space. The infinite series $\sum_{n=1}^{\infty} a_n$ converges to L if the sequence of partial sums $\{A_N\}$ converges to L. Here

$$A_N = \sum_{n=1}^{N} a_n.$$

For infinite-dimensional inner product spaces, the completeness property of Proposition 1.6 need not hold. Its importance suggests that completeness be made part of the definition.

Definition 1.9. An inner product space H is **complete** if, whenever $\{z_n\}$ is a Cauchy sequence in H, then $\lim_{n \to \infty} z_n$ exists.

Definition 1.10. A **Hilbert space** is a complete inner product space.

EXERCISE 1.1. Fix $v \neq 0$. Consider the function $u \mapsto \pi_v(u)$. Verify the following properties:

- $\pi_v(cu) = c\pi_v(u)$ for scalars c.
- $\pi_v(u_1 + u_2) = \pi_v(u_1) + \pi_v(u_2)$.
- $\pi_v(v) = v$.
- $\pi_v^2 = \pi_v$.
- $\langle \pi_v(u), w \rangle = \langle u, \pi_v(w) \rangle$.

EXERCISE 1.2. Show that $\mathbb{C}^n$ is an inner product space, where

$$\langle z, w \rangle = \sum_{j=1}^{n} z_j \overline{w}_j.$$

Explain why the complex conjugation appears.

EXERCISE 1.3. Prove the parallelogram law.

EXERCISE 1.4. Show that $|\sum_{j=1}^{n} z_j|^2 \leq n(\sum_{j=1}^{n} |z_j|^2)$.

EXERCISE 1.5. When does equality hold in the Cauchy–Schwarz inequality?

EXERCISE 1.6. Prove the following inequalities:

$$|\sum_{j=1}^{n} j z_j|^2 \leq \left(\frac{n^3}{3} + \frac{n^2}{2} + \frac{n}{6}\right) \sum_{j=1}^{n} |z_j|^2 \leq n^3 \sum_{j=1}^{n} |z_j|^2.$$

2. Orthonormal expansion

Expansions in terms of orthonormal bases are particularly easy in the finite-dimensional setting. An analogous idea holds in infinite dimensions, but some additional analysis is required. Orthonormal expansions occur throughout physics and engineering. We develop this topic next.

Definition 2.1. A collection of vectors $\{\phi_j\}$ in an inner product space is called an **orthonormal system** if $\langle \phi_j, \phi_k \rangle = 0$ when $j \neq k$ but $\langle \phi_j, \phi_j \rangle = 1$ for each j.

As noted earlier, we can restate the definition using the Kronecker delta notation:

$$\langle \phi_j, \phi_k \rangle = \delta_{jk}.$$

In $\mathbb{C}^n$, we have $z = (z_1, z_2, \cdots, z_n) = z_1 e_1 + \cdots + z_n e_n$. Here the collection of e_j form an orthonormal basis for $\mathbb{C}^n$. We can recover the coefficients z_j easily from z, because

$$z_j = \langle z, e_j \rangle.$$

Given an arbitrary basis $v_1, \ldots, v_n$, one must solve a system of linear equations to write z in terms of this basis. In other words, although $z = \sum a_j v_j$ for some scalars a_j, finding them is annoyingly hard. This simple distinction partially explains why orthonormal bases are so important. The next lemma is also useful.

Lemma 2.2. *Assume that $\{\phi_1, \ldots, \phi_N\}$ is an orthonormal system. For each z,*

$$\left\| \sum_{j=1}^{N} \langle z, \phi_j \rangle \phi_j \right\|^2 = \sum_{j=1}^{N} |\langle z, \phi_j \rangle|^2.$$

PROOF. Expand the squared norm. Using Hermitian symmetry and orthonormality gives

$$\left\| \sum_{j=1}^{N} \langle z, \phi_j \rangle \phi_j \right\|^2 = \sum_{j,k=1}^{N} \langle z, \phi_j \rangle \, \overline{\langle z, \phi_k \rangle} \langle \phi_j, \phi_k \rangle = \sum_{j=1}^{N} |\langle z, \phi_j \rangle|^2.$$

$\square$

Let $\{\phi_j\}$ be a collection of orthonormal vectors in H. A version of the following inequality was discovered (although the terminology necessarily differed) by Bessel in the early 19th century.

Theorem 2.3. *Let H be a Hilbert space. Suppose $\{\phi_j\}$ is an orthonormal system in H. Then, for each positive integer N and each $z \in H$ we have (2). Furthermore, the infinite sum converges.*

$$\left\| \sum_{n=1}^{N} \langle z, \phi_n \rangle \phi_n \right\| \leq \left\| \sum_{n=1}^{\infty} \langle z, \phi_n \rangle \phi_n \right\| \leq \|z\|. \tag{2}$$

PROOF. We first note the following formula:

$$0 \leq \left\| \sum_{n=1}^{N} \langle z, \phi_n \rangle \phi_n - z \right\|^2 = \left\| \sum_{n=1}^{N} \langle z, \phi_n \rangle \phi_n \right\|^2 - 2\mathrm{Re}\langle \sum_{n=1}^{N} \langle z, \phi_n \rangle \phi_n, z \rangle + \|z\|^2.$$

Using the lemma and $\langle \phi_n, z \rangle = \overline{\langle z, \phi_n \rangle}$ we get

$$0 \leq \sum_{n=1}^{N} |\langle z, \phi_n \rangle|^2 - 2\mathrm{Re}\langle \sum_{n=1}^{N} \langle z, \phi_n \rangle \phi_n, z \rangle + \|z\|^2 = - \sum_{n=1}^{N} |\langle z, \phi_n \rangle|^2 + \|z\|^2. \tag{3}$$

Inequality (3) yields, for all N,

$$\sum_{n=1}^{N} |\langle z, \phi_n \rangle|^2 \leq ||z||^2. \tag{3.1}$$

To show that the infinite sum in (2) converges, we show that it is Cauchy. Thus, given $\epsilon > 0$, we want

$$|| \sum_{M+1}^{N} \langle z, \phi_n \rangle \phi_n || < \epsilon$$

for large N, M. It suffices to show that $|| \sum_{M+1}^{N} \langle z, \phi_n \rangle \phi_n ||^2$ tends to zero when M, N do. By Lemma 2.2, it suffices to show that $\sum_{M+1}^{N} |\langle z, \phi_n \rangle|^2$ tends to zero. By (3.1) the sequence $\{t_N\}$ of real numbers defined by $t_N = \sum_{n=1}^{N} |\langle z, \phi_n \rangle|^2$ is bounded and non-decreasing. By Theorem 1.5, $\lim_{N \to \infty} t_N = t$ exists and hence the sequence is Cauchy. Thus the sequence in (2) is also Cauchy and we have $t \leq ||z||^2$. □

Corollary 2.4. *Put* $\xi_N = \sum_{n=1}^{N} \langle z, \phi_n \rangle \phi_n$. *Then* $\{\xi_N\}$ *is Cauchy sequence.*

Definition 2.5. A collection of vectors $\{\phi_n\}$ is called a **complete orthonormal system** in H if

- $||\phi_n||^2 = 1$ for all n,
- $\langle \phi_n, \phi_m \rangle = 0$ for all $n \neq m$,
- $\langle z, \phi_n \rangle = 0$ for all n implies $z = 0$. (The only vector orthogonal to all the ϕ_n is the zero vector.)

The third property provides a test for equality. Vectors z and w are equal if and only if

$$\langle z, \phi_n \rangle = \langle w, \phi_n \rangle.$$

for all n.

Theorem 2.6 (Orthonormal expansion). *Let* $\{\phi_n\}$ *be a complete orthonormal system in a Hilbert space* H. *For all* $z \in H$,

$$z = \sum_{n=1}^{\infty} \langle z, \phi_n \rangle \phi_n. \tag{4}$$

Furthermore, for each z, *Parseval's identity holds:*

$$||z||^2 = \sum_{n=1}^{\infty} |\langle z, \phi_n \rangle|^2.$$

PROOF. We have already established the convergence of the sum. To prove (4), we must show that the value of the sum is z. By the third property from Definition 2.5, if $\{\phi_k\}$ is a complete system, then (4) holds if and only if

$$\langle z, \phi_k \rangle = \langle \sum_{n=1}^{\infty} \langle z, \phi_n \rangle \phi_n, \phi_k \rangle \tag{5}$$

for every k. By the linearity of the inner product and orthonormality, we obtain

$$\langle \sum_{n=1}^{\infty} \langle z, \phi_n \rangle \phi_n, \phi_k \rangle = \sum_{n=1}^{\infty} \langle z, \phi_n \rangle \langle \phi_n, \phi_k \rangle = \langle z, \phi_k \rangle.$$

Parseval's identity follows from (4) after taking squared norms, because the norm is continuous. In other words, whenever $\{w_k\}$ converges to w in H, then $||w_k||^2$ converges to $||w||^2$. □

Let $\phi_1, \phi_2, \cdots$, be an orthonormal system in H and let V be the span of the $\{\phi_j\}$. For $z \in H$ we write $\pi_V(z) = \sum_j \langle z, \phi_j \rangle \phi_j$ for the orthogonal projection of z onto V. For each N, the finite sum $\sum_{n=1}^{N} \langle z, \phi_n \rangle \phi_n$ equals the orthogonal projection of z onto the span of $\phi_1, \phi_2, \cdots, \phi_N$.

One of the major theorems in the basics of Hilbert space theory shows that orthogonal projection provides the solution of an optimization problem.

Theorem 2.7. *Let V be a closed subspace of a Hilbert space H. Given $z \in H$, there is a unique $v \in V$ minimizing $||z - v||$. (If $z \in V$, put $v = z$; in general v is the orthogonal projection of z onto V.)*

PROOF. As noted in the statement, the conclusion is obvious when $z \in V$. Otherwise we consider the collection of numbers $||z - v||$ as v ranges over V. Let $\{v_n\}$ be a sequence in V such that

$$\lim_n ||z - v_n|| = \inf_V ||z - v||.$$

If we can establish that $\{v_n\}$ is Cauchy, then it will have a limit v which is the minimizer. This v turns out to be unique. To first verify uniqueness, assume both v and w are minimizers, and consider their midpoint $u = \frac{v+w}{2}$. Note that u is also in V, by definition of a subspace. The parallelogram law gives

$$||z - v||^2 + ||z - w||^2 = 2||u - z||^2 + 2||u - w||^2 \geq 2||u - z||^2$$

and therefore implies that the midpoint is at least as close to z, and is closer unless $v = w = u$.

The proof that the sequence $\{v_n\}$ is Cauchy also uses the parallelogram law. We write

$$||v_n - v_m||^2 = 2||v_n - z||^2 + 2||v_m - z||^2 - ||(v_n - z) - (z - v_m)||^2 \qquad (6)$$

and note that the subtracted term in (6) is 4 times the squared distance to the midpoint of v_n and v_m. As a result the expression in (6) tends to 0 as m, n tend to infinity. Hence $\{v_n\}$ is Cauchy and therefore converges to some v. Since the subspace V is closed, $v \in V$. $\square$

EXERCISE 2.1. Let π_V be orthogonal projection onto the subspace V. Verify that the properties of Exercise 1.1 hold in this setting.

EXERCISE 2.2. (An easy comparison test) Let $\{\xi_N\}$ be a sequence in a Hilbert space H. Assume that $||\xi_N - \xi_M|| \leq |t_N - t_M|$ for some convergent sequence $\{t_n\}$ of real or complex numbers. Show that $\{\xi_N\}$ converges. Derive from this result a comparison test for an infinite series $\sum_{n=1}^{\infty} v_n$ in a Hilbert space. Compare with the proof of Theorem 2.3.

3. Linear transformations

Let H_1 and H_2 be Hilbert spaces. As in finite dimensions, a map $L : H_1 \to H_2$ is linear if the usual pair of equations hold for all $z, w \in H_1$ and all scalars c:

$$L(z + w) = L(z) + L(w)$$
$$L(cz) = cL(z).$$

As before, the addition and scalar multiplications on the left-hand side are taken in H_1, and on the right-hand side they are taken in H_2. In most of what we do, either $H_2 = H_1$ or $H_2 = \mathbb{C}$. When $H_2 = \mathbb{C}$, the map L is called a *linear functional*. When $H_2 = H_1 = H$, we often call L a linear operator on H.

Definition 3.1. A linear map is *bounded* if there is a constant C such that

$$||L(z)|| \leq C||z|| \qquad (7)$$

for all z in H_1. Hence L is often called a *bounded linear transformation*, or a BLT. The infimum of all C for which (7) holds is called the **operator norm** of L, and written $||L||$.

Inequality (7) is equivalent to saying that L is continuous.

Lemma 3.2. *A linear map between Hilbert spaces is bounded if and only if it is continuous.*

PROOF. We use the ϵ-δ definition to establish continuity, assuming (7) holds. If $\epsilon > 0$ is given, then $||z - w|| < \frac{\epsilon}{C}$ implies

$$||Lz - Lw|| = ||L(z - w)|| \le C||z - w|| < \epsilon.$$

To verify the converse, assume L is continuous. In particular L is continuous at 0. Therefore, given $\epsilon > 0$, there is a $\delta > 0$ such that $||z|| \le \delta$ implies $||Lz|| < \epsilon$. Given an arbitrary non-zero w, the linearity of L and the continuity at 0 guarantee that we can write

$$||L(w)|| = \left|\left|L\left(\frac{w\delta}{||w||}\right)\right|\right| \frac{||w||}{\delta} \le \frac{\epsilon}{\delta}\,||w||.$$

Hence the desired constant C exists. $\qquad\square$

Unbounded linear transformations exist, but they cannot be defined on the entire Hilbert space. We will discuss this matter in Section 7.

Since a bounded linear transformation is continuous, in particular we know that, if $z_n \to z$, then $L(z_n) \to L(z)$. We often use this result without being explicit; for example we might apply L to an infinite sum term-by-term.

The collection $\mathcal{L}(H)$ of bounded linear operators on H has considerable structure. It is a vector space, and furthermore, a non-commutative algebra. In other words, we can compose two linear operators, and the usual algebraic laws hold, except that $LK \ne KL$ in general. As usual, I denotes the identity transformation. Terminology such as invertibility, null space, range, eigenvalues, etc. used in Chapter 1 applies in this setting.

In particular, consider the linear equation $L(z) = w$. All solutions satisfy $z = z_p + v$, where z_p is a particular solution and v is an arbitrary element of the null space of L, as in Theorem 2.5 of Chapter 1. One difference is that the null space could be of infinite dimension.

4. Adjoints

Our next goal is to define the adjoint of a (bounded) linear operator $L : H \to H$. The adjoint $L^* : H \to H$ will be linear and will satisfy $\langle L^*u, v\rangle = \langle u, Lv\rangle$ for all u, v. When $H = \mathbb{C}^n$, and L has matrix (L_{jk}), then L^* has matrix $(\overline{L_{kj}})$, the conjugate transpose of the matrix L.

REMARK. Linear differential operators are not bounded, and the theory is more difficult. We postpone these matters until our study of Sturm–Liouville equations in Section 7. We give the definition of unbounded operator in Definition 7.3 and the definition of self-adjoint (unbounded) operator in Definition 7.4.

A *linear functional* on H is a linear transformation $L : H \to \mathbb{C}$. By Lemma 3.2, a linear functional L is bounded if and only if it is continuous.

Lemma 4.1 (Riesz Lemma). *Let H be a Hilbert space. Let $L : H \to \mathbb{C}$ be a bounded linear functional. Then there is a unique $w \in H$ such that $L(v) = \langle v, w\rangle$.*

The map $v \to \langle v, w\rangle$ is obviously a linear functional. The Riesz lemma asserts the converse. Each continuous linear functional is given by the inner product with some vector. First we prove this statement in finite dimensions, where the conclusion is almost immediate.

In $\mathbb{C}^n$, consider an orthonormal basis e_i, $i = 1, 2, \cdots, n$. Let L be a linear functional. Then L is automatically continuous. We have

$$L(z) = L(\sum_{j=1}^{n} z_j e_j) = \sum_{j=1}^{n} z_j L(e_j) = \sum_{j=1}^{n} z_j \overline{w_j} = \langle z, w \rangle,$$

where we have put $\overline{w_j} = L(e_j)$ and $w = \sum w_k e_k$.

PROOF. We consider the general case. Suppose $L : H \to \mathbb{C}$ is a bounded linear functional. In case $L(v) = 0$ for all v, then $L(v) = \langle v, 0 \rangle$, and the conclusion holds. Hence we may assume that $L(v) \neq 0$ for some v. In other words, $\mathcal{N}(L) \neq H$. Let w_0 be any non-zero vector such that $w_0 \perp \mathcal{N}(L)$.

For $z \in H$, we write $z = z - \alpha w_0 + \alpha w_0$. We want to choose $\alpha \in \mathbb{C}$ to make

$$z - \alpha w_0 \in \mathcal{N}(L).$$

The equation $L(z - \alpha w_0) = 0$ forces $L(z) = \alpha L(w_0)$. Since $w_0 \notin \mathcal{N}(L)$, the complex number $L(w_0)$ is not 0, and we can divide by it. Thus we put

$$\alpha = \frac{L(z)}{L(w_0)}.$$

Then, since w_0 is orthogonal to $\mathcal{N}(L)$,

$$\langle z, w_0 \rangle = \langle z - \alpha w_0, w_0 \rangle + \langle \alpha w_0, w_0 \rangle = \alpha ||w_0||^2 = \frac{L(z)}{L(w_0)} ||w_0||^2. \tag{8}$$

It follows from (8) that

$$L(z) = \frac{L(w_0)}{||w_0||^2} \langle z, w_0 \rangle = \langle z, \frac{\overline{L(w_0)}}{||w_0||^2} w_0 \rangle.$$

We have found the needed w. In fact, w is unique. If

$$Lz = \langle z, w \rangle = \langle z, \zeta \rangle,$$

then $w - \zeta$ is orthogonal to every z, and hence must be 0.   □

The many corollaries of the Riesz lemma include the definition of the adjoint and important results about orthogonal projection onto closed subspaces. We now define the adjoint of a linear transformation.

Let $L : H \to H$ be a bounded linear map. Fix $v \in H$, and consider the map $u \to \langle Lu, v \rangle = \psi(u)$. This map ψ is linear, and it is bounded because

$$|\psi(u)| = |\langle L(u), v \rangle| \leq ||L(u)|| \, ||v|| \leq C ||u|| \, ||v|| = C' ||u||.$$

Since ψ is a bounded linear functional, the Riesz lemma guarantees that $\psi(u) = \langle u, w \rangle$ for some w depending on v. Thus $\langle Lu, v \rangle = \langle u, \text{something} \rangle$. The "something" depends on v, and is denoted $L^* v$. One can easily show that the map L^* is linear.

We give some examples of adjoints.

EXAMPLE 4.1. Let $H = L^2[0,1]$. See Section 6 for the precise definition of L^2. Define an operator T on H by

$$(Tf)(x) = \int_0^x f(t) dt.$$

We claim that T^* is given by $(T^* g)(y) = \int_y^1 g(s) ds$:

$$\langle Tf, g \rangle = \int_0^1 \int_0^x f(t) dt \, \overline{g(x)} dx = \int_0^1 \int_t^1 \overline{g(x)} dx \, f(t) dt = \langle f, T^* g \rangle.$$

EXAMPLE 4.2. A crucial piece of the solution of Example 4.1 was to interchange the order in a double integral. Suppose more generally that a linear map on a space of functions is defined by the rule

$$Tf(x) = \int K(x,y)f(y)dy. \tag{9}$$

Here $K(x,y)$ is called the **integral kernel** of the operator T. Then the adjoint of T has integral kernel $\overline{K(y,x)}$. In other words, $T^*f(x) = \int \overline{K(t,x)}f(t)dt$.

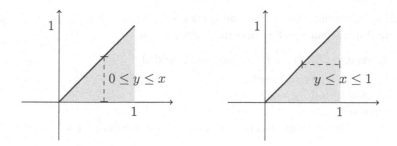

FIGURE 2. Interchanging the order of integration

EXAMPLE 4.3. Suppose $T : \mathbb{C}^n \to \mathbb{C}^n$. We find T^*. Let $e_1, e_2, \cdots, e_n$ be the usual orthonormal basis. Suppose the matrix of T is (c_{kj}). Then we have

$$T(z) = T\left(\sum z_j e_j\right) = \sum_j z_j T(e_j) = \sum_j z_j \sum_k c_{kj} e_k.$$

By switching the roles of the indices, we see that $\langle Tz, w \rangle$ is given by

$$\langle Tz, w \rangle = \sum_j \sum_k c_{kj} z_j \overline{w_k} = \sum_k \sum_j z_k c_{jk} \overline{w_j} = \sum_k z_k \sum_j \overline{\overline{c_{jk}} w_j} = \langle z, T^* w \rangle.$$

Thus the matrix of T^* is the *conjugate transpose* of the matrix of T.

The next example concerns finite-dimensional linear algebra, but it fits nicely here, as evidenced by its nice corollary.

EXAMPLE 4.4. Let V be the complex vector space of n-by-n complex matrices. Given $A, B \in V$, the formula

$$\langle A, B \rangle = \text{trace}(AB^*)$$

defines an inner product on V. See Exercise 4.4.

Corollary 4.2. *Assume that T is a linear functional on the space of n-by-n matrices. Then there is a matrix M such that $T(A) = \text{trace}(AM)$.*

EXERCISE 4.1. Verify that L^* is linear. Verify that $(LM)^* = M^*L^*$.

EXERCISE 4.2. Verify Example 4.2 and use it to show why the formula for T^* from Example 4.1 holds.

EXERCISE 4.3. Evaluate the following integral:

$$\int_0^1 \int_0^1 e^{-\max(x^2, y^2)} dx \, dy.$$

Hint: Break the unit square into two pieces.

EXERCISE 4.4. Prove the assertions in Example 4.4 and Corollary 4.2.

5. Hermitian and unitary operators

Adjoints lead to important classes of bounded operators. We develop the properties of these operators and prove the finite-dimensional spectral theorem in this section.

Definition 5.1. Assume $L : H \to H$ is linear and bounded.

- L is *unitary* if L is invertible and $L^* = L^{-1}$.
- L is *Hermitian* (or self-adjoint) if $L^* = L$.
- L is *positive semi-definite* if $\langle Lz, z \rangle \geq 0$ for all $z \in H$.
- L is *positive definite* if there is a positive constant C such that, for all $z \in H$,

$$\langle Lz, z \rangle \geq C \, ||z||^2. \tag{10}$$

- L is a projection if $L^2 = L$ and an orthogonal projection if $L^* = L = L^2$.

We can express the notions of unitary and Hermitian also in terms of the inner product; L is Hermitian if and only if $\langle Lu, v \rangle = \langle u, Lv \rangle$, for all u, v. We would like to say that L is unitary if and only if $\langle Lu, Lv \rangle = \langle u, v \rangle$, for all u, v, but we need an extra assumption. The correct statement is the following: If L is unitary, then $\langle Lu, Lv \rangle = \langle u, v \rangle$ for all u, v. If $\langle Lu, Lv \rangle = \langle u, v \rangle$ for all u, v, and L is **surjective**, then L is unitary. Equivalently, if L is invertible and preserves the inner product, then L is unitary.

EXAMPLE 5.1. Suppose $H = l^2$ is the collection of sequences $\{z_n\}$ for which $\sum_{n=0}^\infty |a_n|^2 < \infty$. Let L be the shift operator:

$$L(a_0, a_1, a_2, \dots) = (0, a_0, a_1, a_2, \dots).$$

Then $\langle Lu, Lv \rangle = \langle u, v \rangle$ for all u, v, but L is not surjective and hence not invertible. For example, $(1, 0, 0, \dots)$ is not in the range of L. This L preserves the inner product but L is not unitary.

Proposition 5.2. *Assume $L : H \to H$ is linear and bounded. Then $\langle Lu, Lv \rangle = \langle u, v \rangle$ for all u, v if and only if $||Lu||^2 = ||u||^2$ for all u.*

PROOF. One implication is immediate (put $v = u$). The other, called polarization, is not hard, but we omit the proof. See for example [D1]. □

We mention another characterization of unitary transformations which is useful in quantum mechanics.

Theorem 5.3. *A linear operator L on H is unitary if and only if, whenever $\{\phi_n\}$ is a complete orthonormal system in H, then $\{L(\phi_n)\}$ also is.*

PROOF. Suppose L is unitary and $\{\phi_n\}$ is a complete orthonormal system. Then

$$\langle L\phi_m, L\phi_n \rangle = \langle \phi_m, \phi_n \rangle = \delta_{mn}.$$

Hence $\{L\phi_n\}$ is an orthonormal system. To verify completeness, consider z with $\langle L(\phi_n), z \rangle = 0$ for all n. Then $\langle \phi_n, L^{-1}z \rangle = 0$ for all n. By the completeness of $\{\phi_n\}$, we get $L^{-1}(z) = 0$. Hence $z = 0$ as well.

Conversely, suppose L preserves complete orthonormal systems.

We first show L is surjective; in other words, given w we can find a v with $Lv = w$. Let $\{\phi_j\}$ be a complete orthonormal system. Then $\{L(\phi_j)\}$ also is. Then, for scalars c_j we have

$$w = \sum c_j L(\phi_j) = L(\sum c_j \phi_j) = L(v).$$

It remains to show that L preserves inner products. If $v = \sum c_j \phi_j$ and $u = \sum d_k \phi_k$, then

$$\langle v, u \rangle = \sum c_j \overline{d_j}.$$

By linearity and orthonormality we also have

$$\langle Lv, Lu \rangle = \langle \sum c_j \phi_j, \sum d_k \phi_k \rangle = \sum c_j \overline{d_j} = \langle v, u \rangle.$$

Thus L is unitary. $\square$

REMARK. Theorem 5.3 is useful even in finite dimensions. Consider the matrix of a linear map L on $\mathbb{C}^n$ with respect to the standard basis for $\mathbb{C}^n$. Theorem 5.3 implies that the matrix of L is unitary if and only if its column vectors are orthonormal.

Proposition 5.4. *Assume L is unitary and λ is a eigenvalue of L. Then $|\lambda| = 1$.*

PROOF. Assume $v \neq 0$ and $Lv = \lambda v$. Then

$$|\lambda|^2 \langle v, v \rangle = \langle \lambda v, \lambda v \rangle = \langle Lv, Lv \rangle = \langle L^* Lv, v \rangle = \langle v, v \rangle.$$

If $v \neq 0$, then $\langle v, v \rangle = ||v||^2 \neq 0$ and hence $|\lambda|^2 = 1$. $\square$

Proposition 5.5. *Assume $L : H \to H$ is Hermitian and λ is an eigenvalue of L. Then λ is real.*

PROOF. Again suppose that $Lv = \lambda v$ and $v \neq 0$. Then

$$\begin{aligned}
(\lambda - \overline{\lambda})||v||^2 &= \lambda \langle v, v \rangle - \overline{\lambda} \langle v, v \rangle \\
&= \langle \lambda v, v \rangle - \langle v, \lambda v \rangle \\
&= \langle Lv, v \rangle - \langle v, Lv \rangle \\
&= \langle Lv, v \rangle - \langle Lv, v \rangle \\
&= 0.
\end{aligned}$$

Since $||v||^2 \neq 0$, we conclude that $\lambda = \overline{\lambda}$. $\square$

Theorem 5.6. *Assume $L : H \to H$ is Hermitian and u, v are eigenvectors corresponding to distinct eigenvalues. Then u and v are orthogonal.*

PROOF. We are given $u, v \neq 0$ with $Lu = \lambda u$ and $Lv = \eta v$. Hence

$$\begin{aligned}
\lambda \langle u, v \rangle &= \langle \lambda u, v \rangle \\
&= \langle Lu, v \rangle \\
&= \langle u, L^* v \rangle \\
&= \langle u, Lv \rangle \\
&= \langle u, \eta v \rangle \\
&= \overline{\eta} \langle u, v \rangle \\
&= \eta \langle u, v \rangle.
\end{aligned}$$

We have used the reality of η and obtained $(\lambda - \eta)\langle u, v \rangle = 0$. Since $\lambda \neq \eta$, we must have $\langle u, v \rangle = 0$. $\square$

In some sense, Hermitian operators are analogous to real numbers and unitary operators are analogous to points on the unit circle. Figure 8 from Chapter 2 illustrates a linear fractional transformation f mapping the real line to the unit circle. This mapping is called the *Cayley transform*. If L is Hermitian, then $f(L)$ is unitary. Positive semi-definite operators correspond to non-negative real numbers and positive definite operators correspond to positive numbers. These vague correspondences become useful when one considers the spectrum of the operator. Exercise 11.2 offers a nice application of these vague ideas.

REMARK. Inequality (10) defines positive definiteness. In finite dimensions, (10) is equivalent to saying that $\langle Lz, z \rangle > 0$ for $z \neq 0$. In infinite dimensions, these two properties are not equivalent. Consider an infinite diagonal matrix with eigenvalues $\frac{1}{n}$ for each positive integer n. Then there is no positive C such that (10) holds.

These notions are fundamental even in the finite-dimensional setting. We next state and prove the finite-dimensional **spectral theorem**. We then discuss the **singular value decomposition** of a linear transformation whose domain and target are of different dimensions.

We work with matrices here; we can also formulate these theorems in terms of linear maps. A matrix is simply a way to express a linear map in terms of bases. See Section 8 of Chapter 1.

Theorem 5.7 (Spectral theorem, finite-dimensional case). *Let A be an n-by-n Hermitian matrix of complex numbers with eigenvalues λ_j. Let B be a diagonal n-by-n matrix with diagonal entries the eigenvalues of A. Then there is a unitary matrix U such that $A = UBU^* = UBU^{-1}$.*

PROOF. The proof is by induction on the dimension. The result in one dimension is obvious. Suppose it is known in all dimensions up to $n - 1$, and let A be an n-by-n Hermitian matrix. We can solve the characteristic equation $\det(A - \lambda I) = 0$ to find an eigenvalue λ and corresponding eigenspace E_λ. Then E_λ has an orthonormal basis (which consists of eigenvectors for A). If E_λ is the whole space, then we are done. Otherwise, consider the space $E_\lambda^\perp$ of vectors orthogonal to E_λ. We claim that A maps this space to itself. To verify the claim, suppose $\langle w, v \rangle = 0$ whenever $Av = \lambda v$. Then

$$\langle Aw, v \rangle = \langle w, Av \rangle = \lambda \langle w, v \rangle = 0.$$

Thus, if $w \perp E_\lambda$, then $Aw \perp E_\lambda$. The restriction of A to $E_\lambda^\perp$ is Hermitian. By the induction hypothesis, we can find an orthonormal basis of this space consisting of eigenvectors for the restriction of A to $E_\lambda^\perp$. Let U be the matrix whose columns are these two bases. By the remark after Theorem 5.3, a matrix whose columns are orthonormal must be unitary. Then U is unitary and diagonalizes A.   □

Corollary 5.8. *Let A be Hermitian on $\mathbb{C}^n$. Then there are real numbers λ_j (eigenvalues of A) and orthogonal projections π_j such that $\pi_j \pi_k = 0$ for $j \neq k$ and*

$$A = \sum_{j=1}^{k} \lambda_j \pi_j. \tag{11.1}$$

Furthermore, if f is a polynomial (or convergent power series), then

$$f(A) = \sum_{j=1}^{k} f(\lambda_j) \pi_j. \tag{11.2}$$

PROOF. Simply let π_j be the orthogonal projection onto the eigenspace E_{λ_j}. By Theorem 5.6 we must have $\pi_j \pi_k = 0$ for $j \neq k$, and (11.1) simply restates Theorem 5.7. The second statement follows by combining Theorem 5.7 with Lemma 10.1 in Chapter 1; in the calculation we use $\pi_j \pi_k = 0$ for $j \neq k$ and $\pi_j^2 = \pi_j$. See Exercises 5.12 and 5.13.   □

REMARK. Suppose A is an orthogonal projection. Then $I - A$ also is. If A and B are orthogonal projections, then $A + B$ need not be one. But, if $AB = BA = 0$, then

$$(A + B)^2 = A^2 + AB + BA + B^2 = A + B.$$

The proof of Corollary 5.8 uses this fact in a crucial way. See Exercises 5.12 and 5.13 for practice with projections.

REMARK. In Corollary 5.8, the sum of the projections π_j is the identity operator. This collection of orthogonal projections is called a **resolution of the identity**. In more advanced spectral theory, one writes the identity operator as an integral (rather than a sum) of orthogonal projections. See for example [RS].

Thus a Hermitian matrix can be unitarily diagonalized. As we noted in Chapter 1, the column vectors of U are the eigenvectors of A. The conclusions of Theorem 5.7 and Corollary 5.8 also hold for normal matrices; A is **normal** if $AA^* = A^*A$. The proofs are essentially the same. See, for example, [HK].

REMARK. In Theorem 10.2 of this chapter we state the spectral theorem for compact self-adjoint operators on a Hilbert space; its proof is similar to the proof in the finite-dimensional case, although we must work harder to obtain an eigenvalue to get started. Once we do, we again work on the space perpendicular to this eigenspace.

We wish to extend these ideas to linear transformations from $\mathbb{C}^n$ to $\mathbb{C}^m$ when $m \neq n$. Let A be a matrix with complex entries, and with n columns and m rows. The conjugate transpose matrix A^* then has m columns and n rows. The matrix A^*A is then square of size n-by-n. Furthermore A^*A is Hermitian and positive semi-definite:

$$(A^*A)^* = A^*(A^*)^* = A^*A,$$
$$\langle A^*Av, v \rangle = \langle Av, Av \rangle \geq 0.$$

Denote the eigenvalues of A^*A by λ_j. The **singular values** of A are the non-negative numbers $\sigma_j = \sqrt{\lambda_j}$. The following theorem, giving the **singular value decomposition** of A, then holds:

Theorem 5.9. *Let A be an m-by-n matrix of complex numbers with singular values σ_j for $1 \leq j \leq r$. Let B be the m-by-n matrix with $B_{jj} = \sigma_j$ if $1 \leq j \leq r$ and $B_{jk} = 0$ otherwise. Then there are an m-by-m unitary matrix V and an n-by-n unitary matrix U such that $A = UBV^*$.*

PROOF. One applies Theorem 5.7 to both A^*A and AA^*. We leave the proof as an exercise. □

Suppose $A = UBV^*$ for U and V unitary. We then have

$$A^*A = (UBV^*)^*(UBV^*) = (VB^*U^*)(UBV^*) = VB^*BV^*.$$

In particular, the eigenvalues of A^*A must be the squared absolute values of the eigenvalues of B. Note also that the columns of V are the eigenvectors of A^*A and the columns of U are the eigenvectors of AA^*.

We will see matrices of the form A^*A again in Chapter 7 when we discuss least squares regression. Let us now introduce a related useful measurement of the size of a matrix or linear transformation. We have already seen the operator norm $||A||$ in Definition 5.1. Example 4.4 anticipated another possible norm, called the *Frobenius norm*. In the finite-dimensional setting, both the operator norm and the Frobenius norm are always finite. In infinite dimensions, having a finite Frobenius (or Hilbert–Schmidt) norm is a strong restriction on an operator; this property fails even for the identity operator. We let $\text{tr}(L)$ denote the trace of L. Assume $L = A^*A$ for an operator A on a Hilbert space with complete orthonormal system $\{\phi_j\}$. We then define

$$\text{tr}(A^*A) = \sum_{j,k} |\langle A\phi_j, \phi_k \rangle|^2.$$

See also Proposition 5.11 below.

Definition 5.10. The **Frobenius norm** of a linear transformation A in finite dimensions is the non-negative number $||A||_2$ satisfying

$$||A||_2^2 = \mathrm{tr}(A^*A).$$

If A is an operator on Hilbert space, A is called **Hilbert–Schmidt** if $\mathrm{tr}(A^*A)$ is finite, and $\sqrt{\mathrm{tr}(A^*A)}$ is called the **Hilbert–Schmidt** norm of A.

The following simple results in the finite-dimensional case often arise in numerical linear algebra.

Proposition 5.11. *Let A be an m-by-n matrix of complex numbers. Then*

$$||A||_2^2 = \sum_{j=1}^{m} \sum_{k=1}^{n} |a_{jk}|^2.$$

Proposition 5.12. *Let A be an m-by-n matrix of complex numbers with singular values σ_j. Then*

$$\mathrm{tr}(A^*A) = ||A||_2^2 = \sum_j \sigma_j^2.$$

Proposition 5.13. *Let A, B be n-by-n matrices of complex numbers. Then*

$$||BA||_2 \leq ||B||_2 \, ||A||_2.$$

EXAMPLE 5.2. If A is finite-dimensional and Hermitian, then $||A||_2^2 = \sum |\lambda_j|^2$. If A is a compact Hermitian operator on a Hilbert space, then A is Hilbert–Schmidt if $\sum_{j=1}^{\infty} |\lambda_j|^2 < \infty$, and $||A||_2^2 = \sum |\lambda_j|^2$.

We close this section with a brief discussion of **conditioning**. See [St], for example, for additional information. Suppose first that L is an invertible linear map on a finite-dimensional real or complex vector space. When solving the linear equation $Lu = v$ numerically, one needs to bound the relative error in the solution in terms of the relative error in the right-hand side. Assume that u is an exact solution to $Lu = v$. Let δu and δv denote the errors in measuring u and v. These errors are vectors. Then

$$v + \delta v = L(u + \delta u) = Lu + L(\delta u).$$

Hence $\delta u = L^{-1}(\delta v)$. The ratio of errors is then given by

$$\frac{||L^{-1}(\delta v)||}{||\delta v||} \cdot \frac{||v||}{||u||} = \frac{||L^{-1}(\delta v)||}{||\delta v||} \cdot \frac{||Lu||}{||u||}.$$

Therefore the worst possible error is the product

$$\kappa(L) = ||L^{-1}|| \, ||L||$$

of the norms of L^{-1} and L. The number $\kappa(L)$ is the **condition number** of L with respect to the operator norm. In the finite-dimensional setting it is common to use the Frobenius norm instead of the operator norm; in this case the condition number is the ratio

$$\frac{\sigma_{\max}}{\sigma_{\min}}$$

of the largest and smallest singular values of L. When L is Hermitian (or, more generally, normal) this ratio is the same as

$$\frac{|\lambda|_{\max}}{|\lambda|_{\min}}.$$

Here we order the magnitudes of the eigenvalues.

When the condition number is large, the linear map (or its matrix with respect to some basis) is called **ill-conditioned**. Solving ill-conditioned linear equations numerically may lead to inaccurate results. One of the most famous examples of an ill-conditioned matrix is the Hilbert matrix.

EXAMPLE 5.3. For each positive integer N, the N-by-N Hilbert matrix H_N has entries $\frac{1}{j+k+1}$ for $0 \le j, k \le N - 1$. For example,

$$H_4 = \begin{pmatrix} 1 & \frac{1}{2} & \frac{1}{3} & \frac{1}{4} \\ \frac{1}{2} & \frac{1}{3} & \frac{1}{4} & \frac{1}{5} \\ \frac{1}{3} & \frac{1}{4} & \frac{1}{5} & \frac{1}{6} \\ \frac{1}{4} & \frac{1}{5} & \frac{1}{6} & \frac{1}{7} \end{pmatrix}.$$

Each H_N is obviously symmetric and hence Hermitian; it is also positive definite. Hilbert noted that the inverse of H_N has integer entries and found precise formulas for these entries and for the determinant of H_N. When $N = 4$, the eigenvalues of H_4 are approximately

$$1.50021, 0.169141, 0.00673827, 0.0000967023.$$

The determinant of H_4 is $\frac{1}{6048000}$. Even though H_N is invertible, it is ill-conditioned. The Wikipedia page, which seems reliable in this case, contains a considerable amount of interesting information. See also the exercises.

EXERCISE 5.1. Find the singular values of A if

$$A = \begin{pmatrix} -1 & 1 & 1 \\ -1 & 1 & 1 \end{pmatrix}.$$

EXERCISE 5.2. Find the singular values of A if

$$A = \begin{pmatrix} 0 & \frac{3}{\sqrt{2}} & \frac{-3}{\sqrt{2}} \\ -2 & 0 & 0 \end{pmatrix}.$$

Find the U and V guaranteed by Theorem 5.9.

EXERCISE 5.3. Use Theorem 5.7 to prove Theorem 5.9.

EXERCISE 5.4. Prove Propositions 5.11, 5.12, and 5.13.

EXERCISE 5.5. For $\epsilon > 0$, define L by

$$L = \begin{pmatrix} 1 & 1 \\ 1 & 1 + \epsilon \end{pmatrix}.$$

Show that L is Hermitian and positive definite. Find its condition number.

EXERCISE 5.6. When does the condition number of a linear map equal 1?

EXERCISE 5.7. Prove that the Hilbert matrix H_N is positive definite. Suggestion: use

$$\frac{1}{j + k + 1} = \int_0^1 x^{j+k} dx.$$

EXERCISE 5.8. Using row and column operations, prove that

$$\det(H_n) = \frac{((n-1)!)^2}{(2n-1)!} \det(H_{n-1}). \qquad (*)$$

See the online article [Fos]. It follows from (*) that

$$\det(H_n) = \frac{((n-1)!!)^4}{(2n-1)!!}.$$

Here $k!!$ is defined to be $\prod_{j=1}^k j!$.

EXERCISE 5.9 (Hilbert's inequality). Prove the following famous inequality:

$$\sum_{j,k=0}^{\infty} \frac{z_j \overline{z}_k}{j+k+1} \leq \pi \sum_{j=0}^{\infty} |z_j|^2.$$

Suggestion. First prove the following general statement. Let g be continuous on $[0, 2\pi]$ and let $\hat{g}(n)$ denote its n-th Fourier coefficient. Use the Cauchy–Schwarz inequality and Parseval's formula to show that

$$\sum_{j,k=0}^{\infty} \hat{g}(j+k)z_j \overline{z}_k \leq \sup(|g|) \sum_{j=0}^{\infty} |z_j|^2.$$

Then apply this statement with $g(t) = i(\pi - t)e^{-it}$ on the interval $[0, 2\pi]$. See [D3] and its references for additional proofs.

EXERCISE 5.10. Let g be non-negative and continuous on $[0, 2\pi]$ with Fourier coefficients $\hat{g}(n)$. Show that the infinite matrix with entries $\hat{g}(j-k)$ is non-negative definite. A converse result is known as Herglotz's theorem.

EXERCISE 5.11. For $\theta \in \mathbb{C}$, consider the matrix

$$A = \begin{pmatrix} \cos(\theta) & -\sin(\theta) \\ \sin(\theta) & \cos(\theta). \end{pmatrix}$$

Find its eigenvalues and diagonalize it. Also show, if $\theta \in \mathbb{R}$, then A is unitary.

EXERCISE 5.12. Let A be linear. Prove that $A^2 = A$ if and only if $(I - A)^2 = I - A$. Give an example of a projection ($A^2 = A$) that is not orthogonal (that is, $A^* \neq A$).

EXERCISE 5.13. Fill in the details that (11.1) implies (11.2).

EXERCISE 5.14. Consider the polynomials p_j from Theorem 8.2 of Chapter 1. For A as in the spectral theorem, find $p_j(A)$.

6. L^2 spaces

Some of the most important Hilbert spaces are L^2 spaces. One often thinks of $L^2([a, b])$ as the square-integrable functions on the interval $[a, b]$, but doing so is a bit imprecise. In this section we develop some of the analysis needed to make the definition of $L^2([a, b])$ more precise. See [F1] for an excellent treatment of real analysis. See [CV] for an accessible treatment of measure theory at the advanced undergraduate level.

The first required concept is the notion of **measure zero**. Such sets are negligible as far as integration is concerned. A subset S of the real line has **measure zero** if, given any $\epsilon > 0$, we can cover S by a countable union of intervals I_j such that the sum of the lengths of these intervals is at most ϵ. Examples of sets of measure zero include finite sets, countable sets, and uncountable sets called *Cantor sets*. See Example 6.3 for the standard example of a Cantor set. A countable union of sets of measure zero also has measure zero.

When f and g are functions which agree except on a set of measure zero, one usually says that f and g agree **almost everywhere**.

Let us consider some related analytical issues. Let H be an inner product space with inner product $\langle , \rangle$ and squared norm $||v||^2 = \langle v, v \rangle$. Recall that H is complete if each Cauchy sequence in H has a limit, and that a Hilbert space is a complete inner product space.

A subset S of H is called *closed* if, whenever $w_n \in S$ and w_n converges to w in H, then $w \in S$. If $H = \mathbb{C}^n$, then every vector subspace is closed. For Hilbert spaces of infinite dimension, subspaces need not be closed. See Example 6.2. We recall the notion of *dense* set. A subset $S \subset \mathbb{R}$ is called **dense** if, for each $x \in \mathbb{R}$, there is a sequence $\{x_n\} \in S$ converging to x.

EXAMPLE 6.1. The set of rational numbers $\mathbb{Q}$ is dense in $\mathbb{R}$.

Similarly, a subset $\Omega \subset H$ is **dense** if for each $v \in H$, there is a sequence $\{u_n\} \in \Omega$ converging to v. A dense subspace is a dense subset that is also a vector subspace.

EXAMPLE 6.2. Let $C[a,b]$ denote the vector space of continuous functions from the closed interval $[a,b] \to \mathbb{C}$. If f is continuous on $[a,b]$, then (by a basic theorem in real analysis) f is bounded. Thus there exists an M such that $|f(x)| \leq M$ for any $x \in [a,b]$. Hence

$$\int_a^b |f(x)|^2 dx \leq \int_a^b M^2 dx < \infty.$$

Therefore $C[a,b]$ is contained in $L^2([a,b])$. Although it is a subspace, $C[a,b]$ is not closed in $L^2([a,b])$. For simplicity we consider the interval $[0,2]$.

For $0 \leq x \leq 1$, put $f_n(x) = x^n$. For $1 \leq x \leq 2$, put $f_n(x) = 1$. Then $f_n \in C[0,2]$ for all $n \in \mathbb{N}$, and

$$\lim_{n \to \infty} f_n(x) = f(x) = \begin{cases} 0 & \text{if } 0 \leq x < 1 \\ 1 & \text{if } 1 \leq x \leq 2. \end{cases}$$

Thus the limit function is not continuous. Nor does it agree *almost everywhere* with any continuous function. We claim, however, that $\{f_n\}$ converges to f in $L^2([0,2])$. To verify this conclusion, given any $\epsilon > 0$, we must find N such that $n \geq N$ implies $||f_n - f|| < \epsilon$. The function $f_n - f$ is 0 on $[1,2]$, and equals x^n on $[0,1]$. We therefore have

$$||f_n - f||^2 = \int_0^1 x^{2n} dx = \frac{1}{2n+1}.$$

Thus $||f_n - f|| < \epsilon$ if $2n + 1 > \frac{1}{\epsilon^2}$.

This example forces us to confront the precise definition of $L^2([a,b])$.

Definition 6.1. We define

$$L^2([a,b]) = \{\text{equivalence classes of functions } f : [a,b] \to \mathbb{C}, \text{ such that } \int_a^b |f(x)|^2 dx < \infty\}.$$

Here f and g are equivalent if $f(x) = g(x)$ except for x in a set of measure zero.

One of the crowning achievements of the Lebesgue theory of integration is the completeness of L^2, a result whose proof we cannot give here.

Theorem 6.2. $L^2([a,b])$ *is complete.*

REMARK. Things remain rigorous if the reader regards $L^2([a,b])$ as the completion of the space of continuous functions in the L^2 norm. Something is lost, however. Objects in the completion are actually equivalence classes of functions, but this fact requires proof. The same issue applies when we define the real numbers in terms of the rational numbers. If we define a real number to be an equivalence class of Cauchy sequences of rational numbers, then we must prove that a real number has a decimal expansion.

For fun, we define the well-known Cantor set. See [CV], for example, for more information.

EXAMPLE 6.3. Let C be the set of numbers x in $[0,1]$ for which the base three expansion of x contains no two. Then C is uncountable, but nonetheless of measure zero.

Next we mention three different notions of convergence for functions on $[a,b]$.

Definition 6.3. Let $\{f_n\}$ be a sequence of functions from $[a,b]$ to $\mathbb{C}$.

- $f_n \to f$ *pointwise*, if for all x, $f_n(x) \to f(x)$. In other words, for all x, and for all $\epsilon > 0$, there is an N such that $n \geq N$ implies $|f_n(x) - f(x)| < \epsilon$.
- $f_n \to f$ *uniformly*, if for each $\epsilon > 0$, there is an N such that $n \geq N$ implies $|f_n(x) - f(x)| < \epsilon$ for all x. In other words, given ϵ, there is an N that works for all x.
- $f_n \to f$ in L^2, if $||f_n - f|| \to 0$.

The only difference between the definitions of pointwise and uniform convergence is the location of the quantifier "for all x". But the concepts are quite different!

A basic theorem of real analysis states that a uniform limit of continuous functions is itself continuous. In Example 6.2, f_n converges to f both pointwise and in L^2. Since f is not continuous, the convergence is not uniform. Example 6.2 also shows that $C[0, 2]$ is a subspace of $L^2([0, 2])$, but it is not closed in $L^2([0, 2])$. It is, however, dense. Working with closed subspaces is not always possible.

We have defined the Fourier transform of a function in $L^1(\mathbb{R})$, and hence also for a function in the intersection $L^1 \cap L^2$. This space is dense in L^2. There are several ways to extend the definition of the Fourier transform to all of L^2. The following famous theorem is the analogue of the Parseval relation for Fourier series. It implies that $||\hat{f}||_{L^2}^2 = ||f||_{L^2}^2$. See [F1] and [T] for proofs and additional discussion.

Theorem 6.4 (Plancherel). *The Fourier transform extends to be a unitary operator on $L^2(\mathbb{R})$.*

REMARK. In the proof of the Fourier inversion formula, we noted that $\mathcal{F}^*\mathcal{F} = I$, suggesting that $\mathcal{F}$ is unitary. One needs to do a bit more, however, because we defined $\mathcal{F}$ on only a dense subspace of L^2.

EXERCISE 6.1. Suppose that f_n is a sequence of continuous functions on an interval E, and that there are constants M_n such that $|f_n(x)| \leq M_n$ for all x in E. Assume that $\sum_n M_n$ converges. Show that $\sum_n f_n$ converges to a continuous function on E.

EXERCISE 6.2. Use the previous exercise to show that

$$\sum_{n=1}^{\infty} \frac{\sin(nx)}{n^\alpha}$$

converges to a continuous function when $\alpha > 1$.

EXERCISE 6.3. Show that the series

$$\sum_{n=0}^{\infty} e^{-nx}$$

converges for $x > 0$. Find an explicit formula for the sum. Show that the limit function is not continuous at 0. Show that the series converges uniformly on any interval $[a, \infty]$ for $a > 0$. Show that it does not converge uniformly on $(0, \infty)$. See Example 2.3 of Chapter 5.

EXERCISE 6.4. Restate Exercise 4.7 from Chapter 5 using Plancherel's theorem.

EXERCISE 6.5. Use Theorem 6.4 and Example 4.1 of Chapter 5 to evaluate $\int_{-\infty}^{\infty} \frac{\sin^2(x)}{x^2} dx$. This example epitomizes the two ways of computing the total signal energy.

EXERCISE 6.6. Suppose that f_n is continuous on $[a, b]$ and f_n converges uniformly to f. Show that f is continuous. Show furthermore that

$$\lim_{n \to \infty} \int_a^b f_n(x) dx = \int_a^b f(x) dx.$$

Compare with Exercise 4.9 of Chapter 5.

7. Sturm–Liouville equations

Sturm–Liouville theory concerns certain second-order ODE with homogeneous boundary conditions. The ideas lead us to Hilbert spaces and orthonormal expansion, and they generalize Fourier series.

Let $[a,b]$ be a closed interval on $\mathbb{R}$. The Sturm–Liouville problem is to solve the following ODE:

$$(py')' + qy + \lambda wy = 0, \qquad (SL)$$

with given boundary conditions:

$$\alpha_1 y(a) + \alpha_2 y'(a) = 0, \qquad (SL1)$$
$$\beta_1 y(b) + \beta_2 y'(b) = 0. \qquad (SL2)$$

We assume that both equations (SL1) and (SL2) are non-trivial. In other words, at least one of the α_j is not zero, and at least one of the β_j is not zero. We also assume that the functions p, q, w are real-valued, and that p, w are non-negative on the interval $[a, b]$. We write

$$L^2([a,b], w) = \{f : \int_a^b |f(x)|^2 w(x)dx < \infty\}$$

for the square-integrable functions with respect to the weight function w. As in Section 6, to be precise, an element of L^2 is an equivalence class of functions. The inner product here is given by

$$\langle f, g \rangle_w = \int_a^b f(x)\overline{g(x)}w(x)dx.$$

The Sturm–Liouville problem is equivalent to the eigenvalue problem $Ly = \lambda y$ with boundary conditions (SL1) and (SL2). Here the operator L can be written as

$$L = \frac{-1}{w}\left(\frac{\partial}{\partial x}p\frac{\partial}{\partial x} + q\right). \qquad (12)$$

The operator L in (12) is said to be in *divergence form*.

The equation (SL) is more general than it first seems, as we can put any second-order ODE in divergence form. We first illustrate with the Bessel equation.

EXAMPLE 7.1. Consider the Bessel equation:

$$x^2 y'' + xy' + (x^2 - v^2)y = 0 \qquad (13)$$

Although (13) is not in divergence form, it can easily be put in that form, using an *integrating factor*. We multiply both sides by u, where $u(x) = \frac{1}{x}$. For $x \neq 0$, the equation is the same as

$$xy'' + y' + (x - \frac{v^2}{x})y = 0$$

or, equivalently, the Sturm–Liouville form

$$(xy')' + (x - \frac{v^2}{x})y = 0.$$

The integrating factor technique works in general. Given $Py'' + Qy' + Ry = 0$, we can always put it in the form $(py')' + qy = 0$. To find u, note that it must satisfy

$$uPy'' + uQy' - uRy = 0.$$

Therefore $uP = p$, and $uQ = p'$. Hence

$$\frac{p'}{p} = \frac{Q}{P}$$

$$\log(p) = \int \frac{Q}{P}.$$

Then $p = e^{\int \frac{Q}{P}}$ and we obtain

$$u = \frac{p}{P} = \frac{e^{\int \frac{Q}{P}}}{P}.$$

The second-order operator L is not defined on all elements of $L^2([a, b], w)$. There will be a natural domain $D(L)$ on which L is defined and

$$L : D(L) \subseteq L^2([a, b], w) \to L^2([a, b], w).$$

Although L is linear, it is a differential operator and is unbounded. With the homogeneous conditions, however, L turns out to be self-adjoint and there exists a complete orthonormal system of eigenfunctions. The resulting orthonormal expansions generalize Fourier series such as

$$f(x) = \sum_{n=1}^{\infty} c_n \sin(nx), \text{ on } [0, 2\pi].$$

The following crucial theorem summarizes the situation. Its proof is somewhat advanced. We provide most of the details in Sections 8 and 10.

Theorem 7.1. *Let L be the Sturm–Liouville operator (SL), with boundary conditions (SL1) and (SL2):*

(1) *There is a complete orthonormal system for $L^2([a, b], w)$ consisting of of eigenfunctions of L.*
(2) *All the eigenvalues are real and occur with multiplicity one.*
(3) *The eigenfunctions (vectors) ϕ_n corresponding to distinct eigenvalues are orthogonal.*
(4) *If f is a smooth function on $[a, b]$, then*

$$f(x) = \sum_{n=1}^{\infty} \langle f, \phi_n \rangle_w \phi_n(x).$$

We would like to apply the Hilbert space theory we have developed to differential operators, but the following essential difficulty must be overcome.

Lemma 7.2. *The operator $\frac{d}{dx}$ is unbounded on $L^2(S)$ for each non-trivial interval S.*

PROOF. Without loss of generality, suppose $S = [0, 1]$. Put $L = \frac{d}{dx}$. Then L is not defined everywhere. Assume L is defined on $D(L) = \{$differentiable functions$\}$. Note that $L(e^{inx}) = ine^{inx}$. Hence $\|L(e^{inx})\| = n\|e^{inx}\| = n$. Hence there is no C such that $\|L(f)\| \le C\|f\|$ for all f in $D(L)$. $\qquad \square$

We pause to discuss unbounded operators.

Definition 7.3. Let H be a Hilbert space. L is called an *unbounded operator* on H if there is a dense subspace $D(L) \subseteq H$ such that L is linear on $D(L)$, but there is no linear map agreeing with L on all of H.

Since $D(L)$ is dense, each $f \in H$ is the limit of a sequence f_n in $D(L)$. If L is not continuous, then f_n converges but $L(f_n)$ need not converge.

EXAMPLE 7.2. Let V denote the set of continuously differentiable functions $f : \mathbb{R} \to \mathbb{C}$, such that

$$\int_{-\infty}^{\infty} |f(x)|^2 dx < \infty.$$

Then V is a dense subspace of $L^2(\mathbb{R})$. Note that V is not the same space as $C^1(\mathbb{R})$, because not all functions in $C^1(\mathbb{R})$ are square integrable.

Let $D(L)$ be a dense subspace of H, and suppose $L : D(L) \to D(L)$ is linear. Even if L is unbounded, we can still define its adjoint. As before, the functional $u \to \langle Lu, v \rangle$ make sense for $u \in D(L)$. Let

$$D(L^*) = \{ v \in H : u \to \langle Lu, v \rangle \text{ is bounded} \}.$$

For $u \in D(L)$ and $v \in D(L^*)$, we have $\langle Lu, v \rangle = \langle u, L^* v \rangle$.

Definition 7.4. An unbounded operator L is called self-adjoint if $D(L) = D(L^*)$ and $L = L^*$ on $D(L)$.

EXAMPLE 7.3. Consider $H = L^2([0,1])$, and $D(L) = \{\text{continuously differentiable functions in } H\}$, and $L = i\frac{d}{dx}$. We try to compute L^* by using $\langle f', g \rangle = \langle Lf, g \rangle = \langle f, L^* g \rangle$. Using integration by parts, and since $\bar{i} = -i$, we get

$$\langle if', g \rangle = \int_0^1 i f'(x) \overline{g(x)} dx = i f(x) \overline{g(x)}\big|_0^1 + \int_0^1 \overline{ig'(x)} f(x) dx.$$

We need to make the boundary terms drop out. There are several ways to do so. One way is to put $f(0) = f(1) = 0$, with no condition on g. Another way is to put $g(0) = g(1) = 0$ with no condition on f. The symmetric way to make the boundary terms drop out is to assume that f and g are periodic.

We consider two possible definitions of $D(L)$. First let $D(L)$ be the set of continuously differentiable functions vanishing at 0 and 1. If $f \in D(L)$, then $f(0) = f(1) = 0$, and we need no boundary conditions on g. In this case $D(L^*)$ contains the set of continuously differentiable functions, there are no boundary conditions, and $D(L^*)$ is larger than $D(L)$.

For L^* to equal L as operators requires $D(L) = D(L^*)$. The natural definition of $D(L)$ assumes the periodic condition that $f(0) = f(1)$. One also assumes that f is *absolutely continuous*, the condition guaranteeing that f is the integral of its derivative. Thus the operator $i\frac{d}{dx}$ is self-adjoint with this choice of domain.

Let us return to the Sturm–Liouville setting. Put $L = -\frac{1}{w}(\frac{d}{dx}p\frac{d}{dx} + q)$ on $L^2([a,b], w)$, where $w > 0$, $p > 0$ and p is continuously differentiable. Assume the homogeneous boundary conditions. It will then follow that $L^* = L$.

REMARK. Boundary value problems behave differently from initial value problems. Two **boundary** conditions on a second-order ODE do not in general determine the solution uniquely, although two **initial** conditions will.

EXAMPLE 7.4. Consider the eigenvalue problem $y'' + y = -\lambda y$ with $y(0) = y(1) = 0$. We rewrite the equation as $y'' + (1 + \lambda)y = 0$. Assume that $\lambda \geq -1$. We analyze the possibilities.

If $\lambda = -1$, then $y'' = 0$, and thus $y = cx + b$. The conditions $y(0) = y(1) = 0$ imply that $y = 0$.

If $\lambda > -1$, then

$$y(x) = A\cos(\sqrt{1+\lambda}\, x) + B\sin(\sqrt{1+\lambda}\, x).$$

The condition $y(0) = 0$ implies that $A = 0$; therefore $y(x) = B\sin(\sqrt{1+\lambda}\, x)$. If also $y(1) = 0$, then $B\sin(\sqrt{1+\lambda}) = 0$.

If $B \neq 0$, then $\sqrt{1+\lambda} = n\pi$ for some integer n, and $\lambda = n^2\pi^2 - 1$. These numbers are the eigenvalues of this problem. Thus $y(x) = B_n \sin(n\pi x)$ is a solution for each $n \in \mathbb{N}$. By superposition,

$$y(x) = \sum_{n=1}^{\infty} B_n \sin(n\pi x)$$

also is a solution, assuming that the series converges to a twice differentiable function. If $f(0) = f(2\pi) = 0$, and f is continuously differentiable, then f has a Fourier sine series expansion. The functions $\sin(nx)$ form a complete orthonormal system on this subspace of $L^2([0, 2\pi])$. This situation gives the quintessential example of Theorem 7.1.

Theorem 7.5. *The Sturm–Liouville operator $L = -\frac{1}{w}(\frac{d}{dx}p\frac{d}{dx} + q)$ with homogeneous boundary conditions is Hermitian.*

PROOF. We need to show that $L = L^*$; in other words, that

$$\langle Lf, g \rangle = \langle f, L^*g \rangle = \langle f, Lg \rangle$$

on an appropriate dense subspace of $L^2([0, 1], w)$. Using the definition of L we get

$$\langle Lf, g \rangle = \int_a^b -\frac{1}{w}((pf')' + qf)\overline{g}w$$

$$= -\int_a^b ((pf')' + qf)\overline{g}$$

$$= -(pf')\overline{g}|_a^b + \int_a^b pf'\overline{g}' + \int_a^b qf\overline{g}$$

$$= -(pf')\overline{g}|_a^b + fp\overline{g}'|_a^b - \int_a^b (f(p\overline{g}')' + \int_a^b qf\overline{g}.$$

We claim that the homogeneous boundary conditions for both f and g force the boundary terms to be 0. The claim yields

$$\int_a^b f(\overline{Lg})w = \langle f, Lg \rangle$$

and hence $L = L^*$. To verify the claim, note that

$$-(pf')\overline{g}|_a^b + fp\overline{g}'|_a^b = f(b)\overline{g'(b)}p(b) - f(a)\overline{g'(a)}p(a) - f'(b)\overline{g(b)}p(b) + p(a)f'(a)\overline{g(a)}.$$

Assume $\alpha_1 f(a) + \alpha_2 f'(a) = 0$ and $\beta_1 f(b) + \beta_2 f'(b) = 0$. If $\alpha_2 \neq 0$, then $f'(a) = -\frac{\alpha_1}{\alpha_2}f(a)$. If $\beta_2 \neq 0$, then $f'(b) = -\frac{\beta_1}{\beta_2}f(b)$ and similarly with g. Substituting shows that the boundary terms vanish. In case either $\alpha_2 = 0$ or $\beta_2 = 0$, then things simplify and give the needed result easily. $\square$

EXERCISE 7.1. For n a positive number, put $\phi_n(x) = \sin(n\pi x)$. Use complex exponentials to show that the ϕ_n are orthogonal in $L^2([0, 1])$.

EXERCISE 7.2. Put the ODE $xy'' + \alpha y' + \lambda xy = 0$ into divergence form.

EXERCISE 7.3. Consider $L = \frac{d^2}{dx^2}$ on functions on $[0, 1]$ with boundary condition $y(0) = y'(0) = 0$. Find the boundary conditions associated with L^*.

EXERCISE 7.4 (Sines and signs). For each choice of sign, consider the differential equation $y'' = \pm k^2 y$ with boundary conditions $y(0) = y(l) = 0$. With the plus sign, show that the only solution is $y = 0$. With the minus sign, show for certain k that there are infinitely many linearly independent solutions. Determine the condition on k for this situation to arise.

EXERCISE 7.5. For real numbers a, b, compute the following limit:

$$\lim_{l \to \infty} \left(\frac{2}{l} \int_0^l \sin(ax) \, \sin(bx) \, dx \right).$$

(Make sure you treat the case $a = b$ separately.)

8. The Green's function

The Green's function $G(x,t)$ gives a somewhat explicit formula for solving the Sturm–Liouville problem. We will find an integral kernel G such that the operator T defined by

$$Tf(x) = \int_a^b G(x,t)f(t)dt$$

inverts L.

We begin by discussing one of the main uses of orthonormal expansion, at first in the finite-dimensional setting. Let ϕ_j be an orthonormal basis for $\mathbb{C}^n$ and suppose each ϕ_j is an eigenvector for a linear map L. Then we can easily solve the equation $Lz = b$. To do so, put $b = \sum_{j=1}^n \langle b, \phi_j \rangle \phi_j$ and $z = \sum_{j=1}^n \langle z, \phi_j \rangle \phi_j$.

We obtain

$$Lz = \sum_{j=1}^n \langle z, \phi_j \rangle L\phi_j = \sum_{j=1}^n \langle z, \phi_j \rangle \lambda_j \phi_j$$

and hence $\langle b, \phi_j \rangle = \lambda_j \langle z, \phi_j \rangle$ for each j. If $\lambda_j = 0$, we must have $\langle b, \phi_j \rangle = 0$. In other words, for b to be in the range of L, it must be orthogonal to the zero eigenspace. This result is known as the *Fredholm alternative*; see Theorem 11.1. If $\lambda_j \neq 0$, then we must have $\langle z, \phi_j \rangle = \frac{1}{\lambda_j} \langle b, \phi_j \rangle$.

Assume $\{\phi_j\}$ is a complete orthonormal system of eigenvectors for an operator L on a Hilbert space H. When L is a Sturm–Liouville operator, the Green's function $G(x,t)$ solves $Ly = f$ by putting $y(x) = \int_a^b G(x,t)f(t)dt$. We illustrate with the standard example.

EXAMPLE 8.1. Put $L = -\frac{d^2}{dx^2}$ with boundary values $y(0) = 0$ and $y(1) = 0$. We solve $Ly = f$ by finding the Green's function. Assume $y'' = -f$. Then $y' = -\int_0^x f(t)dt + c_1$ and hence

$$y(x) = -\int_0^x \int_0^s f(t)dt\,ds + c_1 x + c_2.$$

The boundary condition $y(0) = 0$ implies $c_2 = 0$. We use the boundary condition $y(1) = 0$ below. First we define the Green's function as follows. Put $G(x,t) = x(1-t)$ if $x \leq t$ and $G(x,t) = t(1-x)$ if $t \leq x$. We show that G does the job. Put $y(x) = \int_0^1 G(x,t)f(t)dt$. Then

$$y(t) = \int_0^1 G(x,t)f(t)dt = (1-x)\int_0^x tf(t)dt + x\int_x^1 (1-t)f(t)dt.$$

Using the fundamental theorem of calculus and the boundary conditions we get

$$y'(x) = -\int_0^x tf(t)dt + (1-x)xf(x) + \int_x^1 (1-t)f(t)dt - x(1-x)f(x) = -\int_0^x tf(t)dt + \int_x^1 (1-t)f(t)dt.$$

Applying the fundamental theorem of calculus again gives

$$y''(x) = -xf(x) - (1-x)f(x) = -f(x).$$

Thus $-y'' = f(x)$.

We summarize the method used to construct the Green's function for the second derivative operator.

- The operator is $L = -\frac{d^2}{dx^2}$ and the boundary conditions are $y(0) = y(1) = 0$.
- We solve $Lu = 0$ with the single boundary condition $u(0) = 0$. We obtain $u(x) = c_1 x$ and $u(0) = 0$. Warning: Here u does **not** equal the Heaviside function.
- We solve $Lv = 0$ with the single boundary condition $v(1) = 0$. We obtain $v(x) = c_2(1-x)$, and $v(1) = 0$.

- Then we merge these two solutions into a function G by putting $G(x,t) = u(x)v(t)$ for $x < t$ and $G(x,t) = u(t)v(x)$ for $x > t$. We want G to be continuous when $x = t$. We obtain

$$u(x) = x(1-t)$$

$$v(x) = t(1-x).$$

We determine the Green's function more generally by a similar method. We first need several results about the Wronskian from ODE. For differentiable functions u, v, we put

$$W(u,v)(x) = u(x)v'(x) - u'(x)v(x).$$

Proposition 8.1. *Let u, v be solutions to $Ly = y'' + qy' + ry = 0$. Then u, v are linearly independent if and only $W(u,v)(x) \neq 0$ for all x. They are linearly dependent if and only if $W(u,v)(x) = 0$ for some x.*

PROOF. Suppose $W(u,v)(x_0) = 0$ for some x_0. Then

$$\det \begin{pmatrix} u(x_0) & v(x_0) \\ u'(x_0) & v'(x_0) \end{pmatrix} = 0$$

and hence there are constants c_1, c_2 such that

$$\begin{pmatrix} u(x_0) & v(x_0) \\ u'(x_0) & v'(x_0) \end{pmatrix} \begin{pmatrix} c_1 \\ c_2 \end{pmatrix} = 0.$$

Equivalently we have the pair of equations

$$c_1 u(x_0) + c_2 v(x_0) = 0$$

$$c_1 u'(x_0) + c_2 v(x_0) = 0.$$

The function $g = c_1 u + c_2 v$ vanishes at x_0 and its derivative vanishes there. Since u, v are solutions and L is linear, $L(g) = 0$ as well. Hence $g''(x_0) = 0$. It follows that g is identically 0. Thus $(c_1 u + c_2 v)(x) = 0$ for all x and hence $W(u,v)(x_0) = 0$, Therefore, there are c_1, c_2 such that $(c_1 u + c_2 v)(x) = 0$, for all x. Hence u, v are linearly dependent.

Conversely, assume $W(u,v)(x) \neq 0$ for all x. We want to show that u, v are independent. Assume $c_1 u + c_2 v = 0$. Differentiating we get

$$c_1 u(x) + c_2 v(x) = 0, \text{ for all } x$$

$$c_1 u'(x) + c_2 v'(x) = 0 \text{ for all } x$$

which (in matrix form) gives $\begin{pmatrix} u(x) & v(x) \\ u'(x) & v'(x) \end{pmatrix} \begin{pmatrix} c_1 \\ c_2 \end{pmatrix} = 0$ for all x. Since the determinant is not 0, the system has only the trivial solution $(c_1, c_2) = (0,0)$. Thus u, v are linearly independent. $\square$

Lemma 8.2. *If f, g satisfy the same Sturm–Liouville problem, including the boundary conditions, then f and g are linearly dependent.*

PROOF. We first write the boundary conditions as a pair of matrix equations:

$$\begin{pmatrix} f(a) & f'(a) \\ g(a) & g'(a) \end{pmatrix} \begin{pmatrix} \alpha_1 \\ \alpha_2 \end{pmatrix} = 0$$

$$\begin{pmatrix} f(b) & f'(b) \\ g(b) & g'(b) \end{pmatrix} \begin{pmatrix} \beta_1 \\ \beta_2 \end{pmatrix} = 0.$$

Each of these matrix equations has a non-trivial solution; hence each determinant vanishes. But the determinants are the Wronskians at a and b; thus $W(f,g)(a) = 0$ and $W(f,g)(b) = 0$. The result now follows from Proposition 8.1. $\qquad\square$

Lemma 8.3. *Let* $L = -\frac{1}{w}(\frac{d}{dx}p\frac{d}{dx} + q)$ *be a Sturm–Liouville operator. Let* $W = W(u,v)$ *be the Wronskian. If* u, v *are linearly independent, and* $Lu = Lv = 0$, *then* $x \mapsto p(x)W(u,v)(x)$ *is constant. Furthermore, if* u *and* v *are linearly independent, then this constant is not zero.*

PROOF. We show that the derivative of this expression is 0.

$$\begin{aligned} \frac{d}{dx}(p(x)W(u,v)(x)) &= \frac{d}{dx}(p(x)u(x)v'(x) - p(x)u'(x)v(x)) \\ &= p'uv' + pu'v' + puv'' - p'u'v - pu''v - pu'v' \\ &= 0. \end{aligned}$$

The second statement follows from Proposition 8.1, since p is not zero. $\qquad\square$

We can now construct the Green's function. Let u be a solution to $Lu = 0$ with boundary condition (SL1) and let v be a solution to $Lv = 0$ with boundary condition (SL2). Let c be the constant $p(x)W(u,v)(x)$. Assume that u and v are linearly independent.

Define the Green's function by

$$G(x,t) = \begin{cases} \frac{1}{c}u(x)v(t), x < t \\ \frac{1}{c}v(x)u(t), x > t \end{cases} . \tag{14}$$

We summarize what we have done in the following theorem:

Theorem 8.4. *Consider the Sturm–Liouville equation* $Ly = f$. *Define* G *by (14) and define* y *by* $y(x) = \int_a^b G(x,t)f(t)dt$. *Then* y *solves the equation* $Ly = f$.

We apply the Green's function to sum an infinite series. The Riemann Zeta function $\zeta(s)$ is defined by the series

$$\zeta(s) = \sum_{n=1}^{\infty} \frac{1}{n^s},$$

which converges if $Re(s) > 1$. One can extend the definition using analytic continuation and a functional equation. The zeta function arises in number theory, physics, probability, applied statistics, and linguistics. The most famous open problem in mathematics, called the *Riemann hypothesis*, concerns the location of the zeroes of (the extended version of) the zeta function. The zeta function is closely connected with the distribution of prime numbers.

It is possible to compute values of $\zeta(2n)$ for positive integers n using Fourier series. For example,

$$\zeta(2) = \sum_{n=1}^{\infty} \frac{1}{n^2} = \frac{\pi^2}{6}.$$

This computation can be done using the Fourier series for $\pi - x$ on an interval.

EXAMPLE 8.2. We compute $\zeta(2)$ using the Green's function. Let

$$G(x,t) = \begin{cases} x(1-t), & x < t \\ t(1-x), & x > t \end{cases} .$$

This G is the Green's function for minus the second derivative operator. Then

$$\int G(x,x)dx = \int_0^1 x(1-x)dx = \frac{1}{6}.$$

What does this integral have to do with $\zeta(2)$? The idea is that we are finding the trace of the infinite-dimensional operator L^{-1}. For finite-dimensional matrices, the trace of (G_{ab}) is defined to be $\sum G_{aa}$. The trace also equals the sum of the eigenvalues.

Hence we compute the eigenvalues for the second derivative. Consider $y'' = 0$ with boundary conditions $y(0) = y(1) = 0$. The solutions are $y(x) = \sin(n\pi x)$. Since $-y'' = -(n\pi)^2 y(x)$, the eigenvalues are $(n\pi)^2$. The eigenvalues for the inverse are the reciprocals. Hence $\sum \frac{1}{n^2\pi^2} = \int_0^1 G(x,x)dx = \frac{1}{6}$. Therefore,

$$\sum_{n=1}^{\infty} \frac{1}{n^2} = \frac{\pi^2}{6}. \tag{15}$$

We mention another way to find $\sum_{n=1}^{\infty} \frac{1}{n^2}$. Consider

$$\int_0^1 \int_0^1 \frac{dx\,dy}{1-xy}. \tag{16}$$

After justifying the interchange of summation and integral, we have

$$\int_0^1 \int_0^1 \frac{dx\,dy}{1-xy} = \iint \sum_0^{\infty} (xy)^n dx\,dy$$

$$= \sum_0^{\infty} \iint (xy)^n dx\,dy$$

$$= \sum_0^{\infty} \frac{1}{(n+1)^2}.$$

Hence, if we can evaluate the integral (16) by another method, we will find $\zeta(2)$. One way uses the change of variables $x = u + v$ and $y = u - v$. We leave the considerable details to the reader.

EXERCISE 8.1. Evaluate (16) by finding the double integral without using series.

EXERCISE 8.2. Find the Fourier series of x^2 on $[-\pi, \pi]$ and use it to evaluate $\sum_{n=1}^{\infty} \frac{(-1)^{n+1}}{n^2}$.

EXERCISE 8.3. Show that $\sum_{n=1}^{\infty} \frac{1}{(2n-1)^2} = \frac{\pi^2}{8}$. Suggestion: relate this sum to the known $\sum_{k=1}^{\infty} \frac{1}{k^2}$.

9. Additional techniques

Our next result is called *Wirtinger's inequality*. We give two proofs. The first proof uses the idea of orthonormal expansion, whereas the second uses the notion of compact operator on Hilbert space; it is sketched in Example 10.2 of the next section. Although different in appearance, the proofs rely on the same ideas from this book.

In this section, we will also discuss a method for finding eigenvalues (of Hermitian operators) that is used even in finite dimensions. First we consider Wirtinger's inequality.

Theorem 9.1. *Assume that f is differentiable, $f(0) = f(1) = 0$, and that $f' \in L^2([0,1])$. Then $||f||^2 \leq \frac{1}{\pi^2}||f'||^2$ and this bound is best possible.*

PROOF. Expand f in a sine series as

$$f(x) = \sum_{n=-\infty}^{\infty} c_n \sin(n\pi x).$$

By hypothesis, $f(0) = 0$, and hence $c_0 = 0$. Extend f to the interval $[-1, 1]$ by putting $f(x) = -f(-x)$ if $x \in [-1, 0]$. This extended f is an odd function, and we have

$$f(x) = 2\sum_{n=1}^{\infty} c_n \sin(n\pi x).$$

Notice that the squared L^2-norm of $\sin(n\pi x)$ on $[-1, 1]$ is 1, and hence on $[0, 1]$ it is $\frac{1}{2}$. By orthonormality, or the Parseval equality, we obtain

$$||f||^2 = \frac{1}{2}\int_{-1}^{1} |f(x)|^2 dx = 2\sum_{n=1}^{\infty} |c_n|^2. \tag{17}$$

Since $f(x) = \sum_{n=-\infty}^{\infty} c_n \sin(n\pi x)$, its derivative satisfies

$$f'(x) = \sum_{n=-\infty}^{\infty} c_n n\pi \cos(n\pi x) = 2\sum_{n=1}^{\infty} c_n n\pi \cos(n\pi x).$$

Computing as before (but with cosine), we obtain

$$||f'||^2 = 2\sum_{n=1}^{\infty} |c_n n\pi|^2.$$

Dividing through by π^2 we get

$$\frac{1}{\pi^2}||f'||^2 = 2\sum_{n=1}^{\infty} |c_n|^2 n^2. \tag{18}$$

Since $1 \leq n^2$, formulas (17) and (18) yield

$$||f||^2 \leq \frac{1}{\pi^2}||f'||^2.$$

Furthermore, the only way to achieve equality is if $c_n = 0$ for $n \geq 2$, in which case $f(x) = \sin(\pi x)$. □

The proof shows that the constant $\frac{1}{\pi^2}$ is achieved by the function $\sin(\pi x)$. In other words, the L^2 norm is maximized by putting all the weight under this one term in the sine series. Figure 3 shows a typical sine series on the left and the function $\sin(\pi x)$, which gives the sharp constant in Theorem 9.1, on the right.

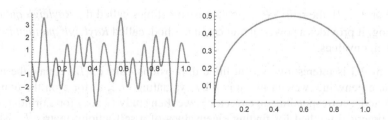

FIGURE 3. Wirtinger inequality

The following easy result looks forward to Theorem 9.3 below.

Lemma 9.2. *Let A be an n-by-n Hermitian matrix with eigenvalues $\lambda_1 \leq \lambda_2 \leq \cdots \leq \lambda_n$. Then, for every non-zero vector z, we have*

$$\lambda_1 \leq \frac{\langle Az, z \rangle}{\langle z, z \rangle} \leq \lambda_n.$$

PROOF. Let $\{\phi_j\}$ be an orthonormal basis of eigenvectors. By orthonormal expansion we have

$$z = \sum \langle z, \phi_j \rangle \phi_j.$$

Plugging in this expression and using orthonormality gives

$$\frac{\langle Az, z \rangle}{\langle z, z \rangle} = \frac{\sum |\langle z, \phi_j \rangle|^2 \lambda_j}{\sum |\langle z, \phi_j \rangle|^2}.$$

This expression obviously lies between the minimum and the maximum of the λ_j.   □

The next result provides a useful method for finding eigenvalues, even in finite dimensions.

Theorem 9.3. *Assume $L : H \to H$ is Hermitian. Then*

$$||L|| = \sup_{f \neq 0} \frac{||Lf||}{||f||} = \sup_{f \neq 0} \frac{|\langle Lf, f \rangle|}{||f||^2}. \tag{19}$$

PROOF. Let us write α for the right-hand side of (19). To show that $||L|| = \alpha$ we establish both inequalities $\alpha \leq ||L||$ and $||L|| \leq \alpha$. The first follows from the Cauchy–Schwarz inequality:

$$\alpha = \sup_{f \neq 0} \frac{|\langle Lf, f \rangle|}{||f||^2} \leq \sup_{f \neq 0} \frac{||L|| \, ||f||^2}{||f||^2} = ||L||.$$

The second is harder. We begin with $4\mathrm{Re}\langle Lf, g \rangle = \langle L(f + g), f + g \rangle - \langle L(f - g), f - g \rangle$. Using the inequality $|\langle Lh, h \rangle| \leq \alpha ||h||^2$ and the parallelogram law, we obtain

$$4\mathrm{Re}\langle Lf, g \rangle \leq \alpha(||f + g||^2 + ||f - g||^2) = 2\alpha(||f||^2 + ||g||^2). \tag{20}$$

Set $g = \frac{Lf}{\alpha}$ in (20). We get

$$4\frac{||Lf||^2}{\alpha} \leq 2\alpha \left(||f||^2 + \frac{||Lf||^2}{\alpha^2} \right).$$

Simple algebra yields $||Lf||^2 \leq \alpha^2 ||f||^2$ and hence

$$\frac{||Lf||}{||f||} \leq \alpha.$$

Taking the supremum over non-zero f gives (19).   □

REMARK. For $f \neq 0$, the ratio $\frac{\langle Lf, f \rangle}{||f||^2}$ in (19) is sometimes called the *Rayleigh quotient*. In the finite-dimensional setting, it provides a powerful numerical method, called *Rayleigh quotient iteration*, for finding eigenvectors and eigenvalues.

Thus Theorem 9.3 is interesting even in finite dimensions. Since $||L||$ equals the maximum value of $|\lambda|$, where λ is an eigenvalue, we can find this λ by evaluating $\langle Lz, z \rangle$ for z on the unit sphere. Once we have found this λ and a corresponding eigenvector v, we then study $\langle Lz, z \rangle$ for z orthogonal to v. Repeating the idea gives a practical method for finding eigenvalues of a self-adjoint matrix L. This method avoids determinants. The restriction to L being self-adjoint is not severe. We can apply the method to the self-adjoint operator L^*L.

Exercise 9.1 uses Lagrange multipliers to find the maximum and minimum of a quadratic form on the unit sphere. Figure 4 illustrates the idea when $n = 2$ and the quadratic form is given by $f(x, y) = xy$. The hyperbolas are level sets of Q; the candidates for minimum and maximum occur where the level sets are tangent to the circle.

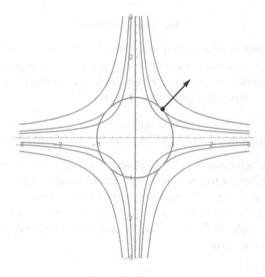

FIGURE 4. Lagrange multipliers

Optimization techniques arise throughout pure and applied mathematics. We next provide a simple example where cleverness enables one to solve a problem simply and directly.

EXAMPLE 9.1. We find the minimum squared distance a certain quadric surface gets to the origin. Let $S = \{(x, y, z) : x + y + z - 1 = xy + xz + yz\}$. We minimize $x^2 + y^2 + z^2$ as follows:

$$x^2 + y^2 + z^2 = (x + y + z)^2 - 2(xy + yz + xz) = (x + y + z)^2 - 2(x + y + z - 1) = (x + y + z - 1)^2 + 1.$$

Therefore the minimum squared distance is 1, and it happens whenever the plane $x + y + z = 1$ intersects S. See Exercise 9.3 for more information.

EXERCISE 9.1. For $x \in \mathbb{R}^n$, and A a symmetric matrix, define f by

$$f(x) = \sum_{j,k=1}^{n} a_{jk} x_j x_k = \langle Ax, x \rangle.$$

Use Lagrange multipliers to find the minimum and maximum of f on the unit sphere. Comment: You can avoid using coordinates by noting that $\nabla f(x) = 2Ax$.

EXERCISE 9.2. Use (19) to show that $\|T^*T\| = \|T\|^2$.

EXERCISE 9.3. Show that the plane $x + y + z = 1$ from Example 9.1 intersects the set S in infinitely many points. In fact, there are infinitely many points with rational coordinates. Show this second statement. Suggestion: Reduce to a quadratic equation, and then use the following fact. The square root of a whole number is either a whole number or is irrational.

EXERCISE 9.4. Assume $x_j \geq 0$ and $a_j > 0$ for $1 \leq j \leq n$. Find the maximum of $\prod_{j=1}^{n} x_j^{a_j}$ given that $\sum x_j = 1$. Comment: In two dimensions, the maximum of $x^a y^b$ on $x + y = 1$ is $\frac{a^a b^b}{(a+b)^{a+b}}$. The answer is $\frac{\alpha^\alpha}{|\alpha|^{|\alpha|}}$ in multi-index notation, which we do not discuss. See page 4 of [T], page 227 of [F1], page 64 of [D1], or many other sources for multi-index notation.

10. Spectral theory

For a complete understanding of the proof of the Sturm–Liouville theorem, one needs to consider *compact operators*. After defining compact operator and giving some examples, we state the spectral theorem for compact Hermitian operators. We use this result to clarify Sturm–Liouville theory. We also sketch a second proof of the Wirtinger inequality using these ideas.

Definition 10.1. Let H be a Hilbert space, and suppose $T : H \to H$ is linear. T is called *compact* if, whenever z_n is a bounded sequence in H, then $T(z_n)$ has a convergent sequence.

We give an intuitive characterization of compactness. First of all, every operator with finite-dimensional range is compact. This conclusion follows from the second statement of Theorem 1.5. An operator on a Hilbert space is compact if and only if its operator norm can be made arbitrarily small outside of a finite-dimensional space. Thus, for any $\epsilon > 0$, we can find a finite-dimensional subspace such that, away from this subspace, $||T|| < \epsilon$. For example, if T were given by an infinite diagonal matrix, then compactness would force the diagonal entries to tend to 0.

EXAMPLE 10.1. Consider

$$
T = \begin{pmatrix}
\lambda_1 & 0 & 0 & \cdots & 0 \\
0 & \lambda_2 & 0 & \cdots & 0 \\
0 & 0 & \lambda_3 & 0 & 0 \\
\cdots & \cdots & \cdots & & \\
0 & \cdots & 0 & \lambda_n & \\
\cdots & \cdots & \cdots & \cdots &
\end{pmatrix}
$$

This operator is compact if and only if $\lambda_n \to 0$.

The primary examples of compact operators are integral operators:

$$
Tf(x) = \int_a^b K(x,t) f(t) \, dt.
$$

Here the function K, the integral kernel of the operator, is assumed continuous in both variables. The Green's function from a Sturm–Liouville problem provides a compelling example.

We therefore state the spectral theorem for compact Hermitian operators; the proof in this setting naturally generalizes the proof in the finite-dimensional setting (Theorem 5.7). See [D3] for a proof for compact Hermitian operators. See [RS] for more general spectral theorems.

Theorem 10.2 (Spectral theorem for compact self-adjoint operators). *Let H be a Hilbert space, and let $T : H \to H$ be a compact, self-adjoint operator. Then:*

- *There is a complete orthonormal system $\{\phi_n\}$ in H consisting of eigenvectors of T.*
- *The eigenvalues are real (by self-adjointness).*
- *Each eigenspace is finite-dimensional (by compactness).*
- *The only possible limit point for a sequence of these eigenvalues is 0 (by compactness).*
- *$||T|| = |\lambda|$, where λ is the eigenvalue of largest magnitude.*

The operator L in the Sturm–Liouville equation is self-adjoint (with the appropriate domain, including the boundary conditions) but it is not compact. Instead we consider $T = (L - \zeta I)^{-1}$, where I is the identity and ζ is real. If $L - \zeta I$ has a continuous Green's function, then T is compact. For ζ sufficiently negative, that holds. Then, by the spectral theorem, T has an orthonormal system of eigenvectors.

Note that $Tf = \lambda f$ if and only if $(L - \zeta I)^{-1} f = \lambda f$, if and only if $f = (L - \zeta I)\lambda f$, and finally if and only if

$$L(f) = \frac{1 + \zeta \lambda}{\lambda} \, f.$$

Thus, as a sequence λ_n of eigenvalues of T tends to 0, the corresponding sequence of eigenvalues of L tends to infinity.

We conclude, when L is a Sturm–Liouville operator, that there is a complete orthonormal system of eigenvectors $\{\psi_n\}$ with $L(\psi_n) = \lambda_n$ and $\lim(\lambda_n) = \infty$. Thus, for $f \in H = L^2([a, b], w)$, we can write

$$f = \sum_n \langle f, \psi_n \rangle_w \, \psi_n.$$

Hence each Sturm–Liouville problem gives rise to an orthonormal expansion similar to a Fourier series. Examples arise throughout mathematical physics. These examples include Legendre polynomials, their generalization called *ultraspherical polynomials*, Hermite polynomials, Chebyshev polynomials, and so on.

EXAMPLE 10.2. Put $Tf(x) = \int_0^x f(t)dt$ on $L^2([0, 1])$. Then

$$(T^*T)(f)(x) = \int_x^1 \int_0^u f(v)dv\,du.$$

It is not difficult to prove that T^*T is compact and self-adjoint. We can easily compute the eigenvalues of T^*T. Taking successive derivatives on both sides of the equation $T^*Tf = \lambda f$, we get first

$$-\int_0^x f(u)du = \lambda f'(x)$$

and then our familiar equation

$$-f''(x) = \frac{1}{\lambda}f(x).$$

If we insist that f vanishes at the endpoints, then f must be a multiple of $\sin(n\pi x)$. The eigenvalues are $\frac{1}{n^2\pi^2}$. The largest of these is $\frac{1}{\pi^2}$. Using the last part of Theorem 9.3, $||T^*T|| = \frac{1}{\pi^2}$. Exercise 9.2 implies that $||T||^2 = ||T^*T||$, and hence $||T|| = \frac{1}{\pi}$. We obtain a second proof of Wirtinger's inequality, Theorem 9.1.

EXERCISE 10.1. Fill in the details of Example 10.2 to obtain a second proof of Theorem 9.1.

EXERCISE 10.2. Give an example of a bounded operator T with $TT^* \neq T^*T$.

EXERCISE 10.3 (Isoperimetric inequality). (Difficult) Use Fourier series to establish the following statement. Let γ be a smooth simple closed curve in the plane. Suppose that the length of γ is L. Prove that the area enclosed by γ is at most $\frac{L^2}{4\pi}$, with equality only when γ is a circle. Suggestion: Parametrize the curve by arc-length (ds) and expand the component functions in Fourier series. Use Green's theorem to show that the area enclosed by γ is

$$\int_\gamma \frac{xy' - yx'}{2} ds.$$

EXERCISE 10.4. The Chebyshev polynomials are defined for $n \geq 0$ by

$$p_n(x) = \cos(n \cos^{-1} x).$$

- Prove that p_n is a polynomial in x.
- Find p_n explicitly for $0 \le n \le 5$.
- Prove that p_n are eigenfunctions for the equation

$$(1 - x^2)y'' - xy' + \lambda y = 0$$

on the interval $(-1, 1)$.

- Explain why these polynomials are orthogonal on $L^2([-1,1], w)$. Here w is the weight function $\frac{1}{\sqrt{1-x^2}}$.

11. Orthonormal expansions in eigenvectors

The purpose of this section is to tie together the linear algebra that began the book with the ideas of orthonormal expansion.

Assume $L : H \to H$ is a linear operator on a Hilbert space, and we wish to solve $Lu = f$. Suppose in addition that $\{\phi_n\}$ is a complete orthonormal system for H with $L(\phi_n) = \lambda_n \phi_n$ for all n. We write

$$f = \sum_n \langle f, \phi_n \rangle \phi_n \tag{21}$$

$$u = \sum_n \langle u, \phi_n \rangle \phi_n. \tag{22}$$

Applying L in (22) and using (21) shows that $Lu = f$ if and only if

$$\sum_n \langle u, \phi_n \rangle \lambda_n \phi_n = \sum_n \langle f, \phi_n \rangle \phi_n. \tag{23}$$

Therefore, by equating coefficients in (23), we obtain

$$\lambda_n \langle u, \phi_n \rangle = \langle f, \phi_n \rangle.$$

Hence, to solve the problem we simply divide by λ_n on each eigenspace. If $\lambda_n = 0$ we see that we cannot solve the equation unless f is orthogonal to the zero eigenspace. See Theorem 11.1 below for the famous generalization of this result.

EXAMPLE 11.1. Consider the two-by-two system $az = c$ and $bw = d$. We want to solve for z and w. (the numbers can be real or complex). The answer is obviously $z = \frac{c}{a}$ and $w = \frac{d}{b}$, as long as we are not dividing by zero. Let us reformulate in the above language.

The vectors $(1, 0)$ and $(0, 1)$ form an orthonormal basis for $\mathbb{C}^2$. We have

$$(z, w) = z(1, 0) + w(0, 1).$$

Also $L((1,0)) = a(1,0)$ and $L((0,1)) = b(0,1)$. Calling $u = (z, w)$ and $f = (c, d)$ we see that we are in the above situation.

Orthonormal expansion generalizes this idea to solving all sorts of (systems of) linear equations, including the second-order differential equations from the Sturm–Liouville problems. Thus solving the linear equation $Lu = f$ is the main unifying theme of this book.

We conclude this chapter by stating and proving the so-called*Fredholm alternative*. Consider a linear equation $Lz = b$, where L is an operator on a Hilbert space H. We want to know whether there is a solution.

Theorem 11.1 (Fredholm alternative). *Assume that $L : H \to H$ is a linear operator and that $\mathcal{R}(L)$ is a closed subspace. Then $Lz = b$ has a solution if and only if b is orthogonal to $\mathcal{N}(L^*)$.*

PROOF. Suppose first that $Lz = b$. Choose $v \in \mathcal{N}(L^*)$. Then

$$\langle b, v \rangle = \langle Lz, v \rangle = \langle z, L^*v \rangle = \langle z, 0 \rangle = 0.$$

Since v is an arbitrary element of the null space, we conclude that b is orthogonal to $\mathcal{N}(L^*)$.

To prove the converse, we assume that $\mathcal{R}(L)$ is closed. We write $b = \pi(b) \oplus w$, where $\pi(b)$ is the orthogonal projection of b onto the range of L, and w is orthogonal to the range of L. Since $0 = \langle w, Lu \rangle$ for all u we have $\langle L^*w, u \rangle = 0$ for all u, and hence $L^*w = 0$. Therefore, by our assumption on b, we have $0 = \langle b, w \rangle$. But

$$0 = \langle b, w \rangle = \langle \pi(b) + w, w \rangle = ||w||^2.$$

Therefore $w = 0$ and $b \in \mathcal{R}(L)$. $\qquad\square$

Corollary 11.2. *In the situation of Theorem 11.1,*

$$H = \mathcal{R}(L) \oplus \mathcal{N}(L^*).$$

EXERCISE 11.1. Put $H = L^2([0, 1])$. Assume $p > 0$. Define an operator L on H by putting

$$Lu(x) = u(x) + \lambda \int_0^1 t^p u(t)\,dt$$

for continuous functions u. Consider the equation $Lu = f$. First find the condition on λ for $\mathcal{N}(L) = \{0\}$. Do the same for L^*. When $\mathcal{N}(L) \neq \{0\}$, use the Fredholm alternative to find the condition on f for a solution to exist. (The range of L is closed here.)

EXERCISE 11.2 (Difficult). Prove that a self-adjoint operator is the sum of two unitary operators. Suggestion. First note that a point in the interval $[-1, 1]$ is the sum of two points on the unit circle. See Figure 5. Then consider the power series expansion for $\sqrt{1 - z^2}$.

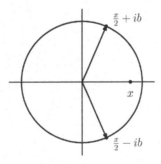

FIGURE 5. Sum of two unitaries

CHAPTER 7

Examples and applications

We consider in this chapter various problems from applied mathematics and engineering which employ the methods we have developed. For example, least squares regression will be understood as orthogonal projection. We discuss enough probability to talk about quantum mechanics, and our discussion of quantum mechanics will be based on our work on Hilbert spaces. We consider some classes of special functions, in addition to complex exponentials (sines and cosines), that form complete orthonormal systems. We offer some of the many applications of complex analysis, such as using conformal mapping to solve the Dirichlet problem and providing techniques for locating roots of polynomials. The last section describes linear time-invariant systems and illustrates the importance of convolution in applied mathematics. The sections are mostly independent of each other, although Section 4 on probability is needed for Section 5 on quantum mechanics. We begin each section with a short comment about prerequisites for reading this section.

1. Least squares

Prerequisites. The ideas used in this section are in Chapter 1 on linear algebra and the finite-dimensional version of Theorem 2.7 from Chapter 6.

Least squares regression is used throughout science, including the social sciences. The method of least squares is essentially Theorem 2.7 from Chapter 6 in disguise!

We are given some data points $(x_1, y_1), \ldots, (x_n, y_n)$. We wish to determine a function f whose graph well approximates these data points. One standard approach is to minimize $\sum_{j=1}^{n} (y_j - f(x_j))^2$, the sum of squared errors, where the minimum is taken over some collection of functions. The simplest example is well known; the class of functions defines lines in the plane. In this case we want to choose constants a and b such that, with $f(x) = a + bx$, the sum of squared errors is minimal. We do not need to assume that f is a linear function; it could be a polynomial of arbitrary degree, or a sine series, or even an arbitrary linear combination of given functions. In Example 1.1 we fit a quadratic to some data. In Example 1.2 we determine the fit with both linear and quadratic functions. In Example 1.3, we consider sums of $\sin(t)$ and $\sin(2t)$. The remarkable thing is that the method of orthogonal projection works in all these cases.

We approach this problem by considering first a Hilbert space H of functions and a linear operator $L : H \to H$. Let $V = \mathcal{R}(L)$. Suppose $z \neq \mathcal{R}(L)$. We want to find the $v \in V$ that minimizes $||z - v||$. Since $\mathcal{R}(L)$ is closed, such a v is guaranteed by Theorem 2.7 of Chapter 6. Furthermore $z - v$ must be perpendicular to $\mathcal{R}(L)$. But, $(z - v) \perp \mathcal{R}(L)$ if and only if $\langle z - v, L\xi \rangle = 0$ for all ξ, and hence if and only if $\langle L^*(z - v), \xi \rangle = 0$ for all ξ. Therefore $z - v$ is in the null space of L^*. Since L^* is linear, the desired condition holds if and only if $L^*(z) = L^*(v)$. Since $v \in \mathcal{R}(L)$, we have $v = Lu$ for some u. We obtain the abstract least squares equation

$$L^*Lu = L^*(z). \tag{1}$$

In (1), the unknown is u.

Let us approach the same problem more concretely. Consider N data points (x_1, y_1) and (x_2, y_2) and so on up to (x_N, y_N). For an unknown function f, consider the vector

$$(y_1 - f(x_1), \ldots, y_N - f(x_N)).$$

If $y = f(x)$ held for all these values of x, then we would have

$$0 = \sum_{j=1}^{N} (y_j - f(x_j))^2. \tag{2}$$

In general, the sum is not zero.

First we consider approximating the data with the values of a polynomial $f(x) = \sum c_j x^j$ of degree K. We choose the coefficients c_j so as to minimize the right-hand side of (2). Thus we regard the c_j as variables, put $G(c) = G(c_0, c_1, c_2, \ldots, c_K)$, and try to minimize G. Setting the gradient of G equal to 0 gives a system of linear equations for the coefficients. This system is precisely the equation

$$M^*M \begin{pmatrix} c_0 \\ c_1 \\ \ldots \\ c_K \end{pmatrix} = M^*y. \tag{3}$$

The column vector of coefficients corresponds to the unknown u in (1). The first column of M consists of all ones; the entry on the j-th row is x_j^0. The second column of M consists of $x_1, x_2, \ldots, x_N$. The third column consists of $(x_1^2, x_2^2, \ldots, x_N^2)$, and so on. The m, n entry of M^*M is then the sum of the $(m+n)$-th powers of the x_j.

We show how to apply the method in several examples, using data made up by the author.

EXAMPLE 1.1. Consider an object moving along the parabola defined by $y(t) = c_0 + c_1 t + c_2 t^2$. We assume that the only significant force is gravity. Here c_0 is the initial position, c_1 is the initial velocity, and $2c_2 = y''(0)$ is the acceleration due to gravity. We want to compute c_2 from some observed data, which we suppose are in feet per seconds.

Suppose we have four data points: $(0, 5.01)$ and $(1, -.99)$ and $(2, -39.5)$ and $(3, -109.2)$. The method above gives

$$M = \begin{pmatrix} 1 & 0 & 0 \\ 1 & 1 & 1 \\ 1 & 2 & 4 \\ 1 & 3 & 9 \end{pmatrix}$$

$$M^* = \begin{pmatrix} 1 & 1 & 1 & 1 \\ 0 & 1 & 2 & 3 \\ 0 & 1 & 4 & 9 \end{pmatrix}.$$

Therefore

$$M^*M = \begin{pmatrix} 4 & 6 & 14 \\ 6 & 14 & 36 \\ 14 & 36 & 98 \end{pmatrix}$$

and this matrix is invertible. We can therefore solve

$$M^*M \begin{pmatrix} c_0 \\ c_1 \\ c_2 \end{pmatrix} = M^*y = M^* \begin{pmatrix} 5.01 \\ -.99 \\ -39.5 \\ -109.2 \end{pmatrix}.$$

We obtain $y(t) = 5.076 + 9.661t - 15.925t^2$ for the best quadratic fitting the data. From these data we would get that the acceleration due to gravity is 31.85 feet per second squared. At the equator, the actual value is about 32.1 feet per second squared.

EXAMPLE 1.2. Consider the five data points

$$(0,-5) \ (1,0) \ (2,5) \ (3,12) \ (4,24).$$

We first find the best line $y = a + bx$ fitting these points, and then we find the best quadratic $y = a + bx + cx^2$ fitting these points. Here *best* is in the sense of minimizing the sum of squared errors.

To find the best line, pictured in Figure 1, we have

$$M = \begin{pmatrix} 1 & 0 \\ 1 & 1 \\ 1 & 2 \\ 1 & 3 \\ 1 & 4 \end{pmatrix}$$

$$M^*M = \begin{pmatrix} 5 & 10 \\ 10 & 30 \end{pmatrix}$$

$$M^*y = \begin{pmatrix} 36 \\ 142 \end{pmatrix}.$$

Solving the system $M^*Mu = M^*y$ yields the best-fitting line to be

$$y = \frac{-34}{5} + 7x.$$

To find the best quadratic, also pictured in Figure 1, we have

$$M = \begin{pmatrix} 1 & 0 & 0 \\ 1 & 1 & 1 \\ 1 & 2 & 4 \\ 1 & 3 & 9 \\ 1 & 4 & 16 \end{pmatrix}$$

$$M^*M = \begin{pmatrix} 5 & 10 & 30 \\ 10 & 30 & 100 \\ 30 & 100 & 354 \end{pmatrix}$$

$$M^*y = \begin{pmatrix} 36 \\ 142 \\ 512 \end{pmatrix}.$$

Solving the system $M^*Mu = M^*y$ yields the best-fitting quadratic to be

$$y = \frac{-158}{35} + \frac{17}{7}x + \frac{8}{7}x^2.$$

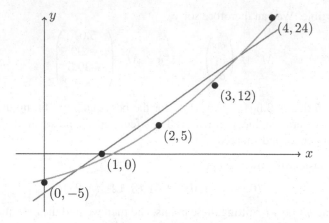

FIGURE 1. The approximations from Example 1.2

We can use the method of least squares to find the best approximation to the span of any finite set of linearly independent functions. Consider functions $f_1, \ldots, f_n$. We evaluate them at points $x_1, x_2, \ldots, x_k$, obtaining values $y_1, y_2, \ldots, y_k$. We want to choose the constants $c_1, \ldots, c_n$ that minimize

$$\sum_{i=1}^{k} \Big(\sum_{j=1}^{n} c_j f_j(x_i) - y_i \Big)^2.$$

As before, we obtain the matrix equation $M^* M c = M^* y$. Here the i, j-th entry of M is $f_j(x_i)$. The column vector y is given by $y = (y_1, \ldots, y_k)$, and the unknown column vector c is given by $c = (c_1, \ldots, c_n)$.

EXAMPLE 1.3. We are given a curve which is fixed to be of height 0 at the endpoints 0 and π of an interval. We believe that the position $f(t)$ of the curve at time t satisfies the equation

$$f(t) = c_1 \sin(t) + c_2 \sin(2t) \tag{4}$$

fo appropriate constants c_1 and c_2. We obtain the following measurements:

$$f(\tfrac{\pi}{2}) = 5.1, \quad f(\tfrac{\pi}{3}) = 7.8, \quad f(\tfrac{\pi}{4}) = 7.5.$$

We wish to find the values of c_1 and c_2 that minimize the squared error. Put $f_1(t) = \sin(t)$ and $f_2(t) = \sin(2t)$. The matrix M satisfies

$$M = \begin{pmatrix} f_1(x_1) & f_2(x_1) \\ f_1(x_2) & f_2(x_2) \\ f_1(x_3) & f_2(x_3) \end{pmatrix}.$$

Hence

$$M = \begin{pmatrix} 1 & 0 \\ \frac{\sqrt{3}}{2} & \frac{\sqrt{3}}{2} \\ \frac{\sqrt{2}}{2} & 1 \end{pmatrix}$$

$$M^* = \begin{pmatrix} 1 & \frac{\sqrt{3}}{2} & \frac{\sqrt{2}}{2} \\ 0 & \frac{\sqrt{3}}{2} & 1 \end{pmatrix}$$

$$M^*M = \begin{pmatrix} \frac{9}{4} & \frac{3+2\sqrt{2}}{4} \\ \frac{3+2\sqrt{2}}{4} & \frac{7}{4} \end{pmatrix}.$$

To find the best values of c_1 and c_2, we must solve the equation

$$M^*M \begin{pmatrix} c_1 \\ c_2 \end{pmatrix} = M^* \begin{pmatrix} 5.1 \\ 7.8 \\ 7.5 \end{pmatrix}.$$

The inverse of M^*M is given approximately by

$$(M^*M)^{-1} = \begin{pmatrix} 0.964538 & -0.803106 \\ -0.803106 & 1.24012 \end{pmatrix}.$$

We obtain the values $c_1 = 5.102$ and $c_2 = 3.898$.

With these coefficients, the values of the function in (4) at the points $\frac{\pi}{2}$, $\frac{\pi}{3}$, and $\frac{\pi}{4}$ are 5.102, 7.793, and 7.505. It is thus quite plausible that this curve can be represented by these two sine functions.

We close this section by re-emphasizing the power of orthogonal projection. Consider any finite set of functions. We ask whether a set of measurements is consistent with some linear combination of these functions. If not, we seek the linear combination that minimizes the sum of squared errors. To find the coefficients, we solve the equation $M^*Mc = M^*y$. The matrix M is determined by the points x_j at which we evaluate, and the vector y is the set of values we obtained by evaluating.

We close with a simple lemma explaining when M^*M is invertible. In this case, there is a unique solution to the least squares problem.

Proposition 1.1. *Let $M : V \to W$ be a linear map between inner product spaces with adjoint M^* : $W \to V$. Then $\mathcal{N}(M^*M) = \mathcal{N}(M)$. In particular, in the finite-dimensional setting, M^*M is invertible if and only if M has full rank.*

PROOF. Suppose $v \in \mathcal{N}(M)$. Then $M^*Mv = M^*(0) = 0$ and thus $\mathcal{N}(M) \subseteq \mathcal{N}(M^*M)$. Conversely, suppose $v \in \mathcal{N}(M^*M)$. Then we have

$$0 = \langle M^*Mv, v \rangle = \langle Mv, Mv \rangle = ||Mv||^2.$$

Hence $Mv = 0$, and thus $\mathcal{N}(M^*M) \subseteq \mathcal{N}(M)$. Suppose V is finite-dimensional. Since $M^*M : V \to V$, it is invertible if and only if it has trivial null space. But $\mathcal{N}(M^*M) = \mathcal{N}(M)$. Recall that

$$\dim(V) = \dim \mathcal{N}(M) + \dim \mathcal{R}(M).$$

Since the rank of M is the dimension of its range, the result follows. $\qquad \square$

If we want to approximate data by a polynomial of degree d, then the matrix M is a Vandermonde matrix whenever the x_j are distinct points. Thus M will have maximum rank as long as the points x_j are distinct and there are at least $d + 1$ of them.

REMARK. Let us return to the basic question of finding a function that *best fits* some data. Minimizing the sum of squared errors is not the only possible way to proceed. One could, for example, minimize the sum of any fixed positive power α of the (absolute) errors. Using squared errors ($\alpha = 2$) is the most common method; it enables us to use the Hilbert space methods we have developed and is the most symmetric. Using large values of α controls the effect of the maximum error; using $\alpha = 1$ controls the average error.

EXERCISE 1.1. Find the least squares line fitting the data

$$(0, -3), \ (1, 3.1), \ (2, 8.9), \ (3, 16), \ (4, 20.8).$$

EXERCISE 1.2. Find the matrix M^*M in case we want to find the least squares line and the data are evaluated at the points $2, 3, 5, 7$. What is the matrix M^*M in case we want to find the least squares parabola?

EXERCISE 1.3. We believe that certain data fit the formula $\sum_{j=1}^{3} c_j \sin(jt)$ for some constants c_j. Suppose we evaluate at $\frac{\pi}{2}, \frac{\pi}{3}, \frac{\pi}{4}, \frac{\pi}{6}$. What is the matrix M^*M in this case?

EXERCISE 1.4. Develop equations to minimize the sum of the absolute errors rather than the sum of squared errors. Compare the two methods.

EXERCISE 1.5. Suppose $c, d > 0$. Find the limit, as $\alpha \to \infty$, of

$$(c^\alpha + d^\alpha)^{\frac{1}{\alpha}}.$$

How does the result relate to the above remark?

2. The wave equation

Prerequisites. This section should be accessible to all readers who have taken a basic course in differential equations. The ideas here motivated Fourier series and orthonormal expansion.

We first discuss the 1-dimensional wave equation, the PDE that governs the motion of a vibrating guitar string. Given the initial position and velocity of the string, we solve the equation by a classical method dating to D'Alembert in the 18th century. Reconsidering the equation using the method of separation of variables leads us back to Fourier series. We then discuss separation of variables for several other PDEs.

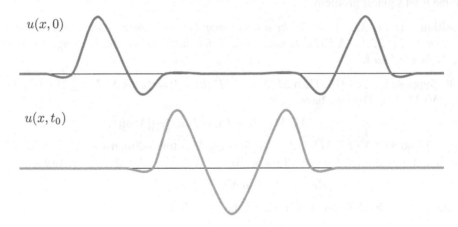

$u(x,0)$

$u(x,t_0)$

FIGURE 2. Vibrating string

As indicated in Figure 2, let $u(x,t)$ denote the displacement of a vibrating string at position x along the string and at time t. By elementary physics, one can derive the PDE

$$u_{xx}(x,t) = \frac{1}{c^2} u_{tt}(x,t) \tag{5}$$

where c is the speed of propagation. Assume the string is fixed at the ends, which for convenience we take as 0 and π. Thus $u(0,t) = 0$ for all t and $u(\pi,t) = 0$ for all t. We also assume the initial position of the string is given by $u(x,0) = f(x)$ and the initial velocity is given by $u_t(x,0) = g(x)$.

Our first result is the D'Alembert solution.

Theorem 2.1. *If u satisfies the wave equation (5) and u is twice continuously differentiable, then there are twice differentiable functions F and G such that*

$$u(x,t) = F(x+ct) + G(x-ct). \tag{6}$$

Conversely, if u satisfies (6) for twice differentiable functions, then u satisfies the wave equation (5).

Furthermore, if the initial position of the string is given by $u(x,0) = f(x)$ and the initial velocity is given by $u_t(x,0) = g(x)$, then

$$u(x,t) = \frac{1}{2}(f(x+ct) + f(x-ct)) + \frac{1}{2c}\int_{x-ct}^{x+ct} g(\alpha)d\alpha.$$

PROOF. The converse assertion follows from the chain rule: if $u(x,t) = F(x+ct) + G(x-ct)$, then $u_{xx} = F'' + G''$ and $u_{tt} = c^2 F'' + (-c)^2 G''$. Therefore $u_{xx}(x,t) = \frac{1}{c^2}u_{tt}(x,t)$.

Next assume u satisfies (5). We change coordinates in the (x,t) plane by putting

$$(x_1, x_2) = (x + ct, x - ct).$$

Equivalently we have

$$(x,t) = \left(\frac{x_1 + x_2}{2}, \frac{x_1 - x_2}{2c}\right).$$

Define ϕ by $\phi(x_1, x_2) = u(x,t)$. Then $\phi_{x_1} = \frac{1}{2}u_x + \frac{1}{2c}u_t$ and hence

$$\phi_{x_1 x_2} = \frac{1}{4}u_{xx} - \frac{1}{4c}u_{xt} + \frac{1}{4c}u_{tx} + \frac{1}{4c^2}u_{tt} = \frac{1}{4}\left(u_{xx} - \frac{1}{c^2}u_{tt}\right) = 0.$$

We have used the equality of mixed partial derivatives, Theorem 2.2 of Chapter 3. Since $\phi_{x_1 x_2} = 0$, it follows that ϕ_{x_1} is independent of x_2 and ϕ_{x_2} is independent of x_1. Hence there are twice differentiable functions A and B for which

$$\phi(x_1, x_2) = A(x_1) + B(x_2).$$

Therefore $u(x,t) = A(x+ct) + B(x-ct)$ as desired.

Next we express the solution in terms of the initial conditions. Assume $u(x,t) = A(x+ct)+B(x-ct)$. The initial conditions yield

$$u(x,0) = A(x) + B(x) = f(x)$$

$$u_t(x,0) = cA'(x) - cB'(x) = g(x).$$

Differentiating the first equation gives $A' + B' = f'$. We obtain a linear system of equations:

$$\begin{pmatrix} f' \\ g \end{pmatrix} = \begin{pmatrix} 1 & 1 \\ c & -c \end{pmatrix} \begin{pmatrix} A' \\ B' \end{pmatrix}.$$

Solving this system yields

$$A' = \frac{1}{2}\left(f' + \frac{1}{c}g\right)$$

$$B' = \frac{1}{2}\left(f' - \frac{1}{c}g\right).$$

We integrate to finally obtain the desired conclusion:

$$u(x,t) = \frac{1}{2}\left(f(x+ct) + f(x-ct)\right) + \frac{1}{2c}\int_{x-ct}^{x+ct} g(\alpha)d\alpha. \tag{7}$$

$\square$

We check the answer. Setting $t = 0$ in (7) gives $u(x, 0) = f(x)$. Computing u_t gives

$$u_t(x, t) = \frac{1}{2}(cf(x + ct) - cf(x - ct)) + \frac{1}{2c}(cg(x + ct) - (-c)g(x - ct)).$$

Setting $t = 0$ again yields $u_t(x, 0) = g(x)$. Thus (7) gives the solution to the wave equation with the given initial conditions. The proof of Theorem 2.1 derived (7) from (5).

We next approach the wave equation via the standard technique called **separation of variables**. Assume u satisfies the wave equation and that $u(x, t) = A(x)B(t)$. Then

$$u_{xx}(x, t) = A''(x)B(t)$$

$$u_{tt}(x, t) = A(x)B''(t).$$

The equation $u_{xx}(x, t) = \frac{1}{c^2}u_{tt}(x, t)$ implies

$$A''(x)B(t) = \frac{1}{c^2}A(x)B''(t). \tag{8}$$

But (8) implies

$$\frac{A''}{A} = \frac{1}{c^2}\frac{B''}{B}. \tag{9}$$

The expression $\frac{A''}{A}$ is independent of t and the expression $\frac{B''}{B}$ is independent of x. The left-hand side of (9) depends only on x; the right-hand side depends only on t. Hence each side is a constant λ.

The solutions to (9) depend on the sign of λ. To get oscillation we assume $\lambda < 0$. Then

$$A(x) = c_1 \cos(\sqrt{|\lambda|}x) + c_2 \sin(\sqrt{|\lambda|}x)$$

$$B(t) = b_1 \cos(c\sqrt{|\lambda|}t) + b_2 \sin(c\sqrt{|\lambda|}t).$$

The condition $u(0, t) = 0$ for all t implies $c_1 = 0$ and $u(\pi, t) = 0$ for all t implies $c_2 \sin(\sqrt{|\lambda|}\pi) = 0$. Therefore $\sqrt{|\lambda|} = n$, for some integer n. We obtain solutions

$$A_n(x) = a_n \sin(nx)$$

$$B_n(t) = b_{n1} \cos(cn\pi t) + b_{n2} \sin(cn\pi t).$$

Since sine is an odd function, it is convenient to extend the solution to $[-\pi, \pi]$ using $\sin(-x) = -\sin(x)$.

The wave equation (5) is linear. We may therefore superimpose solutions. We therefore try

$$u(x, t) = \sum_{n=1}^{\infty} a_n \sin(nx)B_n(t).$$

We must check the initial conditions. To make $u(x, 0) = f(x)$, we require

$$f(x) = \sum_{n=1}^{\infty} d_n \sin(nx) \tag{10}$$

for some numbers d_n. A similar result applies to make $u_t(x, 0) = g(x)$. To solve the 1-dimensional wave equation with $u(x, 0) = f(x)$, we must be able to write f as the sine series (10).

By Theorem 2.1 we can solve (5) for given twice differentiable f and g. Fourier therefore claimed that if f is any smooth odd function on $[-\pi, \pi]$, then f can be expanded in a sine series as in (10). Since

$$\int_{-\pi}^{\pi} \sin(nx) \sin(mx)dx = 0$$

for $n \neq m$, we regard (10) as an orthonormal expansion. If $f(x) = \sum_{n=1}^{\infty} d_n \sin(nx)$, then we can recover the coefficients d_m by the inner product:

$$\langle f(x), \sin(mx) \rangle = \frac{1}{2\pi} \int_{-\pi}^{\pi} f(t) \sin(mt) dt$$

$$= \frac{1}{2\pi} \int_{-\pi}^{\pi} \sum_{n=1}^{\infty} d_n \sin(nt) \sin(mt) dt$$

$$= \sum_{n=1}^{\infty} d_n \frac{1}{2\pi} \int_{-\pi}^{\pi} \sin(nt) \sin(mt) dt$$

$$= d_m \| \sin(mt) \|^2$$

To clarify what is going on, let us briefly write s_m for the function $\sin(mx)$. We have just shown that

$$\langle f, s_m \rangle = d_m \| s_m \|^2.$$

Had we used $\phi_m = \frac{s_m}{\|s_m\|}$, then (10) would be $f = \sum D_n \phi_n$, the coefficient D_n would be given by $D_n = \langle f, \phi_n \rangle$, and (10) would become

$$f = \sum_n \langle f, \phi_n \rangle \phi_n.$$

Thus solving the wave equation leads us to Fourier sine series. As we have seen, Fourier series is but one of many examples of orthonormal expansion. Whenever $\{\phi_n\}$ are a complete orthonormal system in a Hilbert space H, we can write

$$f(x) = \sum_n \langle f, \phi_n \rangle \phi_n(x).$$

The series converges in H, but not necessarily at each x. Thus the Hilbert space theory of Chapter 6 has its roots in mathematics from more than two hundred years ago.

Separation of variables in many linear PDE leads to similar ideas and methods. In Section 5 we consider the Schrödinger equation. Here we consider the heat equation in two space variables and one time variable.

EXAMPLE 2.1. Consider the heat equation $u_t = u_{xx} + u_{yy}$. First we write $u(x, y, t) = A(x, y)B(t)$. The technique of separation of variables and using Δ for the Laplacian in the plane gives us two equations:

$$\Delta A = \lambda A$$

$$B' = \lambda B.$$

The second equation is easy to integrate: $B(t) = e^{\lambda t}$. To ensure that the heat dissipates, we assume the constant λ to be negative. To solve the first equation we use separation of variables again, in polar coordinates. The Laplacian in polar coordinates can be written

$$\Delta A = A_{rr} + \frac{A_r}{r} + \frac{A_{\theta\theta}}{r^2}. \tag{11}$$

See Exercise 2.2, or Exercise 8.5 of Chapter 3, for a derivation. We want $\Delta A = \lambda A$. Write $A(r, \theta) = f(r)g(\theta)$ in (11), obtaining

$$f''g + \frac{f'}{r}g + f\frac{g''}{r^2} = \lambda fg.$$

Divide both sides by fg and then collect appropriate terms in a manner similar to the above. We get

$$r^2 \frac{f''}{f} + r\frac{f'}{f} - \lambda r^2 = \frac{-g''}{g}.$$

The left-hand side depends on r, and the right-hand side depends on θ. Hence we obtain, for a new constant μ, the pair of equations

$$g'' = -\mu g$$

$$r^2 f'' + rf' - (\lambda r^2 + \mu)f = 0.$$

Recall that $\lambda < 0$. To make g periodic, we want $\mu > 0$. Writing $\mu = \kappa^2$ gives

$$r^2 f''(r) + rf'(r) + (|\lambda|r^2 - \kappa^2)f(r) = 0.$$

This equation becomes a Bessel equation after a change of variables. Techniques from Sturm–Liouville theory yield orthonormal expansions in terms of Bessel functions.

EXERCISE 2.1. Solve the wave equation by factoring the operator $D_x^2 - \frac{1}{c^2}D_t^2$.

EXERCISE 2.2. Prove formula (11) for the Laplacian in polar coordinates.

EXERCISE 2.3. Find a formula for the Laplacian in spherical coordinates in three-dimensional space. Use it to solve the following problem. Suppose that the temperature distribution $f(\phi)$ is prescribed on the unit sphere in a manner independent of θ. Assume the temperature inside is $T(\rho, \phi)$, where T satisfies the heat equation. Assume by separation of variables that $T(\rho, \phi) = A(\rho)B(\phi)$. Find the second-order ODE satisfied by $A(\rho)$.

3. Legendre polynomials

Prerequisites. This section can be read on its own, but works best in conjunction with Section 7 from Chapter 6 on Sturm–Liouville equations.

Legendre polynomials provide one of the most interesting examples of a complete orthonormal system. We define them by way of a Sturm–Liouville equation.

Consider the second-order ODE

$$\frac{d}{dx}\left((1-x^2)\frac{d}{dx}P_n(x)\right) + n(n+1)P_n(x) = 0. \tag{12}$$

When n is a non-negative integer, (12) admits a polynomial solution P_n of degree n. The Legendre polynomials are polynomial solutions to (12). The Rodrigues formula

$$P_n(x) = \frac{1}{2^n n!}\frac{d^n}{dx^n}\left((x^2-1)^n\right)$$

gives one of several nice expressions for these polynomials.

Equation (12) has the equivalent form

$$(1-x^2)y'' - 2xy' + n(n+1)y = 0 \tag{13}$$

which yields (12) when we put the equation into divergence form. We also observe that the equation degenerates at the endpoints ± 1 of the interval.

Using the Sturm–Liouville notation, we have $p(x) = 1 - x^2$ and $\lambda = n(n+1)$. Thus we want to solve $L(y) = n(n+1)y$ where $L(y) = -((1-x^2)y')'$. Here are examples of solutions with the normalization $P_n(1) = 1$.

- $P_0(x) = 1$

- $P_1(x) = x$

- $P_2(x) = \frac{1}{2}(3x^2 - 1)$

- $P_3(x) = \frac{1}{2}(5x^3 - 3x)$

- $P_4(x) = \frac{1}{8}(35x^4 - 30x^3 + 3)$

- $P_5(x) = \frac{1}{8}(63x^5 - 70x^3 + 15x)$.

Figures 3 and 4 illustrate the first few Legendre polynomials. Notice that $P_n(1) = 1$ for all n and $P_n(-1) = (-1)^n$ for all n. As the degree increases, the graphs become more like waves. The graphs suggest (but do not prove) that the Legendre polynomials form a complete system for $L^2([-1,1])$.

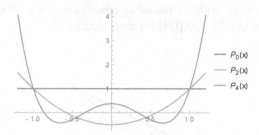

FIGURE 3. Even degree Legendre polynomials

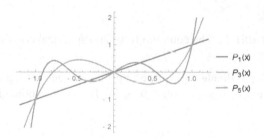

FIGURE 4. Odd degree Legendre polynomials

The general results from Sturm–Liouville theory apply in this setting. It follows that the functions $P_n(x)$ for $n \geq 0$ form a complete orthogonal system on $L^2[-1,1]$. One can compute, using the generating function in (14), a simple formula for the L^2 norms: $||P_n||^2 = \frac{2}{2n+1}$.

The Legendre polynomials have an elementary generating function:

$$\frac{1}{\sqrt{1 - 2xt + t^2}} = \sum_{n=0}^{\infty} P_n(x)t^n. \tag{14}$$

The function $\frac{1}{\sqrt{1-2xt+t^2}}$ appears when one puts the Coulomb potential into spherical coordinates, and hence the Legendre polynomials are related to spherical harmonics.

The Coulomb potential $f(\mathbf{x})$ at a point $\mathbf{x}$ in $\mathbb{R}^3$ due to a point charge at $\mathbf{p}$ is a constant times the reciprocal of the distance from $\mathbf{x}$ to $\mathbf{p}$. Thus

$$f(\mathbf{x}) = \frac{c}{||\mathbf{x} - \mathbf{p}||}. \tag{15}$$

The electric field due to this point charge is $-\nabla f$. See the discussion in Section 5 of Chapter 3.

Here it is convenient to place the charge at $\mathbf{p} = (0, 0, 1)$. We use spherical coordinates to write

$$\mathbf{x} = (\rho \cos(\theta) \sin(\phi), \rho \sin(\theta) \sin(\phi), \rho \cos(\phi)). \tag{16}$$

Substitute (16) into (15). An easy simplification then expresses the potential function as

$$\frac{c}{\sqrt{1 - 2\rho \cos(\phi) + \rho^2}}. \tag{17}$$

We obtain the left-hand side of (14) with x replaced by $\cos(\phi)$ and t by ρ. Thus $P_n(\cos(\phi))$ is the coefficient of ρ^n in the Taylor expansion of (17). See [D1] for more details.

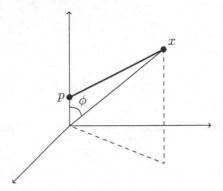

FIGURE 5. The coulomb force and spherical coordinates

EXAMPLE 3.1. Consider a **dipole**. We place opposite charges (of equal magnitude q) at $\mathbf{p}$ and $-\mathbf{p}$. Assume $\mathbf{p}$ is as in Figure 5, but with coordinates $(0, 0, \frac{a}{2})$. The electric potential energy due to these charges at a point $\mathbf{x}$ is then

$$\frac{q}{||\mathbf{x} - \mathbf{p}||} - \frac{q}{||\mathbf{x} + \mathbf{p}||}. \tag{17.1}$$

The number a is called the **dipole bond length**. We can then expand the potential energy as a function of a. The Taylor series contains only terms for odd powers of a, as (17.1) defines an odd function of a. Again, the coefficients are Legendre polynomials in ϕ times powers of ρ.

EXERCISE 3.1. Let $\{f_n\}$ be elements of a Hilbert space. Show that $\langle f_j, f_k \rangle = \delta_{jk}$ (Kronecker delta) if and only if

$$|| \sum_{n=0}^{\infty} f_n t^n ||^2 = \frac{1}{1 - |t|^2}.$$

Use this formula and (14) to show that $||P_n||^2 = \frac{2}{2n+1}$.

EXERCISE 3.2. Verify Bonnet's recursion formula:

$$(n + 1)P_{n+1}(x) = (2n + 1)xP_n(x) - nP_{n-1}(x).$$

EXERCISE 3.3. Find numbers $c_{n,k}$ such that

$$P_n(x) = \sum c_{n,k}(1+x)^{n-k}(1-x)^k.$$

EXERCISE 3.4. Find the coefficient of a^{2n+1} in the Taylor series expansion of (17.1) in powers of a.

4. Probability

Prerequisites. This section is self-contained; the only prerequisites are calculus and some familiarity with the Gaussian integral discussed also in Chapter 5.

We give a rapid discussion of some aspects of probability to prepare for the next section on quantum mechanics. At the end of the section, we discuss the usual abstract axioms for probability.

We begin, as is customary, with the situation when there are finitely many possible distinct outcomes $s_1, \ldots, s_N$. We write $\mathbf{X}$ to denote what is called a (numerical) *random variable*; we will be more precise about this term a bit later. Let p_j denote the probability that a random experiment results in an outcome with value j. We write

$$\mathbf{Prob}(\mathbf{X} = j) = p_j.$$

Then we have $0 \le p_j \le 1$ and $\sum_{j=1}^{N} p_j = 1$. The *expectation* or *mean* of $\mathbf{X}$ is defined by

$$\mathbf{E}(\mathbf{X}) = \mu_{\mathbf{X}} = \sum_{j=1}^{N} j p_j.$$

For a nice function f of $\mathbf{X}$, we also define expectations by

$$\mathbf{E}(f(\mathbf{X})) = \sum_{j=1}^{N} f(j) p_j.$$

Expectation is linear for functions of a given random variable:

$$\mathbf{E}((f+g)(\mathbf{X})) = \sum_{j=1}^{N}(f+g)(j)p_j = \sum_{j=1}^{N}(f(j)+g(j))p_j = \mathbf{E}(f(\mathbf{X})) + \mathbf{E}(g(\mathbf{X})).$$

$$\mathbf{E}(cf)(\mathbf{X}) = c\mathbf{E}(f(\mathbf{X})).$$

The *variance* $\sigma_{\mathbf{X}}^2$ of $\mathbf{X}$ is defined by

$$\sigma_{\mathbf{X}}^2 = \mathbf{E}((\mathbf{X} - \mu_{\mathbf{X}})^2).$$

When the situation is given and the random variable $\mathbf{X}$ is understood, we drop the $\mathbf{X}$ as a subscript on the mean and variance. The mean is the usual average, and the variance is the sum of squared deviations from the mean.

The following well-known result gives a second formula for the variance; we have seen the proof in several previous contexts, such as the parallel axis theorem. After we consider continuous probability densities, we will discuss it again.

Proposition 4.1. *Let* $\mathbf{X}$ *be a numerical random variable with variance* σ^2. *Then*

$$\sigma^2 = \mathbf{E}(\mathbf{X}^2) - \left(\mathbf{E}(\mathbf{X})\right)^2.$$

PROOF. The result follows easily from the linearity of expectation:

$$\mathbf{E}((\mathbf{X} - \mu)^2) = \mathbf{E}(\mathbf{X}^2) - \mathbf{E}(2\mu\mathbf{X}) + \mathbf{E}(\mu^2) = \mathbf{E}(\mathbf{X}^2) - 2\mu^2 + \mu^2\mathbf{E}(1). \tag{18}$$

The expectation of the constant function 1 is itself equal to 1, because it computes the sum of the probabilities of all possible outcomes. Hence the last two terms on the right-hand side of (18) combine to form $-\mu^2$. □

Probability becomes more interesting when one allows for continuous random variables. For simplicity we work on the real line.

Definition 4.2. A probability density function is a piecewise continuous function $f : \mathbb{R} \to [0, 1]$ such that $\int_{-\infty}^{\infty} p(x)dx = 1$. Given such a function p we define a random variable $\mathbf{X}$ by the rule

$$\mathbf{Prob}(\mathbf{X} \leq t) = \int_{-\infty}^{t} p(x)dx. \tag{19}$$

For $s \leq t$ it follows from (19) that

$$\mathbf{Prob}(s \leq \mathbf{X} \leq t) = \int_{s}^{t} p(x)dx. \tag{20}$$

In particular, when $s = t$, formula (20) gives 0. Thus the probability of any specific value is 0.

Definition 4.3. Let $\mathbf{X}$ be the random variable in (19). The mean μ and variance σ^2 are defined by

$$\mu = \int_{-\infty}^{\infty} xp(x)dx$$

$$\sigma^2 = \int_{-\infty}^{\infty} (x - \mu)^2 p(x)dx.$$

For the mean to be defined, the integral defining it must be absolutely convergent. Thus we require

$$\int_{-\infty}^{\infty} |x|p(x)dx < \infty.$$

Even if the mean exists, the integral defining the variance might diverge. Often one must hypothesize that this integral converges; one therefore often sees the words "with finite variance" in the hypotheses of theorems from probability. Another way to state that assumption is that $x - \mu$ is square-integrable with respect to the weight function p. Exercise 4.3 provides an interesting example.

As before, we define the expectation of functions of $\mathbf{X}$ by

$$\mathbf{E}(f(\mathbf{X})) = \int_{-\infty}^{\infty} f(x)p(x)dx.$$

Expectation is linear and we have $\sigma^2 = \mathbf{E}(\mathbf{X}^2) - (\mathbf{E}(\mathbf{X}))^2$.

EXAMPLE 4.1 (The uniform distribution). Put $p(x) = \frac{1}{b-a}$ for $a \leq x \leq b$ and $p(x) = 0$ otherwise. For $a \leq s \leq t \leq b$ we have

$$\mathbf{Prob}(s \leq \mathbf{X} \leq t) = \frac{1}{b-a} \int_{s}^{t} dx = \frac{t-s}{b-a}.$$

Routine computations show that the mean is $\frac{a+b}{2}$ and the variance is $\frac{(b-a)^2}{12}$. See Exercise 4.2.

We also wish to mention probability density functions on domains in higher dimensions. Let Ω be a domain in $\mathbb{R}^n$. Assume that p is a continuous density function on Ω; thus

$$\int_\Omega p(x)dV(x) = 1.$$

We do not attempt to define *nice subset*. We imagine that there is a collection of nice subsets on which the integrals are defined. We assume that Ω itself is nice, that the collection is closed under taking complements, and that the collection is closed under countable unions and countable intersections. If E is a nice subset of Ω, then we define

$$\mathbf{Prob}(\mathbf{X} \in E) = \int_E p(x)dV(x).$$

EXAMPLE 4.2. The uniform distribution on a rectangle. Consider the rectangle R defined by $a \le x \le b$ and $c \le y \le d$. Its area is $(b-a)(d-c)$. We define the uniform random variable on this rectangle by decreeing that

$$\mathbf{Prob}(\mathbf{X} \in E) = \frac{\text{Area}(E)}{\text{Area}(R)}$$

for nice subsets E.

We next turn to perhaps the most important probability density function.

EXAMPLE 4.3 (Gaussians). See also Example 4.2 of Chapter 5. Put

$$p(x) = \frac{1}{\sqrt{2\pi}}e^{\frac{-x^2}{2}}.$$

Then p is a probability density function on the real line. The corresponding random variable is called the *standard normal* or *standard Gaussian*. The mean is 0 and the variance is 1. The normal random variable with mean μ and variance σ^2 has probability density function given by

$$p(x) = \frac{1}{\sqrt{2\pi}\sigma}e^{\frac{-(x-\mu)^2}{2\sigma^2}}. \tag{21}$$

We end this section with an axiomatic description of probability. Consider an experiment whose outcome is not certain. The *sample space* S consists of all possible outcomes; an *event* is a subset of S. In some situations, all subsets are events. In general, however, it is not possible to make this assumption. The basic axioms for events are easy to state. The sample space itself is an event. Whenever E_j is a sequence of events, then the union of these events is also an event. This union is the event that at least one of the E_j occurs. Also the intersection of these events is an event. The intersection is the event that all of the E_j occur. For each event, its complement is also an event. In particular, since S is an event, the empty set is an event. If it makes sense to say that an event E occurs, it is only natural to assume that it makes sense for this event **not** to occur.

For each event E, we assign a probability $\mathbf{Prob}(E)$; we always have

$$0 \le \mathbf{Prob}(E) \le 1.$$

We assume $\mathbf{Prob}(S) = 1$. If E_1 and E_2 are mutually exclusive events, then

$$\mathbf{Prob}(E_1 \cup E_2) = \mathbf{Prob}(E_1) + \mathbf{Prob}(E_2). \tag{22}$$

Formula (22) is not completely adequate; one must replace it with a countable version. Namely, suppose for each positive integer n, there is an event E_n. Suppose in addition that $E_j \cap E_k$ is empty if $j \ne k$. Then

$$\mathbf{Prob}(E_1 \cup E_2 \cup \dots) = \sum_{n=1}^\infty \mathbf{Prob}(E_n). \tag{23}$$

Formula (23) is called *countable additivity*.

Several intuitive results are consequences of these axioms. For example, the probability of the empty set is 0. It then easily follows, for each event E, that the complement of E has probability $1 - \mathbf{Prob}(E)$. If E_1 is a subset of E_2, then $\mathbf{Prob}(E_1) \leq \mathbf{Prob}(E_2)$. Slightly more difficult, but still easy, is the simple law

$$\mathbf{Prob}(E_1) + \mathbf{Prob}(E_2) = \mathbf{Prob}(E_1 \cup E_2) + \mathbf{Prob}(E_1 \cap E_2). \tag{24}$$

Formula (24) is easily understood by drawing a Venn diagram. For example, if we interpret probability as area, then (24) says that the area of the union of two sets is the sum of their areas minus the part that gets double counted, namely the area of the intersection.

By induction, one can generalize (24) to find the probability of the union of n events. The resulting formula is sometimes known as the **inclusion-exclusion principle**. We write down the result for three events, called E, F, G. We write $\mathbf{P}$ instead of $\mathbf{Prob}$ to shorten the formula.

$$\mathbf{P}(E \cup F \cup G) = \mathbf{P}(E) + \mathbf{P}(F) + \mathbf{P}(G) - \mathbf{P}(E \cap F) - \mathbf{P}(E \cap G) - \mathbf{P}(F \cap G) + \mathbf{P}(E \cap F \cap G).$$

Definition 4.4. A *random variable* is a real-valued function defined on the possible outcomes of an experiment. A random variable is *finite* if it takes on only a finite set of values. Let Ω be a subset of $\mathbb{R}^n$. A random variable $\mathbf{X}$ is a real-valued function on Ω. It is called *continuous* if there is a non-negative continuous real-valued function f, called its probability density function, such that

- For each event E,

$$\mathbf{Prob}(\mathbf{X} \in E) = \int_E f(x)dV(x).$$

  Here dV is the n-dimensional volume form.
- $\int_\Omega f(x)dV(x) = 1$.

When more than one random variable appears in the same setting, one denotes the density function corresponding to $\mathbf{X}$ by $f_{\mathbf{X}}$.

In case Ω is the real line, intervals are always nice sets, and

$$\mathbf{Prob}(s \leq \mathbf{X} \leq t) = \int_s^t f(x)dx.$$

The function $t \to F(t) = \int_{-\infty}^t f(x)dx$ is called the *cumulative distribution function* of $\mathbf{X}$.

We develop these ideas in the exercises. The results of Exercises 4.13 and 4.14 might seem at first a bit counter-intuitive. But, any sports fan has seen examples where A defeats B, B defeats C, but C defeats A. Nonetheless, it is somewhat surprising that such a result can happen with dice. Such dice are called *non-transitive*.

EXAMPLE 4.4. A famous example of non-transitive dice is called *Schwenk's dice*. Each random variable is the roll of a die. A has sides $1, 1, 1, 13, 13, 13$. B has sides $0, 3, 3, 12, 12, 12$. C has sides $2, 2, 2, 11, 11, 14$. Each die totals 42. See Figure 6. In each case, assume all six sides are equally likely to occur. In this example, the probability that $A > B$ equals $\frac{7}{12}$, the probability that $B > C$ equals $\frac{7}{12}$, and the probability that $C > A$ is also $\frac{7}{12}$.

EXERCISE 4.1. Roll two fair dice, and let $\mathbf{X}$ denote the sum of the two numbers shown. Find the probability p_j of each of the 11 possible outcomes E_j. Then compute the mean and variance.

EXERCISE 4.2. Compute the mean and variance of the uniform distribution.

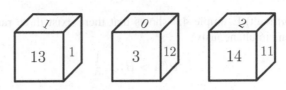

FIGURE 6. Non-transitive dice

EXERCISE 4.3 (The Cauchy distribution). Consider the probability density function $\frac{1}{\pi}\frac{1}{1+x^2}$ on the real line. Show that the mean and variance do not exist. Comment: One might want to say that the mean is 0 by symmetry, using the principal value integral. The definition of an improper integral requires letting the limits of integration tend to $-\infty$ and $+\infty$ separately. (Compare with Example 8.3 of Chapter 4 and its alternative derivation in Section 1 of Chapter 5.)

EXERCISE 4.4. Break a stick at random in two places. Determine the probability that the three pieces form a triangle. Suggestion: assume that the stick has length 1, and use the uniform distribution on the unit square. Graph $\{(x,y)\}$ such that choosing the points x and y in $[0,1]$ leads to the needed inequalities.

EXERCISE 4.5. Flip a fair coin $2n$ times. What is the probability of getting exactly n heads?

EXERCISE 4.6. Flip a fair coin 1000 times. What is the probability that, sometime during your experiment, you get 10 consecutive heads?

EXERCISE 4.7. Two equal teams play a series of games until one team wins 4 games. What is the expected number of games played? Answer the question when one team wins each game with probability p.

EXERCISE 4.8. Verify that: the function p in (21) integrates to 1, the mean is μ, and the variance is σ^2.

EXERCISE 4.9. Let $\mathbf{X}$ be the standard normal. Find $\mathbf{E}(\mathbf{X}^p)$ for $p \geq 1$. Suggestion: Use the Gamma function.

EXERCISE 4.10. By consulting tables (easily found online), find the probability that a standard normal random variable takes on a value between 0 and 2.

EXERCISE 4.11. Suppose that X is normal with mean 100 and variance 225. Find $\mathbf{P}(\mathbf{X} \geq 120)$.

EXERCISE 4.12. Recall that a permutation of a set S is a bijection from S to itself. Let S have n elements. Count the number t_n of permutations with no fixed points. Suggestion: use inclusion-exclusion. Find $\lim_{n\to\infty}\frac{t_n}{n!}$. The answer is $\frac{1}{e}$.

EXERCISE 4.13. Verify the claims about Schwenk's dice. Then show that, although $\mathbf{P}(A > B) > \frac{1}{2}$, if we roll each die *twice* and consider the sums A_s and B_s, then $\mathbf{P}(A_s > B_s) < \frac{1}{2}$.

EXERCISE 4.14. Interpret the sum of two random variables in terms of convolution.

EXERCISE 4.15. The Gamma distribution with parameter a is defined for $x \geq 0$ by the density

$$\frac{1}{\Gamma(a)}e^{-x}x^{a-1}.$$

Suppose $\mathbf{X}$ has the Gamma distribution with parameter a and $\mathbf{Y}$ has the Gamma distribution with parameter b. For $t \in \mathbb{R}$, find the probability that $\mathbf{X} + \mathbf{Y}$ is at most t.

EXERCISE 4.16. (Difficult) Example 4.4 shows that there exist three random variables A, B, C for which all three can happen simultaneously:

$$\mathbf{P}(A > B) > \frac{1}{2}$$

$$\mathbf{P}(B > C) > \frac{1}{2}$$

$$\mathbf{P}(C > A) > \frac{1}{2}.$$

Assign probabilities p_j to each of the six permutations of the letters A, B, C. Of course $0 \leq p_j$ for each j and $\sum p_j = 1$. Express the three conditions leading to non-transitivity as a set of inequalities on the p_j. Find the volume of the region where all three conditions occur. Exercise 4.15 might be useful.

5. Quantum mechanics

Prerequisites. The ideas used in this section require the previous section on probability and much of the material on Hilbert spaces from Chapter 6. Some material from Chapter 3 on vector analysis and basic things about complex numbers from Chapter 2 arise as well.

Quantum mechanics differs from classical physics in several ways. We do not attempt to discuss the physical implications of the theory; instead we formulate the basics of quantum mechanics, as is now standard, in terms of Hilbert spaces. We give two formulations of the Heisenberg uncertainty principle (Theorems 5.3 and 5.4). We also clarify (in Theorem 5.1) a hypothesis needed to guarantee that the norm of a wave function is constant in time. We also apply separation of variables to study the Schrödinger equation.

The starting point is a Hilbert space $\mathcal{H}$ of states. We use the notation $\mathcal{H}$ in this section, reserving H for (our brief mention of) the Hamiltonian. Let $\langle \, , \, \rangle$ denote the inner product and $\| \ \|$ the norm. Often this Hilbert space is $L^2(\mathbb{R}^3)$. In this section $|x|$ will denote the Euclidean length of a vector in $\mathbb{R}^3$, to avoid confusion with the Hilbert space norm. The **wave function** $\boldsymbol{\Psi}$ of a particle is a solution of Schrödinger's equation, discussed below. For now $\boldsymbol{\Psi}$ will be a function of space and time; at each time t we have $\boldsymbol{\Psi}(x, t) \in \mathcal{H}$ and

$$\|\boldsymbol{\Psi}(x, t)\|^2 = 1.$$

Thus the **state** of a particle is a unit vector in $\mathcal{H}$. An **observable** is a self-adjoint operator A on $\mathcal{H}$. The **expectation** of A on the state $\boldsymbol{\Psi}$ is given by

$$\mathbf{E}(A) = \langle A(\boldsymbol{\Psi}), \boldsymbol{\Psi} \rangle.$$

This definition is equivalent to the definition of expectation in the section on probability, because we can think of $|\boldsymbol{\Psi}|^2$ as a probability density.

Since A is self-adjoint, its eigenvalues are real. If we do a measurement, then (by definition) we must obtain an eigenvalue of A.

We first therefore assume that the eigenfunctions ϕ_n of A form a complete orthonormal system for $\mathcal{H}$. Let λ_n denote the eigenvalue corresponding to the eigenfunction ϕ_n. The probability $|c_n|^2$ of obtaining a specific eigenvalue λ_n is given by

$$|c_n|^2 = |\langle \boldsymbol{\Psi}, \phi_n \rangle|^2. \tag{25}$$

Orthonormal expansion provides the explanation!

Expand $\boldsymbol{\Psi} = \sum \langle \boldsymbol{\Psi}, \phi_n \rangle \phi_n$ and plug into $\langle A\boldsymbol{\Psi}, \boldsymbol{\Psi} \rangle$. Using orthonormality, the result is

$$\langle A\boldsymbol{\Psi}, \boldsymbol{\Psi} \rangle = \sum_n |\langle \boldsymbol{\Psi}, \phi_n \rangle|^2 \lambda_n.$$

Since $||\mathbf{\Psi}||^2 = 1$, we have $\sum_n |\langle \mathbf{\Psi}, \phi_n \rangle|^2 = 1$. Thus, the probability of obtaining the eigenvalue λ_n is given by (25).

To see where all these ideas arise, we first consider the expression for energy as the sum of kinetic and potential energy. Suppose a particle of mass m has velocity vector $\mathbf{v}$. Then the kinetic energy is $\frac{1}{2}m||\mathbf{v}||^2$. Let the potential energy be denoted by V. Thus the total energy E is given by

$$E = \frac{1}{2}m||\mathbf{v}||^2 + V. \tag{26}$$

The momentum vector $\mathbf{p}$ equals $m\mathbf{v}$. Let $\hbar$ denote Planck's constant divided by 2π. This extremely small number is approximately 10^{-34} Joule-seconds.

Quantization consists in substituting $\frac{\hbar}{i}\nabla$ for the momentum. The energy in (26) becomes

$$H = -\frac{\hbar^2}{2m}\nabla^2 + V.$$

The operator H is called the (standard) **Hamiltonian**. It is a real operator. The (time-dependent) Schrödinger equation is

$$i\hbar\mathbf{\Psi}_t = H\,\mathbf{\Psi}. \tag{27}$$

In (27), the subscript t means the partial derivative with respect to time. Without the factor of i, (27) would be a diffusion equation.

The solution $\mathbf{\Psi}$ defines a probability density function. Let us work in three dimensions. Thus $\mathbf{x}$ denotes position. Let Ω denote a nice subset of $\mathbb{R}^3$. The probability that a particle is found in the set Ω at time t is given by

$$\mathbf{Prob}(\mathbf{X} \in \Omega) = \int_\Omega |\mathbf{\Psi}(\mathbf{x},t)|^2 dV(\mathbf{x}). \tag{28}$$

For (28) to make sense, we need to have $||\mathbf{\Psi}(.,t)||^2 = 1$ for all t, where the L^2 norm is taken over all of $\mathbb{R}^3$. It is asserted in [Gr] that, if $||\mathbf{\Psi}(.,t)||^2 = 1$ for some time t, and $\mathbf{\Psi}$ satisfies the Schrödinger equation, then $||\mathbf{\Psi}(.,t)||^2 = 1$ for all t. The incomplete proof there requires an additional assumption on $\mathbf{\Psi}$ to be correct. Any assumption that guarantees that $\mathbf{\Psi}$ vanishes at infinity suffices. See Exercise 5.2.

We make the following extra assumption about wave functions. Not only does $\mathbf{\Psi}$ satisfy equation (27), but also we assume $\nabla\mathbf{\Psi}$ satisfies the following property for all t:

$$\lim_{R\to\infty}\int_{|x|=R}\overline{\mathbf{\Psi}}\nabla\mathbf{\Psi}\cdot d\mathbf{S} = 0. \tag{29}$$

Notice in (29) that $\nabla\mathbf{\Psi}(x,t)$ is a vector in $\mathbb{R}^3$ and that the flux integral is taken over a sphere of radius R. Assumption (29) is one of several ways one can guarantee that Theorem 5.1 holds.

Assumption (29) and the divergence theorem imply the following fundamental fact.

Theorem 5.1. *Suppose $\mathbf{\Psi}$ satisfies the Schrödinger equation in $\mathbb{R}^3$ and (29) holds. If $||\mathbf{\Psi}(.,t)||^2 = 1$ for some time t, then $||\mathbf{\Psi}(.,t)||^2 = 1$ for all t.*

PROOF. We start with

$$n(t) = ||\mathbf{\Psi}(.,t)||^2 = \langle \mathbf{\Psi}(.,t), \mathbf{\Psi}(.,t) \rangle$$

and differentiate with respect to t. We obtain

$$n'(t) = 2\text{Re}(\langle \mathbf{\Psi}_t, \mathbf{\Psi} \rangle). \tag{30}$$

Using the Schrödinger equation, we replace the time derivative $\mathbf{\Psi}_t$ in (30) with $\frac{1}{i\hbar}H\mathbf{\Psi}$. Thus for real constants c_1 and c_2 we have

$$n'(t) = 2\text{Re}(\langle \frac{1}{i\hbar}H\mathbf{\Psi}, \mathbf{\Psi} \rangle) = 2\text{Re}(i\langle c_1\nabla^2\mathbf{\Psi}, \mathbf{\Psi}\rangle + ic_2\langle V\mathbf{\Psi}, \mathbf{\Psi}\rangle). \tag{31}$$

We note the simple identity:

$$\nabla^2(\boldsymbol{\Psi})\,\overline{\boldsymbol{\Psi}} = \mathrm{div}(\overline{\boldsymbol{\Psi}}\nabla\boldsymbol{\Psi}) - |\nabla\boldsymbol{\Psi}|^2. \tag{32}$$

Then we integrate (32) over $\mathbb{R}^3$ to obtain

$$\langle\nabla^2\boldsymbol{\Psi},\boldsymbol{\Psi}\rangle = \int \mathrm{div}(\overline{\boldsymbol{\Psi}}\nabla\boldsymbol{\Psi})dV - \int |\nabla\boldsymbol{\Psi}|^2 dV. \tag{33}$$

Plug (33) into (31). The terms in $n'(t)$ involving $|\nabla\boldsymbol{\Psi}|^2$ and $\langle V\boldsymbol{\Psi},\boldsymbol{\Psi}\rangle$ each get multiplied by i, and hence the real part of the corresponding terms vanish. The only remaining term in $n'(t)$ is given for some constant c by

$$n'(t) = c\int \mathrm{div}(\overline{\boldsymbol{\Psi}}\nabla\boldsymbol{\Psi})dV = c\lim_{R\to\infty}\int_{|x|\le R} \mathrm{div}(\overline{\boldsymbol{\Psi}}\nabla\boldsymbol{\Psi})dV.$$

For any ball of radius R centered at 0, we apply the divergence theorem to obtain an integral over the sphere of radius R:

$$n'(t) = c\lim_{R\to\infty}\int_{|x|=R} \overline{\boldsymbol{\Psi}}\nabla\boldsymbol{\Psi}\cdot d\mathbf{S}.$$

Assumption (29) implies that the limit must be 0, and thus $n'(t) = 0$. Hence $n(t)$ is a constant. $\square$

REMARK. Just because a function is square-integrable does not force it or its derivative to vanish at infinity. Here is an interesting example in one dimension. Put

$$f(x) = \frac{\sin(\lambda x^2)}{\sqrt{x^2+1}}.$$

Then $|f(x)|^2 \le \frac{1}{x^2+1}$ (no matter what λ is) and hence f is square-integrable. Its derivative, however, behaves like $2\lambda\cos(\lambda x^2)$ for large x, and in particular does not tend to 0. See also Exercises 5.2 and 5.11.

Definition 5.2. Let A, B be linear maps on a vector space V. Their commutator $[A, B]$ is the linear map $AB - BA$. A linear map C is called a *commutator* if there exist A and B such that $C = [A, B]$.

Consider the special case of quantum mechanics on the real line. Then multiplication by x denotes the position operator A and $-i\hbar\frac{d}{dx}$ denotes the momentum operator B. We compute their commutator:

$$[A,B](f) = (AB - BA)(f) = -ix\hbar f' - \left(-i\hbar\frac{d}{dx}(xf)\right) = i\hbar f.$$

Thus $[A, B]$ is a tiny imaginary multiple of the identity. This formula implies that we cannot simultaneously measure both position and momentum. Since the commutator is not zero, it matters which measurement is done first. This result is known as the *Heisenberg uncertainty principle*. A general result holds for self-adjoint operators on a Hilbert space. Later in the section we give a simpler version.

Theorem 5.3 (General Heisenberg uncertainty principle). *Assume that A and B are self-adjoint operators on a Hilbert space $\mathcal{H}$. Let $[A, B]$ denote their commutator $AB - BA$. Let σ_A^2 and σ_B^2 denote the variances with respect to the probability density function $|\Psi|^2$. Then*

$$\sigma_A^2\sigma_B^2 \ge \frac{1}{4}(\mathbf{E}([A,B]))^2.$$

PROOF. We estimate the product of the variances using the Cauchy–Schwarz inequality as follows: By definition of variance and expectation, we have

$$\sigma_A^2\sigma_B^2 = \mathbf{E}((A-\mu_A)^2))\,\mathbf{E}((B-\mu_B)^2) = \langle(A-\mu_A)^2\boldsymbol{\Psi},\boldsymbol{\Psi}\rangle\,\langle(B-\mu_B)^2\boldsymbol{\Psi},\boldsymbol{\Psi}\rangle. \tag{34}$$

Since A and B are self-adjoint, their expectations are real, and hence both $A - \mu_A$ and $B - \mu_B$ are also self-adjoint. We can therefore rewrite (34) as

$$\sigma_A^2 \sigma_B^2 = ||(A - \mu_A)\boldsymbol{\Psi}||^2 \, ||(B - \mu_B)\boldsymbol{\Psi}||^2. \tag{35}$$

By the Cauchy–Schwarz inequality, we have

$$||(A - \mu_A)\boldsymbol{\Psi}||^2 \, ||(B - \mu_B)\boldsymbol{\Psi}||^2 \geq |\langle(A - \mu_A)\boldsymbol{\Psi}, (B - \mu_B)\boldsymbol{\Psi}\rangle|^2. \tag{36}$$

For any complex number z we have

$$|z|^2 \geq (\mathrm{Im}(z))^2 = (\frac{1}{2i}(z - \overline{z}))^2 = \frac{-1}{4}(z - \overline{z})^2.$$

Using this inequality with z equal to the inner product in (36) gives

$$\sigma_A^2 \sigma_B^2 = \mathbf{E}((A - \mu_A)^2)) \, \mathbf{E}((B - \mu_B)^2) \geq |\langle(A - \mu_A)\boldsymbol{\Psi}, (B - \mu_B)\boldsymbol{\Psi}\rangle|^2$$
$$\geq \frac{-1}{4}\langle(A - \mu_A)\boldsymbol{\Psi}, (B - \mu_B)\boldsymbol{\Psi}\rangle - \langle(B - \mu_B)\boldsymbol{\Psi}, (A - \mu_A)\boldsymbol{\Psi}\rangle)^2. \tag{37}$$

Now move the self-adjoint operator $B - \mu_B$ over in the first term in (37) and the self-adjoint operator $A - \mu_A$ over in the second term to obtain $\frac{-\langle C\boldsymbol{\Psi}, \boldsymbol{\Psi}\rangle}{4}$. Here C satisfies

$$C = (B - \mu_B)(A - \mu_A) - (A - \mu_A)(B - \mu_B) = BA - AB.$$

Thus

$$\sigma_A^2 \sigma_B^2 \geq \frac{1}{4}\langle(AB - BA)\boldsymbol{\Psi}, \boldsymbol{\Psi}\rangle = \frac{1}{4}(\mathbf{E}(AB - BA))^2.$$

$\square$

The main point of Theorem 5.3 is that the product of the variances is bounded below by a positive quantity, unless A and B commute. Hence the result of performing two measurements depends upon the order in which the two are done. We noted above, in the one-dimensional case, that the position and momentum operators do not commute. For position, A is multiplication by x. For momentum, B is $-i\hbar\frac{d}{dx}$. Their commutator is multiplication by $i\hbar$. Note that B is unbounded; the next remark provides a partial explanation.

REMARK. If A, B are finite-dimensional matrices, then $[A, B]$ cannot be a non-zero multiple of the identity. To see why, take the trace of both sides. The trace of a commutator is always 0, whereas the trace of a multiple of the identity is that multiple of the dimension. It is also true that the commutator of bounded operators on a Hilbert space cannot be a multiple of the identity. It happens in quantum mechanics because the momentum operator involves differentiation, and hence is an unbounded operator on the Hilbert space.

Next we give a version of the Heisenberg uncertainty principle for position and momentum on the real line. This inequality helps clarify why we assume the first space derivative of a wave function is square-integrable. Recall that multiplication by x corresponds to position and differentiation corresponds to (a constant times) momentum.

Theorem 5.4. *Assume that both $x\boldsymbol{\Psi}$ and $\boldsymbol{\Psi}'$ are square integrable on $\mathbb{R}$. Then*

$$||\boldsymbol{\Psi}||^4 \leq 4||x\boldsymbol{\Psi}||^2 \, ||\boldsymbol{\Psi}'||^2.$$

PROOF. We start by differentiating $x\boldsymbol{\Psi}\overline{\boldsymbol{\Psi}}$ to get

$$|\boldsymbol{\Psi}|^2 = (x\boldsymbol{\Psi}\overline{\boldsymbol{\Psi}})' - 2\mathrm{Re}(x\boldsymbol{\Psi}\boldsymbol{\Psi}'). \tag{38}$$

Integrating (38) gives

$$||\boldsymbol{\Psi}||^2 = -2\mathrm{Re}(\langle x\boldsymbol{\Psi}, \boldsymbol{\Psi}'\rangle), \tag{39}$$

because the hypotheses imply that $|x|\,|\Psi|^2$ tends to 0 at $\pm\infty$. Applying the Cauchy–Schwarz inequality to (39) then gives

$$||\Psi||^2 \leq 2||x\Psi||\,||\Psi'||. \tag{40}$$

Squaring (40) gives the conclusion.

$\square$

In our section on the wave equation, we saw the effectiveness of separation of variables in analyzing a PDE. Let us now use separation of variables on the Schrödinger equation. For simplicity, we work in one space dimension. We also assume that the potential function V is independent of time.

Let us assume that the wave function Ψ satisfies

$$\Psi(x,t) = A(x)B(t).$$

Plug this expression into the Schrödinger equation and collect the terms depending on time. The result is

$$i\hbar\frac{B'(t)}{B(t)} = \frac{-\hbar^2}{2m}\frac{A''(x)}{A(x)} + V(x). \tag{41}$$

The left-hand side of (41) depends on time but not on x; the right-hand side depends on x but not on time. Hence each is a constant λ. We conclude that

$$B(t) = e^{\frac{-i\lambda t}{\hbar}} \tag{42}$$

and that A satisfies the second-order ODE

$$\frac{-\hbar^2}{2m}A''(x) + (V(x) - \lambda)A(x) = 0. \tag{43}$$

The equation (43) is often called the **time-independent Schrödinger equation**. It is a second order ODE of the type we have studied in Chapter 6. In terms of the Hamiltonian, we can rewrite (43) in the form

$$HA = \lambda A.$$

For each possible eigenvalue λ_n of the problem we obtain a solution A_n. Such a λ_n is called an **allowed energy**. The discrete nature of quantum mechanics arises because not all energy levels are possible.

We can expand solutions of the general Schrödinger equation in the form

$$\Psi(x,t) = \sum_{n=1}^{\infty} c_n e^{\frac{-i\lambda_n t}{\hbar}} A_n(x).$$

Once again we observe the significance of eigenvalues and expansions in terms of eigenfunctions.

Situations in mathematical physics where the spectrum of a differential operator is more complicated than simply a discrete set of points also arise. We therefore briefly discuss the notion of continuous spectrum. Consider a bounded operator L on a Hilbert space. Based on finite-dimensional intuition, we might expect the spectrum of L to be a discrete set of eigenvalues. Doing so does not capture the right concept. Rather than defining the spectrum $\sigma(L)$ to be the set of eigenvalues of L, we define it to the set of λ for which $(L - \lambda I)$ does not have an inverse. Then $\sigma(L)$ is a closed subset of $\mathbb{C}$. In the finite-dimensional case, these two concepts are identical. In the infinite-dimensional situation, they are not the same.

In the infinite-dimensional setting, there are two ways that an operator can fail to be invertible. It could have a non-trivial null space, that is, fail to be injective, or it could fail to be surjective. We give a beautiful example illustrating some of these issues.

EXAMPLE 5.1. Let the Hilbert space $\mathcal{H}$ consist of all square-integrable complex-analytic functions in the unit disk. Thus $\iint |f(z)|^2 dx\, dy < \infty$, where the integral is taken over the unit disk. We define $L : \mathcal{H} \to \mathcal{H}$ by $Lf(z) = \frac{f(z)-f(0)}{z}$. We find the eigenvalues of L. Setting $Lf = \lambda f$ gives

$$\frac{f(z) - f(0)}{z} = \lambda f(z).$$

Solving for f gives $f(z) = \frac{f(0)}{1-\lambda z}$. Hence, the function $\frac{1}{1-\lambda z}$ is an eigenvector as long as it is square integrable, which happens for $|\lambda| < 1$. Thus every point in the open unit disk is an eigenvalue! Since the spectrum of an operator is a closed set, the spectrum here is the entire closed unit disk. The boundary circle is part of the spectrum but does not consist of eigenvalues.

EXERCISE 5.1. Let V be a finite-dimensional vector space. Show that the trace of a commutator is 0. (Difficult) Suppose that the trace of C is 0. Prove that C is a commutator. Suggestion: Proceed by induction on the dimension of V.

EXERCISE 5.2. Sketch the graph of a smooth non-negative function f on $\mathbb{R}$ such that $\int_{-\infty}^{\infty} f(x)dx = 1$ but $\lim_{x\to\infty} f(x)$ does not exist. Note that the derivative of such a function must be often large. Explain why such a function would create a counterexample to Theorem 5.1 if we omitted assumption (29).

EXERCISE 5.3. Verify the identity (32).

EXERCISE 5.4. Assume that A and B are self-adjoint. Show that AB is self-adjoint if and only if A and B commute.

EXERCISE 5.5. Show that $[[A,B],C] + [[C,A],B] + [[B,C],A] = 0$.

EXERCISE 5.6. Suppose $[A,B] = I$. Show by induction that $[A,B^n] = nB^{n-1}$.

EXERCISE 5.7. Use the previous exercise to show that there are no bounded operators whose commutator is the identity.

EXERCISE 5.8. Find a complete orthonormal system $\{\phi_n\}$ for the Hilbert space in Example 5.1. Then prove that

$$\sum_{n=1}^{\infty} |\phi_n(z)|^2 = \frac{1}{\pi}\frac{1}{(1-|z|^2)^2}. \tag{44}$$

The function in (44) is known as the **Bergman kernel** function of the unit disk.

EXERCISE 5.9. The Hamiltonian for the harmonic oscillator of frequency ω is given by

$$H = \frac{1}{2m}\left((-i\hbar\frac{d}{dx})^2 + (m\omega x)^2\right).$$

Here m is the mass. By factoring the operator inside the parentheses, find operators $a_{\pm}$ such that

$$H = a_- a_+ - \frac{\hbar\omega}{2}$$

$$H = a_+ a_- + \frac{\hbar\omega}{2}.$$

Determine the commutator $[a_+, a_-]$. These operators are called **ladder operators**. The operator a_+ raises the energy level and a_- lowers it. Determine the relationship of these operators to the stationary states of the harmonic oscillator, expressed in terms of the Hermite polynomials from Exercise 4.15 of Chapter 5.

EXERCISE 5.10. The quantum mechanical analogue of charge density is $|\Psi|^2$. The quantum mechanical analogue of current density is the **probability current** given by

$$\mathbf{J} = \frac{\hbar}{2im}\left(\overline{\Psi}\nabla\Psi - \Psi\nabla\overline{\Psi}\right).$$

Using the Schrödinger equation, verify the continuity equation

$$0 = \text{div}(\mathbf{J}) + \frac{\partial\rho}{\partial t}.$$

Compare with Maxwell's equations.

EXERCISE 5.11. For $m > 0$ put $f(x) = \frac{\sin(x^m)}{\sqrt{1+x^2}}$. Show that $f \in L^2(\mathbb{R})$. Compute $f'(x)$ and determine the condition on m that ensures $f(x)f'(x)$ tends to 0 at infinity.

6. The Dirichlet problem and conformal mapping

Prerequisites. The ideas used in this section rely on Chapters 2 and 4 on Complex Analysis. Fourier series makes a brief appearance as well.

Complex variable theory provides a powerful classical approach to one of the most important applied partial differential equations. Given a region Ω in $\mathbb{C}$ with reasonably nice boundary $b\Omega$, and a nice function g on $b\Omega$, the **Dirichlet problem** is to find a function u that is harmonic on Ω and agrees with g on $b\Omega$. Several physical models lead to this problem. For example, u could be the steady state temperature distribution resulting from maintaining the temperature on the boundary to be the function g. Electric potential energy provides a second example. For a region Ω in the plane, we can think of g as giving the distribution of charge on $b\Omega$; then u gives the electric potential on Ω. Ideal fluid flow (both incompressible and irrotational) provides a third model.

Because of its applied importance, the Dirichlet problem has been studied extensively. We limit our discussion to situations where complex variable theory provides great insight, with the primary purpose being to unify many of the ideas of the book. Fourier series and power series appear. Corollary 2.6 of Chapter 4, stating that the real and imaginary parts of complex analytic functions are harmonic, is of crucial importance. Lemma 6.5 and Corollary 6.7 below link conformal maps with the Dirichlet problem.

Our first Dirichlet problem concerns the unit disk. Because of the Riemann mapping theorem, which is beyond the scope of this book, this special case turns out to be remarkably general. The upper half-plane also provides a useful model.

EXAMPLE 6.1. Let Ω be the unit disk, and consider a trig polynomial

$$g(\theta) = \sum_{n=-d}^{d} a_n \cos(n\theta).$$

To make g real-valued, we assume $a_{-n} = \overline{a_n}$. In particular a_0 is real. Put $z = e^{i\theta}$. Recall that

$$\cos(n\theta) = \text{Re}(e^{in\theta}) = \text{Re}(z^n).$$

With $z = x + iy$ we define a function u by

$$u(x,y) = a_0 + 2\text{Re}\left(\sum_{n=1}^{d} a_n z^n\right).$$

Then u is the real part of a complex analytic polynomial, and hence u is harmonic. Furthermore, $u = g$ on the circle. We have solved a Dirichlet problem by inspection.

In case g is continuous on the circle, with Fourier series

$$g(\theta) = \sum_{-\infty}^{\infty} a_n e^{in\theta},$$

the same approach solves the Dirichlet problem. One must however be a bit careful about convergence issues. Thus the Dirichlet problem for the unit disk is closely connected to Fourier series. See Theorem 6.3.

Theorem 6.3 solves the Dirichlet problem by using the Poisson kernel as an approximate identity.

Definition 6.1 (Poisson kernel). For $z = re^{i\theta}$ and $r < 1$, put

$$P_r(\theta) = \frac{1 - |z|^2}{|1 - z|^2}.$$

Lemma 6.2. *For* $0 \le r < 1$ *we have*

$$P_r(\theta) = \sum_{n=-\infty}^{\infty} r^{|n|} e^{in\theta}.$$

PROOF. Using the geometric series, the sum can be rewritten as

$$1 + 2\text{Re} \sum_{n=1}^{\infty} r^n e^{in\theta} = 1 + 2\text{Re} \sum_{n=1}^{\infty} z^n = 1 + \frac{z}{1-z} + \frac{\overline{z}}{1-\overline{z}} = \frac{1 - |z|^2}{|1-z|^2}.$$

$\square$

Theorem 6.3. *Assume* g *is continuous on the unit circle. Let* u *be the convolution:*

$$u(re^{i\theta}) = (P_r * g)(\theta).$$

Then u *is infinitely differentiable,* u *is harmonic, and* $u = g$ *on the circle.*

PROOF. We sketch the proof, omitting some analysis required to be rigorous. We assume g is given by the Fourier series

$$g(\theta) = \sum_{n=-\infty}^{\infty} \hat{g}(n) e^{in\theta}.$$

A simple computation and Lemma 6.2 then give

$$(P_r * g)(\theta) = \sum_{n=-\infty}^{\infty} \hat{g}(n) r^{|n|} e^{in\theta}. \tag{45}$$

As r increases to 1, the sum in (45) is Abel summable (See Exercise 2.12 of Chapter 5) to the Fourier series of g. Since g is real-valued, it follows that $\hat{g}(-n) = \overline{\hat{g}(n)}$. Hence the sum in (45) is $\hat{g}(0)$ plus twice the real part of a convergent power series in z. Since the real part of a complex analytic function is infinitely differentiable and harmonic, the result follows. $\square$

The technique of conformal mapping enables one to solve the Dirichlet problem for *simply connected* domains other than the whole plane. Roughly speaking, a domain is simply connected if it has no holes. An annulus is not simply connected, whereas the upper half-plane is simply connected. The complement of a branch cut is also simply connected. Here is the precise definition.

Definition 6.4. An open connected subset Ω of $\mathbb{C}$ is **simply connected** if each simple closed curve in Ω encloses no point in the complement of Ω.

When a domain Ω is simply connected, but not the entire space $\mathbb{C}$, the Riemann mapping theorem states that there is a conformal map from Ω to the unit disk. By Corollary 6.7, solving the Dirichlet problem on Ω reduces to solving it on the unit disk. Exercise 6.10 asks why the whole plane is excluded; the crucial issue is Liouville's theorem.

We define and discuss conformal mappings next. Recall that multiplication by a non-zero complex number amounts to a rotation of the angle. In other words, the argument of a product ζw is the sum of the arguments of ζ and w. How does this idea connect with differentiation? A function f is complex analytic near a point z_0 if, for each complex number h with sufficiently small magnitude, we have

$$f(z_0 + h) - f(z_0) = f'(z_0)h + \text{error}.$$

In other words, changing z_0 by adding h (infinitesimally) changes the argument of h by adding the argument of $f'(z)$. A complex analytic function with non-vanishing derivative therefore infinitesimally preserves angles. The preservation of angles motivates the following basic definition.

Definition 6.5. Let Ω_1 and Ω_2 be open subsets of $\mathbb{C}$. A complex analytic function $f : \Omega_1 \to \Omega_2$ is called a **conformal map** from Ω_1 to Ω_2 if f is injective, surjective, and f^{-1} is also complex analytic. When the domains are understood, we say simply f is **conformal**.

REMARK. The domains Ω_1 and Ω_2 are a crucial part of the definition. For example, the map $z \mapsto z^2$ is conformal from the open first quadrant to the open upper half-plane but it is not conformal from $\mathbb{C}$ to itself.

REMARK. The composition of conformal maps is a conformal map. Given domains Ω and U, one often finds a conformal map between them by finding many intermediate conformal maps. In other words, if $f : \Omega \to \Omega_1$ is conformal, $g : \Omega_1 \to \Omega_2$ is conformal, and $h : \Omega_2 \to U$ is conformal, then $h \circ g \circ f : \Omega \to U$ is conformal.

REMARK. One can test for conformality using the derivative, but some care is required. If f is conformal, then $f' \neq 0$ on Ω. See Exercise 6.12. If f' is never 0 on Ω, then f is locally injective, but not necessarily injective. As usual, e^z provides a compelling example.

To picture a conformal map, it is helpful to consider two copies of $\mathbb{C}$, the z-plane and the w-plane. Linear fractional transformations are conformal mappings. In Section 6 of Chapter 2 we saw how to conformally map disks to disks or to half-planes.

EXAMPLE 6.2. Figure 8 of Chapter 2 shows a conformal map from the upper half-plane to the unit disk. Figure 9 of Chapter 2 depicts a conformal map from a disk to another disk, and Figure 10 of Chapter 2 illustrates a conformal map from the unit disk minus the origin to the exterior of the unit disk.

EXAMPLE 6.3. Assume $|a| < 1$. Put $f(z) = \frac{a-z}{1-\overline{a}z}$. Then f is a conformal map from the unit disk to itself. This remarkable map sends a to 0 and 0 to a, but yet it is injective, surjective, and preserves angles. See Exercise 6.2 of Chapter 2 for one use of this map. See also Exercise 6.13. This map is also important in non-Euclidean geometry. See [D2] for an accessible introduction.

Conformal maps do not necessarily preserve angles on the boundary. We give a standard example where a right angle becomes a straight angle.

EXAMPLE 6.4. Let Ω_1 be the open set defined by $\frac{-\pi}{2} < \text{Re}(z) < \frac{\pi}{2}$ and $\text{Im}(z) > 0$. Put $f(z) = \sin(z)$. Then f is a conformal map from Ω_1 to the upper half-plane. Notice that the derivative equals 0 at the points $z = \pm\frac{\pi}{2}$. These points are on the boundary of Ω_1, and angles are not preserved there.

We have discussed Einstein's principle of choosing coordinates to make a problem easy. Conformal maps amount to changing coordinates. To apply this idea to the Dirichlet problem we need the following lemma and corollary.

Lemma 6.6. *Let Ω_1 and Ω_2 be open subsets of $\mathbb{C}$. Assume that $A : \Omega_2 \to \Omega_1$ is complex analytic, and u_1 is harmonic on Ω_1. Put $u_2 = u_1 \circ A$. Then u_2 is harmonic on Ω_2.*

PROOF. By Exercise 2.4 of Chapter 4, it suffices to prove that $(u_2)_{z\bar{z}} = 0$. Since A is complex analytic, $A_{\bar{z}} = 0$. Since u_1 is harmonic, $(u_1)_{z\bar{z}} = 0$. The chain rule therefore gives $(u_2)_z = (u_1)_z A'$ and then

$$(u_2)_{z\bar{z}} = (u_1)_{z\bar{z}}|A'|^2 = 0.$$

$\square$

We will apply Lemma 6.5 when F is a conformal map from Ω_1 to Ω_2, and $A = F^{-1}$.

Corollary 6.7. *Assume $f : \Omega_1 \to \Omega_2$ is a conformal map, and f extends to be bijective from $b\Omega_1$ to $b\Omega_2$. Let g_2 be continuous on $b\Omega_2$. Put $g_1 = g_2 \circ f$ and suppose u_1 solves the Dirichlet problem with boundary data g_1. Put $u_2 = u_1 \circ f^{-1}$. Then u_2 solves the Dirichlet problem with boundary data g_2.*

PROOF. Since f^{-1} is complex analytic, u_2 is harmonic by Lemma 6.5. On the boundary $b\Omega_2$ we have $u_2 = u_1 \circ f^{-1}$ and hence $g_2 = g_1 \circ f^{-1} = u_1 \circ f^{-1} = u_2$ there. $\square$

EXAMPLE 6.5. We seek a harmonic function u in the upper half-plane such that $u(x,0) = 1$ for $|x| < 1$ and $u(x,0) = 0$ for $|x| > 1$. (We do not define u at ± 1.) This problem is equivalent to the much simpler problem of finding a harmonic function in the strip $0 < \text{Im}(\zeta) < \pi$ whose value on the real axis is 0 and whose value on the line $\text{Im}(\zeta) = \pi$ is 1. This second problem is instantly solved by putting $h(x,y) = \frac{\text{Im}(\zeta)}{\pi}$. All we need to do is find a conformal map from the upper half-plane to the strip and keep track of the boundary condition. The desired map is

$$f(z) = \log\left(\frac{z-1}{z+1}\right).$$

Exercise 6.4 asks for the details. In particular, the solution is given by

$$\frac{1}{\pi}\text{Im}(f(z)) = \frac{1}{\pi}\tan^{-1}\left(\frac{2y}{x^2+y^2-1}\right).$$

EXAMPLE 6.6. This example uses conformal mapping to analyze fringe lines at the end of a parallel plate capacitor. We begin with infinitely long parallel plates, which we regard as the lines in $\mathbb{C}$ given by $\text{Im}(z) = \pm\pi$. Let Ω denote the strip defined by $-\pi < \text{Im}(z) < \pi$. Put $w = f(z) = z + e^z$. We claim that f defines a conformal map from Ω to a certain region. This image region is the complement of two half lines: $\text{Im}(w) = \pm\pi$ with $-\infty < \text{Re}(w) \le -1$. See Figure 6. The lines of equal electric potential energy are the lines $\text{Im}(z) = c$ in the z-plane. The images of such lines under f are the equipotential curves in the w-plane. We analyze the function f on the lines $\text{Im}(z) = \pm\pi$. Since $e^{i\pi} = -1$, the real part is the function $z - e^z$. This function maps the real line to the interval $(\infty, 1]$. By calculus it has a maximum of -1 at the point $x = 0$. The imaginary part is $\pm i\pi$. Parallel lines in the strip in the z-plane get mapped to curves that smoothly "turn around" when $x > -1$. Since $f'(\pm i\pi) = 0$, we see that f is not conformal at these points, explaining why the lines defining the boundary of the strip sharply turn around. We can also consider this example in the context of ideal fluid flow. The streamlines $\text{Im}(z) = c$, for $-\pi < c < \pi$ in the z-plane get mapped into complicated curves describing the flow in the w-plane.

EXAMPLE 6.7. We use a conformal map from the unit disk Ω_1 to the upper half-plane Ω_2 to compare corresponding Dirichlet problems. The function $h_1(x,y) = x = \text{Re}(z)$ is harmonic in Ω_1 and has boundary values $g_1(\theta) = \cos(\theta)$. What happens in Ω_2? The map $f : \Omega_1 \to \Omega_2$ given by

$$w = f(z) = -i\frac{z-1}{z+1}$$

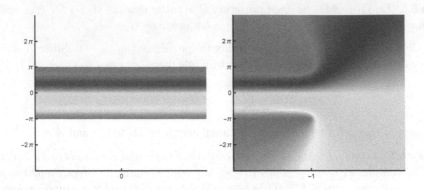

FIGURE 7. Capacitor fringe lines

is conformal. Its inverse is given by

$$z = f^{-1}(w) = \frac{i - w}{i + w}.$$

Put $w = u + iv$. According to Corollary 6.7, the harmonic function h_2 is given by $h_1 \circ f^{-1}$, and hence

$$h_2(u, v) = \mathrm{Re}\left(\frac{i - w}{i + w}\right) = \mathrm{Re}\left(\frac{(i - w)(-i + \overline{w})}{|w + i|^2}\right) = \frac{1 - u^2 - v^2}{u^2 + (v + 1)^2}.$$

As the real part of an analytic function, h_2 is harmonic. (Verifying the harmonicity by computing the Laplacian, however, is a nuisance.) The boundary of Ω_2 is the real line, where $v = 0$. Hence h_2 solves the Dirichlet problem in Ω_2 with boundary values $\frac{1 - u^2}{1 + u^2}$.

Had we begun with the boundary value $g_1 = \sin(\theta)$, we would obtain h_2 as

$$h_2(u, v) = \mathrm{Im}\left(\frac{i - w}{i + w}\right) = \mathrm{Im}\left(\frac{(i - w)(-i + \overline{w})}{|w + i|^2}\right) = \frac{2u}{u^2 + (v + 1)^2}.$$

If we now evaluate along $v = 0$, we obtain the boundary values $\frac{2u}{1 + u^2}$. Combining the two boundary values gives us the map

$$u \to \left(\frac{1 - u^2}{1 + u^2}, \frac{2u}{1 + u^2}\right).$$

We have obtained a rational parametrization of the unit circle.

EXERCISE 6.1. For an appropriate branch of $\tan^{-1}$, and for $x \neq 0$, the function $\tan^{-1}\left(\frac{y}{x}\right)$ is harmonic. Verify this result first by computing the appropriate second derivatives. Then give a proof without any computation at all. Then read Section 8 on differential forms from Chapter 3 and explain what is going on.

EXERCISE 6.2. What harmonic function agrees with $\sin(\theta)$ on the circle? What harmonic function agrees with $\sin^2(\theta)$ on the unit circle? Write the answers in terms of (x, y).

EXERCISE 6.3. Check the last step in the proof of Lemma 6.2.

EXERCISE 6.4. Fill in the details in Example 6.5. In particular, show that the argument of $\frac{z-1}{z+1}$ is $\tan^{-1}\left(\frac{2y}{x^2 + y^2 - 1}\right)$.

EXERCISE 6.5. Put $f(z) = z + \frac{1}{z}$. Find a domain Ω_1 such that f is a conformal map from Ω_1 to the upper half-plane. Find a domain Ω such that f is a conformal map from Ω to the complement of the closed interval $[-2, 2]$. Suggestion: first put $z = e^{i\theta}$ and find $f(z)$.

EXERCISE 6.6. As in Example 6.6, put $w = f(z) = z + e^z$. Show that f maps the strip onto the complement of the two half-lines.

EXERCISE 6.7. Find a conformal map from the upper half-plane to the first quadrant. Find a conformal map from the upper half-plane to the sector in the first quadrant given by $0 < \theta < \beta < \frac{\pi}{2}$.

EXERCISE 6.8 (Poisson kernel). Given a continuous function g on the real line, verify that the following integral formula extends g to be harmonic in the upper half-plane.

$$Tg(x, y) = \frac{1}{\pi} \int_{-\infty}^{\infty} \frac{yg(t)}{(x - t)^2 + y^2} dt.$$

EXERCISE 6.9. Show that $w = \sin(z)$ defines a conformal map from the set Ω to the upper half-plane. Here Ω is the set where $\frac{-\pi}{2} < x < \frac{\pi}{2}$ and $y > 0$. Why are angles not preserved on the boundary?

EXERCISE 6.10. There is no conformal map from $\mathbb{C}$ to the unit disk. Why not?

EXERCISE 6.11. Verify the statements in the first two remarks after Definition 6.5.

EXERCISE 6.12. Suppose $f : \Omega_1 \to \Omega_2$ is conformal. Show that $f' \neq 0$ on Ω_1. Give an example of a complex analytic f and an Ω such that $f' \neq 0$ on Ω but f is not injective.

EXERCISE 6.13. Denote the mapping in Example 6.3 by ϕ_a. What is its inverse mapping? Prove that each conformal map from the unit disk to itself can be expressed as

$$f(z) = e^{i\theta}\phi_a(z)$$

for some θ and some a in the unit disk. Given arbitrary points a and b in the unit disk, find a conformal map that sends a to b.

7. Root finding

Prerequisites. This chapter relies on Chapters 2 and 4 on Complex Analysis. It also makes connections with the Laplace transform from Chapter 5.

In many applications one needs to find roots of polynomial equations or to locate them approximately. In this section we will assume that *polynomial* means polynomial with complex coefficients in a complex variable z. The fundamental theorem of algebra (Theorem 5.5 of Chapter 4) guarantees that a non-constant polynomial has (possibly complex) roots. For polynomials of degree two, the quadratic formula (Exercise 1.1 of Chapter 2) provides formulas for these roots in terms of the coefficients. For polynomials of degree three and four, rather complicated formulas exist for the roots in terms of the coefficients, but for polynomials of degree at least five, it can be proved that no such formula can exist. One must therefore resort to numerical methods. We briefly consider several such methods in this section.

We begin with an elementary approach for finding rational roots for polynomials with rational coefficients. Observe first, after clearing denominators, we may assume when seeking roots that such a polynomial has integer coefficients. We have the following result from elementary mathematics. The reader should have no trouble writing a proof.

Proposition 7.1 (Rational zeroes theorem). *Let* $p(x) = a_0 + \cdots + a_d x^d$ *be a polynomial with integer coefficients. Suppose* $c = \frac{m}{n}$ *is rational and reduced to lowest terms. If* $p(c) = 0$, *then* m *divides* a_0 *and* n *divides* a_d.

EXAMPLE 7.1. Put $p(x) = x^5 - 15x + 2$. If $p(\frac{m}{n}) = 0$, then m must be ± 1 or ± 2 and n must be ± 1. Thus the only possible rational roots are ± 1 and ± 2. Checking them all, we find that $p(-2) = 0$. The other values are not roots. Hence p has only one rational root.

We continue with a rather naive method for finding real roots, the method of bisection. Consider a polynomial $p(x)$ with real coefficients. Suppose that $p(a)$ and $p(b)$ have opposite signs. Let I_0 denote the interval $[a, b]$. Since p is continuous, the intermediate value theorem guarantees that p has a root between a and b. One then evaluates p at the midpoint $\frac{a+b}{2}$. If p is zero there, we are done. If not, we obtain an interval I_1, whose length is half of that of I_0, on which p has opposite signs at the endpoints. Continuing in this fashion we either find a root in finitely many steps or we obtain a sequence I_k of nested closed intervals. The limiting interval (See [R] for example) is a single point, which must be a root. We obtain a sequence of guesses converging to a root.

EXAMPLE 7.2. Again put $p(x) = x^5 - 15x + 2$. Since $p(0) > 0$ and $p(1) < 0$, there is a root in $(0, 1)$. Since $p(\frac{1}{2}) < 0$, there is a root in the interval $(0, \frac{1}{2})$. Since $p(\frac{1}{4}) < 0$, there is a root in the interval $(0, \frac{1}{4})$. The next step tells us that there is a root in $(\frac{1}{8}, \frac{1}{4})$. We can regard this method as finding the binary expansion of the root. Each step gains one place in the binary expansion. For this polynomial there is one root in the interval $(0, 1)$; its base ten value is approximately 0.133336. Since $p(1) < 0$ and $p(2) > 0$, there is another root in the interval $(1, 2)$. It turns out to be approximately 1.93314.

Next we consider Newton's method for root finding of complex polynomials. Given a differentiable function f, we define a new function g by

$$g(z) = z - \frac{f(z)}{f'(z)}.$$

Assuming that $f'(z) \neq 0$, we note that $g(z) = z$ if and only if $f(z) = 0$. Therefore we seek fixed points for the function g. We begin with an initial guess z_0. We define a sequence recursively by $z_{n+1} = g(z_n)$. For most initial guesses, this procedure converges rapidly to a fixed point for g and hence a root for f. It is quite interesting to note the connection with fractals. Many Web sites have beautiful pictures obtained via Newton's method. The reader can google "Newton's method fractals" and find many such sites. Suppose, for example, that a polynomial has four complex roots a, b, c, d. We color the plane in four colors, say A, B, C, D, as follows. If an initial guess leads to the root a, we color that point A, and so on. The boundaries of the four sets are complicated fractals, even for rather simple polynomials. These boundaries are those points for which the method fails. This set is typically uncountable, but nonetheless small. See [Fa] and [MH] and their references for much more information.

EXAMPLE 7.3. We again consider the polynomial $z^5 - 15z + 2$. The corresponding function g is

$$g(z) = \frac{2 - 4z^5}{15 - 5z^4}.$$

We try the initial guess $z = 1$. Then $g(1) = \frac{-1}{5}$ and $g(\frac{-1}{5}) = \frac{3127}{23425}$. Plugging this value into g gives a rational number λ whose numerator is 21 decimal digits and whose denominator is 23 decimal digits. The decimal expansion of λ, to ten places, is 0.1333361429. Furthermore $p(\lambda)$ is approximately 5.6×10^{-10}.

Another approach, though not usually as efficient as Newton's method, uses Rouche's theorem to try to locate the roots of p. To apply the theorem, ones chooses a closed curve γ and a function f for which the number of roots inside γ is obvious. An appropriate inequality on γ then guarantees that f and p have the same number of roots inside. Below we consider several examples and applications. See [A], [B], [D2], and [MH] for additional references and discussion. Before stating Rouche's theorem, we discuss a more general result, called the *argument principle*, that allows us to locate both zeroes and poles.

Theorem 7.2 (The argument principle). *Suppose f is complex analytic in a simply connected region Ω except for finitely many poles. Let γ be a positively oriented simple closed curve in Ω such that $f(z)$ is neither 0 nor singular along γ. Let N denote the number of zeroes of f (with multiplicity counted) inside γ and let P be the number of poles (with multiplicity counted as well). Then*

$$\frac{1}{2\pi i}\int_\gamma \frac{f'(z)}{f(z)}dz = N - P. \tag{46}$$

PROOF. The idea behind the proof is easy. The expression $\frac{f'}{f}$ is the derivative of $\log(f)$. Since f is not zero along γ, it is not identically 0 and hence has at most finitely many zeros inside γ. By assumption it has finitely many poles. Call the zeroes z_j, and repeat to account for multiplicity. Call the poles p_j and again allow repeats if necessary. Hence we can write

$$f(z) = \frac{\prod_{j=1}^{N}(z - z_j)}{\prod_{j=1}^{P}(z - p_j)}u(z),$$

where u is analytic and not zero on and inside γ. A simple computation (motivated by the logarithmic derivative) gives us

$$\frac{f'(z)}{f(z)} = \sum_{j=1}^{N}\frac{1}{z - z_j} - \sum_{j=1}^{P}\frac{1}{z - p_j} + \frac{u'(z)}{u(z)}. \tag{47}$$

Since $\frac{u'}{u}$ is analytic, the last term in (47) integrates to zero by Cauchy's theorem. Since $\int_\gamma \frac{dz}{z-p} = 2\pi i$ when γ encloses p, each of the terms in the first sum in (47) contributes 1 to the integral in (46) and each of the terms in the second sum contributes -1. The result follows. $\qquad\square$

REMARK. The argument principle can be stated in the following way. Put $w = f(z)$. Consider the image $f(\gamma)$ of the curve γ in the w-plane. Then $f(\gamma)$ winds around the origin m times, where $m = N - P$.

EXAMPLE 7.4. Put $f(z) = 2z^2 - z$. Then f has two zeroes and no poles inside the unit circle. The image of the unit circle under f wraps twice around the origin. For a second example, put

$$g(z) = \frac{24z^3 - 26z^2 + 9z - 1}{5z - 1}.$$

Then g has three zeroes and one pole inside. Again the image of the unit circle wraps twice around the origin. See Figure 8.

EXAMPLE 7.5. Suppose $p(z) = (z - a)(z - b)(z - c)$. Then

$$\frac{p'(z)}{p(z)} = \frac{1}{z - a} + \frac{1}{z - b} + \frac{1}{z - c}.$$

Assume $|a| < |b| < |c|$. A circle about 0 whose radius is larger than $|c|$ encloses all three points; three roots lie inside such a circle. A circle whose radius r satisfies $|b| < r < |c|$ encloses two roots, and so on.

EXAMPLE 7.6. Suppose f has a pole of order m at p and we wish to determine m. We find a sufficiently small closed disk about p with boundary circle γ on which $f(z) = \frac{u(z)}{(z-p)^m}$, where u is analytic and not zero. By the argument principle, $\frac{1}{2\pi i}\int_\gamma \frac{f'(z)}{f(z)}dz = -m$. We could check this result directly, because $\frac{f'(z)}{f(z)} = \frac{-m}{z-p} + \frac{u'(z)}{u(z)}$. Since $u(z) \neq 0$, the function $\frac{u'}{u}$ is analytic and its integral is thus 0.

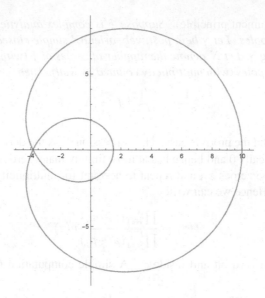

FIGURE 8. Argument principle for Example 7.4

Theorem 7.3 (Rouche's theorem). *Suppose f and g are complex analytic on and inside a simple closed curve γ. Suppose the inequality $|f - g| < |f|$ holds at points of γ. Then f and g have the same number of roots (with multiplicity counted) inside γ.*

PROOF. The hypothesis $|f - g| < |f|$ on γ is equivalent to $|1 - \frac{g}{f}| < 1$ on γ. Hence the function $h = \frac{g}{f}$ has no zeroes or poles on γ. Furthermore, the image of γ under h is at distance less than 1 from the point 1, and hence cannot wind around 0. Applying the argument principle to h then yields

$$0 = \int_\gamma \frac{h'(z)}{h(z)} dz.$$

Expressing $\frac{h'}{h}$ in terms of f and g yields

$$\frac{h'}{h} = \frac{g'}{g} - \frac{f'}{f}.$$

The argument principle now implies the conclusion of Rouche's theorem. Here is a simple explanation of the idea behind the proof. Consider a person walking a dog around a tree. If the person walks around the tree, never getting closer to the tree than the length of the leash, and the person and dog return to their starting locations, then the person and the dog wind the same number of times about the tree. □

REMARK. We offer a second way to express the idea in the proof. We use a new letter ζ for $f - g$ to avoid confusion. The hypothesis becomes $|\zeta| < |f|$ on γ and the conclusion becomes f and $f - \zeta$ have the same number of zeroes inside. Subtracting ζ from f is a small perturbation and hence should not change the number of zeroes. To make this idea precise, consider $H_t = f - t\zeta$ for $0 \le t \le 1$. The number $Z(t)$ of zeroes of H_t inside γ is given by (46):

$$Z(t) = \frac{1}{2\pi i} \int_\gamma \frac{H_t'(z)}{H_t(z)} dz.$$

The formula shows that $Z(t)$ is continuous in t; but $Z(t)$ is also an integer. The only possibility is that $Z(t)$ is constant, and hence $Z(0) = Z(1)$. Thus f and $f - \zeta = g$ have the same number of zeroes inside γ.

EXAMPLE 7.7. Put $p(z) = z^5 - 15z + 2$ and put $f(z) = z^5$. Then f has five roots (with multiplicity counted) and

$$|p(z) - f(z)| = |2 - 15z| < 32 + 15\epsilon < (2 + \epsilon)^5 = |f(z)|$$

on the curve $|z| = 2 + \epsilon$ for each $\epsilon > 0$. Hence all the roots of p lie in the set $\{z : |z| \leq 2\}$. Here are the roots of p to four decimal places:

$$-2, 1.33336, 1.9331, .0332 \pm i1.969.$$

Conformal mapping can be used to help in root finding.

Lemma 7.4. *For $z \in \mathbb{C}$, we have $\mathrm{Re}(z) < 0$ if and only if $\left|\frac{1+z}{1-z}\right| < 1$.*

PROOF. The second inequality is equivalent to $|1 + z|^2 < |1 - z|^2$ which, upon expanding, is equivalent to $4\,\mathrm{Re}(z) < 0$. $\qquad\square$

An alternative way to state the lemma is that the map $z \mapsto \frac{1+z}{1-z} = w$ is a conformal map from the left half-plane to the unit disk. A simple computation shows that the inverse map is given by $z = \frac{w-1}{w+1}$. In many applied problems the criterion for stability is that all the poles of a given rational function lie in the left half-plane. The condition $\mathrm{Re}(z) < 0$ becomes the condition $|w| < 1$. We can thus express stability in terms of the unit disk.

EXAMPLE 7.8. The polynomial $f(s) = s^2 - 2s + 2$ has roots $1 \pm i$, both of which have positive real part. Replacing s by $\frac{w-1}{w+1}$ yields the rational function

$$\frac{(1 - w)^2 - 2(w - 1)(1 + w) + 2(1 + w)^2}{(w + 1)^2} = \frac{w^2 + 2w + 5}{(w + 1)^2}.$$

The roots of this rational function are at $-1 \pm 2i$. Thus the two roots of $f(s)$ in the right half-plane get transformed into two roots outside the unit circle; the point at infinity gets transformed into a pole of order two at -1 on the unit circle.

EXAMPLE 7.9. The polynomial $g(z) = 17z^5 + 31z^4 + 198z^3 + 82z^2 + 49z - 1$ has three roots inside the unit disk and two outside. We use Rouche's theorem to verify this statement. We compare g with $f(z) = 198z^3$, which has (multiplicity counts!) three roots in the unit disk. Furthermore, on the unit circle,

$$|f(z) - g(z)| = |17z^5 + 31z^4 + 82z^2 + 49z - 1| \leq 17 + 31 + 82 + 49 + 1 = 180 < 198 = |f(z)|.$$

By Rouche's theorem, g also has three zeroes inside the disk. If we replace z by $\frac{1+w}{1-w}$, then we obtain a rational function in w which has three roots in the left half-plane. That function is

$$\frac{-8(19w^5 + 15w^4 - 2w^3 - 30w^2 + 19w + 47)}{(w - 1)^5}. \tag{48}$$

In other words, we can establish a stability criterion for the rational function in (48) by applying Rouche's theorem to the function g.

REMARK. The **Jury stability criterion** is a method for determining whether a polynomial has its roots inside the unit disk. This criterion provides an algorithm, somewhat akin to elementary row operations, for deciding this information. We next establish this criterion using Rouche's theorem.

Definition 7.5. Let $p(z)$ be a polynomial of degree d. Define its *reflected polynomial q*, or simply its *reflection*, by

$$q(z) = z^d \overline{p}\left(\frac{1}{z}\right).$$

In this formula, $\overline{p}$ is the polynomial whose coefficients are the complex conjugates of those of p.

EXAMPLE 7.10. Put $p(z) = (1 - az)(1 - bz) = 1 - (a + b)z + abz^2$. Then

$$q(z) = z^2 \overline{p}\left(\frac{1}{z}\right) = z^2 \left(1 - \frac{\overline{a}}{z}\right)\left(1 - \frac{\overline{b}}{z}\right) = (z - \overline{a})(z - \overline{b}) = \overline{ab} - (\overline{a + b})z + z^2.$$

Lemma 7.6. *If q is the reflection of p, then $|p(z)| = |q(z)|$ when $|z| = 1$.*

PROOF. Since $z\overline{z} = 1$ on the circle, and $|\overline{w}| = |w|$, we have

$$|q(z)| = \left|z^d \overline{p}\left(\frac{1}{z}\right)\right| = |z^d \overline{p}(\overline{z})| = |\overline{p}(\overline{z})| = |p(z)|. \tag{49}$$

$\square$

REMARK. Why do we call q the reflection of p? Suppose p has k roots (multiplicity counted) **inside** the unit disk, l roots on the unit circle, and m roots outside the unit disk. Then q has k roots **outside**, l on the circle, and m inside. This assertion has a simple proof. If $p(z) = z - a$, then $q(z) = 1 - \overline{a}z$. Each root a of p is reflected across the circle to the root $\frac{1}{\overline{a}}$ of q. The product of these two roots has modulus 1. Since each polynomial factors into linear factors, the general statement follows.

The coefficients of q are the complex conjugates of those in p, but in reverse order. In particular, if the coefficients are real, then q is obtained from p simply by reversing the order of the monomials. Lemma 7.6 is the key fact behind the Jury stability criterion, which is equivalent to the following proposition.

Proposition 7.7. *Let p be a polynomial with reflection $q(z)$. For $c \in \mathbb{C}$ put $f(z) = p(z) - cq(z)$. If $|c| < 1$, then f has the same number of zeroes in the unit disk as p does.*

PROOF. We compare f and p on the circle using Rouche's theorem. Using Lemma 7.6 and the hypothesis, we obtain

$$|f(z) - p(z)| = |c\,q(z)| = |c|\,|p(z)| < |p(z)|.$$

By Rouche's theorem, f and p have the same number of zeroes inside. $\square$

Corollary 7.8 (Stability algorithm). *Let $p(z) = a_0 z^n + a_1 z^{n-1} + \cdots + a_n$ be a polynomial with reflection $q(z)$. Put $f(z) = p(z) - (\frac{a_n}{\overline{a_0}})q(z)$. If $|a_n| < |a_0|$, then f has the same number of zeroes in the unit disk as p does. If $|a_n| > |a_0|$, we first interchange the roles of p and q, and then obtain the corresponding conclusion.*

Let us elaborate on Proposition 7.7; $f = p - cq$ has the same number of zeroes inside the circle as p whenever $|c| < 1$. We choose the coefficient $\frac{a_n}{\overline{a_0}}$ in order to ensure that $f(0) = 0$. We can then divide f by z and at the next step reduce the degree of the polynomial considered. When all the roots of p are inside the circle, we must have $|a_n| < |a_0|$. If at some stage this ratio has modulus larger than 1, we therefore know that there is a root outside. In some settings we can halt here. In case we want to know about additional zeroes, we can interchange the roles of p and q, and then continue with a polynomial which has a zero inside. Given a polynomial with at least one root inside, the process then creates a new polynomial f, one of whose roots is at 0. The remaining roots get moved around, but never change their positions relative to the unit circle. The algorithm as stated breaks down when all the roots are on the unit circle. See Exercise 7.10 for some examples.

What happens if the ratio $\frac{a_n}{a_0}$ equals 1? Either all the roots are on the unit circle, or there are roots both inside and outside the circle. One must refine the algorithm in case there are roots *on* the circle.

EXAMPLE 7.11. Put $p(z) = 2z^3 - z^2 + 8z - 4$. Then $q(z) = -4z^3 + 8z^2 - z + 2$. Then $c = -2$ and there is at least one root of p outside. This information might be adequate. If not, we interchange p and q. Thus we consider

$$f(z) = -4z^3 + 8z^2 - z + 2 - (\frac{2}{-4})(2z^3 - z^2 + 8z - 4) = -3z^3 + \frac{15}{2}z^2 + 3z = z(-3z^2 + \frac{15}{2}z + 3).$$

This polynomial has the same number of zeroes inside as q does. One of them is at 0. Divide out the z, and for convenience multiply by 2. We need to study the zeroes of $h(z) = -6z^2 + 15z + 6$. Note that the coefficient ratio for h has modulus 1. We instead apply the quadratic formula to see that h has one zero inside and one outside. Hence q has two zeroes inside and one zero outside. One can easily check that the roots of q are 2 and $\frac{\pm i}{2}$.

EXAMPLE 7.12. Put $p(z) = (1 - 2z)(1 - 3z)(1 - 4z) = 1 - 9z + 26z^2 - 24z^3$. Its reflection q is given by $q(z) = -24 + 26z - 9z^2 + z^3$. The polynomial f from Proposition 7.7 is

$$p(z) + \frac{1}{24}q(z) = \frac{-5}{24}(38z - 123z^2 + 115z^3).$$

The roots of p are $\frac{1}{2}, \frac{1}{3}, \frac{1}{4}$. The roots of f are at 0 and $\frac{123 \pm i\sqrt{2351}}{230} \approx 0.53 \pm .21i$.

REMARK. If one writes the coefficients of p as the first row of a matrix, and the coefficients of q as the second row, then the process in Proposition 7.7 amounts to adding a multiple of the second row to the first row to make one element 0. Hence we are simply doing row operations!

REMARK. When applying the algorithm from Proposition 7.7 using row operations, one can of course multiply a row (corresponding to p) by a non-zero constant; of course one must then multiply the row corresponding to its reflection by the same constant.

We briefly discuss connections between the topics in this section and other ideas in this book. Why do half-planes arise? Consider first the simple situation where $f(t) = t^\alpha$ for $\alpha > -1$. We saw in Chapter 5 that the Laplace transform $F(s)$ of f exists when $\text{Re}(s) > 0$, and we have $F(s) = \frac{\Gamma(p+1)}{s^{p+1}}$. Here the region of convergence is the right half-plane. For another example, the Laplace transform of $e^{\lambda t}$ exists for $\text{Re}(s) > \lambda$.

Consider next a constant coefficient homogeneous differential equation with characteristic polynomial

$$p(D) = (D - \lambda_1)(D - \lambda_2) = D^2 + bD + c.$$

The functions $e^{\lambda_1 t}$ and $e^{\lambda_2 t}$ form a basis for the solution space. If the real part of either λ_j is positive, then that solution blows up as t tends to infinity. When the real part is 0 we get oscillation, as shown for example by $\sin(\lambda t)$. Only when both real parts are negative can we conclude that the limit at $t = \infty$ exists. Solving the inhomogeneous equation $p(D)y = f$ via Laplace transforms, including the initial conditions, yields

$$p(s)Y(s) = F(s) + y'(0) + (s + b)y(0).$$

Dividing as usual by the characteristic polynomial shows that

$$Y(s) = \frac{F(s) + y'(0) + (s + b)y(0)}{s^2 + bs + c}.$$

The Laplace transform has poles at the roots of the characteristic equation $p(s) = s^2 + bs + c = 0$. To ensure stability, we need the roots to be in the left half-plane. In fact, many authors define a polynomial to be **stable** if all of its zeroes lie in the left half-plane.

We have seen throughout the book that finding eigenvalues of a linear map is fundamental to math and science. In finite-dimensional situations, eigenvalues are roots of the characteristic polynomial equation $p_A(z) = \det(A - zI)$. In general, however, computing determinants is difficult and one does not use this approach for finding eigenvalues. Instead one develops algorithms based on Theorem 9.3 of Chapter 6.

EXERCISE 7.1. Assuming Rouche's theorem, prove the fundamental theorem of algebra. Suggestion: If $p(z) = z^d + \cdots$, where $\cdots$ denotes lower order terms, compare p with z^d on a circle of large radius.

EXERCISE 7.2. How many roots of $z^5 - 20z + 2$ lie in $\{z : |z| < 3\}$? How many lie in the unit disk?

EXERCISE 7.3. How many zeroes of $z^4 - 7z - 1$ lie in $\{z : |z| < 2\}$? How many in $\{z : |z| < 1\}$?

EXERCISE 7.4. In engineering it is often important to know that the poles of a rational function lie in the left half-plane. Devise a test for deciding this matter. Suggestion: Consider a closed contour consisting of a large interval along the imaginary axis and a semi-circle in the right half-plane. In case there are poles on the imaginary axis, avoid them by using indented contours (rotate the contour used in Figure 7 of Chapter 4). Apply the argument principle.

EXERCISE 7.5. Suppose you choose an initial guess in Newton's method that happens to be a zero of the derivative. What happens? What other things can go wrong?

EXERCISE 7.6. Put $p(x) = (x - 1)(x - 2)(x - 3)(x - 4)$. Show that there are infinitely many initial guesses that fail in Newton's method.

EXERCISE 7.7. Assume each a_j is in the unit disk. Put

$$f(z) = \prod_{j=1}^{N} \frac{z - a_j}{1 - \overline{a_j}z}.$$

Show that f maps the unit circle to itself. Evaluate (in your head!)

$$\frac{1}{2\pi i} \int_{|z|=1} \frac{f'(z)}{f(z)} dz.$$

EXERCISE 7.8. Put $p(z) = 6z^3 - 35z^2 + 26z - 5$. Apply the algorithm in Proposition 7.7 twice to show that p has two zeroes in the unit disc and one zero outside. Apply Proposition 7.1 to find the roots exactly.

EXERCISE 7.9. Under what circumstances does p equal its reflection q?

EXERCISE 7.10. Put $p(z) = (z^4 + 1)(z^3 + 1)$. What happens when we apply Proposition 7.7? More generally suppose that $p(z) = \prod_{j=1}^{d}(z - \lambda_j)$, and each λ_j is on the circle. What happens in Proposition 7.7?

EXERCISE 7.11. Suppose that f and γ are as in the statement of the argument principle. Assume h is analytic on and inside γ. Evaluate the following:

$$\frac{1}{2\pi i} \int_{\gamma} \frac{f'(z)}{f(z)} h(z)dz.$$

Suggestion: multiply both sides of (47) by $h(z)$ and then integrate.

EXERCISE 7.12. Suppose you wanted to use the argument principle to show that a polynomial had no zeroes in the first quadrant. What curve γ might you try?

EXERCISE 7.13. In Definition 5.1 of Chapter 1 we defined *field*. We mentioned there that the set of integers modulo a prime defines a field. Show that the fundamental theorem of algebra fails in such fields; in other words, given a prime number p, find a polynomial f such that the coefficients of f are in the set $\{0, 1, \ldots, p-1\}$ and such that $f(n) \neq 0$ for n in this set. (That is, $f(n)$ is never congruent to 0 modulo p.) For example, the polynomial $x^2 + x + 1$ has no roots in the field of two elements. More generally, show that the fundamental theorem of algebra fails in every field with finitely many elements.

EXERCISE 7.14 (The open mapping theorem). Suppose h is complex analytic on an open connected set but not a constant. Then the image of h is an open set. Follow the following outline to derive an even stronger result from Rouche's theorem. Pick some $z_0 \in \Omega$. First consider the function $z \mapsto h(z) - h(z_0) = f(z)$. Since $f(z_0) = 0$ but f is not constant, there is an integer m such that $f(z) = (z - z_0)^m u(z)$, where $u(z) \neq 0$ for z near z_0. For w near $f(z_0)$, consider now the two functions $f(z)$ and $h(z) - w$. Apply Rouche's theorem to a circle of radius δ about z_0 inside of which f has m zeroes but has none on the circle. Then there is a positive ϵ such that $|f| \geq \epsilon$ on this circle. By the continuity of h,

$$|f(z) - (h(z) - w)| = |w - h(z_0)| < \epsilon \leq |f(z)|$$

when w is close enough to $f(z_0)$. Hence $h(z) - w$ has m zeroes inside, because $f(z) = h(z) - h(z_0)$ does. In particular w is in the image of h.

EXERCISE 7.15 (The maximum principle). Suppose f is complex analytic and not constant on an open connected set. Then $|f|$ cannot attain a maximum. Derive this important result from the previous exercise. Suggestion: The image must contain an open disk about each $f(z_0)$; hence $|f(z_0)|$ cannot be at a maximum.

EXERCISE 7.16. Use the Cauchy integral formula to give a different proof of the maximum principle.

EXERCISE 7.17. Derive the maximum principle for harmonic functions (in two dimensions) from the maximum principle for complex analytic functions. Comment: The maximum principle for harmonic functions holds in all dimensions.

8. Linear time-invariant systems

Prerequisites. This material relies on differential equations, complex exponentials, linear transformations, convolutions, and the Dirac delta function. It is enhanced by understanding the Fourier and Laplace transforms. The Green's function also makes a brief appearance in this section.

Recall from Chapter 1 that the current in an RLC circuit is governed by a second-order, constant-coefficient, linear ordinary differential equation. We write $\mathcal{I}(t)$ for the current in a circuit at time t, and let $E(t)$ be the voltage drop at time t. Let R denote the resistance, L denote the inductance, and C denote the capacitance. We have three relationships:

$$E_{\text{resistor}}(t) = R\,\mathcal{I}(t).$$

$$E_{\text{inductor}}(t) = L\,\mathcal{I}'(t).$$

$$E'_{\text{capacitor}}(t) = \frac{1}{C}\,\mathcal{I}(t).$$

These lead to the second-order ODE

$$L\mathcal{I}''(t) + R\mathcal{I}'(t) + \frac{1}{C}\mathcal{I}(t) = E'(t), \tag{50}$$

expressing the current in terms of the voltage drop, the inductance, the resistance, and the capacitance.

We can connect many of the topics in this book by viewing these ideas in terms of **linear time-invariant (LTI)** systems.

The starting point is a vector space V whose elements are functions $f : \mathbb{R} \to \mathbb{C}$. For some purposes we need to put some restrictions on these functions, such as how smooth they are or how they behave at infinity. For now, however, just imagine V being a very large space of functions.

We think of the variable in $\mathbb{R}$ as time, and we think of elements of V as signals. A **linear system** is simply a linear function $G : V \to V$. Thus we have the usual properties from Chapter 1:

$$G(f_1 + f_2) = G(f_1) + G(f_2)$$

$$G(cf) = c\, G(f).$$

Given an **input signal** f, we think of $G(f)$ as an **output signal**. One example of such a G is the shift S_a defined by

$$(S_a f)(t) = f(t + a).$$

A linear system G is called **time-invariant** if, for each $a \in \mathbb{R}$,

$$S_a \circ G = G \circ S_a.$$

In other words, for each function f and each time t we have

$$(G \circ f)(t + a) = G(t \mapsto f(t + a)).$$

In particular, given a delayed input signal, the output signal is delayed by the same amount. A typical circuit defines a time-invariant system, but it would not be time-invariant if, for example, the resistance changed with time.

Since differentiation and integration are linear, differential equations determine various linear systems. But one must be careful of initial conditions.

EXAMPLE 8.1. For f continuous, put

$$G(f)(t) = \int_0^t f(x)\, dx.$$

Then G defines a linear time-invariant system. If, however, we consider

$$G(f)(t) = 1 + \int_0^t f(x)\, dx,$$

then G is not a linear map.

EXAMPLE 8.2. Consider a capacitor and an inductor. The relationship $\mathcal{I}(t) = CE'(t)$ implies that

$$E(t) - E(t_0) = \frac{1}{C} \int_{t_0}^t \mathcal{I}(\tau)\, d\tau. \tag{51.1}$$

The relationship $E(t) = L\mathcal{I}'(t)$ implies that

$$\mathcal{I}(t) - \mathcal{I}(t_0) = \frac{1}{L} \int_{t_0}^t E(\tau)\, d\tau. \tag{51.2}$$

If we assume that $E(t_0) = 0$ in (51.1), then we can regard (51.1) as a linear system. The input signal is the current and the output signal is the voltage. If we assume that $\mathcal{I}(t_0) = 0$ in (51.2), then we can regard (51.2) as a linear system. Now the input signal is the voltage and the output signal is the current. In both cases the system is time-invariant.

Complex exponentials provide important examples of signals. The next example illustrates a crucial property of such signals.

EXAMPLE 8.3. Consider an input signal of the form $t \mapsto e^{st}$. Here we allow s to be a complex number. Thus, if $s = x + iy$, then the input signal is $e^{xt}\left(\cos(yt) + i\sin(yt)\right)$. Suppose G is a linear time-invariant system. We claim that the output signal has the form $F(s)e^{st}$ for some function F. Thus the exponential is an eigenfunction of G. To verify this statement, note by the basic property of the exponential and then by linearity of the map G, we have

$$G\left(t \mapsto e^{s(t+a)}\right) = G\left(t \mapsto e^{sa}e^{st}\right) = e^{sa}G\left(t \mapsto e^{st}\right). \tag{52.1}$$

By time-invariance, the left side of (52.1) is the same as $G\left(t \mapsto e^{st}\right)(t + a)$. Therefore

$$G\left(t \mapsto e^{st}\right)(t + a) = e^{sa}G\left(t \mapsto e^{st}\right). \tag{52.2}$$

To understand (52.2), we denote $G(t \mapsto e^{st})$ by $u_s(t)$. Then (52.2) says that

$$u_s(t + a) = e^{sa}u_s(t). \tag{52.3}$$

Setting $t = 0$ in (52.3) gives us, for some function F of s (namely $u_s(0)$)

$$u_s(a) = e^{sa}u_s(0) = e^{sa}F(s). \tag{52.4}$$

But a is arbitrary, and hence we can regard it as a variable. The claim follows.

Time-invariant systems are given by convolution. Let G be a time-invariant system. Consider the Dirac delta function δ as an input signal. To be mathematically precise, one must use an approximation. Following standard practice, we will first imagine that δ is a valid input. Let h denote the output signal $G(\delta)$. This function h is called the **impulse response**.

The square pulse provides a way to view the delta function rigorously. See Exercise 4.10 of Chapter 5. Other approximate identities (defined in Section 4 of Chapter 5) can also be used. We have formula (53), where the middle term is an abbreviation for the limit:

$$f(0) = \int_{-\infty}^{\infty} f(x)\delta(x)\,dx = \lim_{\epsilon \to 0} \frac{1}{2\epsilon} \int_{-\epsilon}^{\epsilon} f(x)\,dx. \tag{53}$$

We have seen in Chapter 5, Sections 1 and 4, that the delta function serves as the identity for convolution:

$$f(t) = (f * \delta)(t) = \int_{-\infty}^{\infty} f(\tau)\delta(t - \tau)\,d\tau. \tag{54}$$

Let G be a time-invariant linear system. Apply G to both sides of (54). We may regard the integral as a limit of sums. Therefore, assuming a continuity property of G, we may move G past the integral sign, obtaining

$$(Gf)(t) = \int_{-\infty}^{\infty} f(\tau)G\left(\delta(t - \tau)\right)\,d\tau. \tag{55}$$

Recall that $G(\delta) = h$. Since G is time-invariant, $G(\delta(t - \tau)) = h(t - \tau)$. Plugging into (55) gives

$$(Gf)(t) = \int_{-\infty}^{\infty} f(\tau)h(t - \tau)\,d\tau = (h * f)(t). \tag{56}$$

Equation (56) is remarkable; it tells us that knowing what G does to the delta function tells us what G does to every function for which the convolution is defined. We summarize in simple notation:

$$G(f) = G(f * \delta) = f * G(\delta) = f * h = h * f. \tag{57}$$

(The last step uses the commutivity of convolution.) The derivation of (57) shows that we must impose some conditions on the space V of functions and on the map G. Because of (57), it is reasonable to **define** a time-invariant linear system to be the map

$$f \mapsto h * f = G(f),$$

as the **impulse response** h completely determines G. We can for example assume that both h and the input signals are absolutely integrable; this condition guarantees that $h * f$ makes sense, and as we see below, that the system is stable.

Both Fourier and Laplace transforms convert convolution into multiplication. Taking Fourier transforms, we have

$$\mathcal{F}(G(f)) = \mathcal{F}(h)\mathcal{F}(f) = HF$$

and hence we have an alternative way to compute $G(f)$. We start with the impulse response h and take its Fourier transform $\mathcal{F}(h)$ which we call H. This function in the frequency domain is called the **frequency response**. We multiply H by F, the Fourier transform of f. We then take the inverse Fourier transform of the result. One can proceed in a similar way with the Laplace transform. The Laplace transform of h is called the **transfer function**.

The advantage of the transfer function over the frequency response is that it can exist even for certain unstable systems. It is important to realize that the definition $G(f) = h * f$ can make sense even when neither the Laplace nor Fourier transform of h exists (the system is unstable). See [KM] for considerable information about stable systems. One natural stability criterion is that the impulse function h be absolutely integrable:

$$||h||_{L^1} = \int_{-\infty}^{\infty} |h(t)|dt < \infty.$$

This condition implies that the Fourier transform of h exists. See Theorem 8.2 below for more information on stability.

Most realistic systems are causal; outputs depend only upon the past and present values of the signal f. This condition is equivalent to demanding that $h(t) = 0$ for $t < 0$. In this setting, Laplace transforms and Fourier transforms are closely related:

$$\mathcal{F}h(\omega) = \frac{1}{\sqrt{2\pi}} \int_0^{\infty} e^{-i\omega t} h(t) \, dt = \frac{1}{\sqrt{2\pi}} \mathcal{L}(h)(i\omega).$$

EXAMPLE 8.4. Consider the function e^{-t}. It is not absolutely integrable as t tends to $-\infty$. By contrast the function given by e^{-t} for $t \geq 0$ and 0 otherwise *is* absolutely integrable.

In the following definition of stability, we consider *locally integrable* functions. A (measurable) function on $\mathbb{R}$ is **locally integrable** if, for each $R > 0$, we have

$$\int_{-R}^{R} |f(t)| \, dt < \infty.$$

Every continuous function is locally integrable. A locally integrable function need not be integrable on all of $\mathbb{R}$; the most obvious example is the constant function 1. Another example is the sinc function.

The acronym BIBO in Definition 8.1 stands for *bounded-input, bounded-output*.

Definition 8.1. Let V be the space of locally integrable functions from $\mathbb{R}$ to $\mathbb{C}$. Consider the linear time-invariant system $G : V \to V$ defined by $G(f) = h * f$ for some real-valued continuous impulse function h. The system is called **BIBO stable** if f bounded implies $h * f$ bounded.

The proof that $h \in L^1$ implies BIBO stability amounts to a proof of an easy case of Young's inequality, stated in Section 4 of Chapter 5.

Theorem 8.2. *The system $f \mapsto h * f$ is BIBO stable if and only if h is absolutely integrable:*

$$\|h\|_{L^1} = \int_{-\infty}^{\infty} |h(t)| \, dt < \infty. \tag{58}$$

PROOF. Suppose first that (58) holds and f is bounded; say $|f(x)| \le M$. Then

$$|(h * f)(t)| = \left| \int_{-\infty}^{\infty} f(t - \tau) h(\tau) \, d\tau \right| \le \int_{-\infty}^{\infty} |f(t - \tau)| \, |h(\tau)| \, d\tau \le M \, \|h\|_{L^1} < \infty.$$

The bound $M \, \|h\|_{L^1}$ works for all time t and thus $f * h$ is bounded.

Conversely, suppose that h is not absolutely integrable. We need a bounded input f for which $h * f$ is the unbounded output. The idea is to define $f(t) = 1$ where $h(-t) > 0$ and $f(t) = -1$ where $h(-t) < 0$. We can put $f(t) = 0$ if $h(t) = 0$. Then f is bounded, and

$$(h * f)(0) = (f * h)(0) = \int_{-\infty}^{\infty} f(-\tau) h(\tau) \, d\tau = \int_{-\infty}^{\infty} |h(\tau)| \, d\tau. \tag{59}$$

Thus f is locally integrable and bounded, but $(h * f)(0)$ is infinite. Hence $h * f$ is unbounded and BIBO stability fails. $\qquad\square$

Recall that the system is causal if and only if $h(t) = 0$ for $t < 0$. In this setting, the integrals in the proof of Theorem 8.2 are taken over the non-negative real axis. The condition of BIBO stability can then be expressed in terms of the poles of the transfer function, which is the Laplace transform of h. This condition is that all the poles lie in the left half-plane. See [KM] for example.

One can compose linear time-invariant systems in several ways. A *cascade* configuration of such systems arises by defining the impulse function to be the convolution product of the impulse functions of the subsystems. When each system is regarded as a linear map G_j, we are simply composing these linear maps:

$$(G_2)(G_1(f)) = G_2(h_1 * f) = h_2 * (h_1 * f) - (h_2 * h_1) * f.$$

We can of course consider any finite number of such systems.

A *parallel* configuration arises by defining the impulse function to be the sum of the impulse functions of the subsystems. *Feedback* configurations also arise throughout applied mathematics. For example, let f be an input and y the desired output. We call $f - y$ the error signal, which we then use as an input. We want $G(f - y) = y$. By the linearity of G, this equation is equivalent to $y + Gy = Gf$, which leads to $y = (I + G)^{-1} Gf$. See Exercises 8.12 and 8.13.

Closely related to these ideas is the notion of *fundamental solution* of a differential equation. Here we discuss constant coefficient linear ODEs, such as the equation of an LRC circuit. For concreteness, we will illustrate the idea via first and second-order equations, but our methods apply more generally, including to higher order linear ODE and to linear PDE. See for example [T].

We start with a polynomial $p(x)$, which we assume is monic. Thus

$$p(x) = x^n + c_{n-1} x^{n-1} + \cdots + c_1 x + c_0. \tag{60}$$

As in Chapter 1, we define $p(D)$ by substituting the operator $\frac{d}{dt}$ for x in (60). We wish to find a particular solution to the ODE

$$p(D)y = f. \tag{61}$$

To do so, we first solve the ODE

$$p(D)\phi = \delta,$$

where δ is the Dirac delta. We call ϕ the **fundamental solution** of the equation (61). We then have

$$p(D)(f * \phi) = f * p(D)\phi = f * \delta = f, \tag{62}$$

and thus $\phi * f$ solves (61). Note the similarity between (57) and (62). In both cases we apply the operator to the delta function in the convolution.

How do we compute fundamental solutions?

EXAMPLE 8.5. Consider the simplest case, where $P(D) = D = \frac{d}{dt}$. We put $\phi(t) = 1$ for $t > 0$ and $\phi(t) = 0$ for $t < 0$. Here ϕ is the Heaviside function, discussed in Chapter 5. Then, by the definition of convolution, the definition of ϕ, and the fundamental theorem of calculus, we get

$$p(D)(\phi * f)(t) = \frac{d}{dt} \int_{-\infty}^{\infty} \phi(t - \tau) f(\tau)\, d\tau = \frac{d}{dt} \int_{-\infty}^{t} f(\tau)\, d\tau = f(t).$$

Therefore ϕ is the fundamental solution of differentiation.

EXAMPLE 8.6. Consider the second-order equation where $p(D) = D^2 + aD + bI$. We want to construct the fundamental solution ϕ. Following the ideas from the Sturm–Liouville section in Chapter 6, we first find g and h such that $p(D)g = p(D)h = 0$ and where $h(0) = g(0)$ but $h'(0) = 1 + g'(0)$. The recipe from Chapter 6 for finding the Green's function suggests trying

$$\phi(t) = g(t) \text{ if } t \leq 0$$

$$\phi(t) = h(t) \text{ if } t \geq 0.$$

We claim that ϕ does the job.

We must verify that $p(D)(\phi * f) = f$. First we note that

$$(\phi * f)(t) = \int_{-\infty}^{\infty} \phi(t - \tau) f(\tau)\, d\tau = \int_{-\infty}^{t} h(t - \tau) f(\tau)\, d\tau + \int_{t}^{\infty} g(t - \tau) f(\tau)\, d\tau. \tag{63}$$

Applying $p(D)$ to (63) is somewhat painful, but things get easier if we remember that $p(D)g = p(D)h = 0$. By the fundamental theorem of calculus (see Exercise 8.4), we get two kinds of terms when we differentiate once the right side of (63):

$$\frac{d}{dt} \int_{-\infty}^{t} h(t - \tau) f(\tau) d\tau = h(0) f(t) + \int_{-\infty}^{t} h'(t - \tau) f(\tau)\, d\tau. \tag{64.1}$$

$$\frac{d}{dt} \int_{t}^{\infty} g(t - \tau) f(\tau) d\tau = -g(0) f(t) + \int_{t}^{\infty} g'(t - \tau) f(\tau)\, d\tau. \tag{64.2}$$

Since $g(0) = h(0)$, two of these terms cancel when we add (64.1) and (64.2). When we differentiate again, we get terms in which derivatives are applied to g and h, which will cancel out when we include everything and use $p(D)g = p(D)h = 0$. The new terms arising from differentiating (64.1) and (64.2) are $h'(0)f(t)$ and $-g'(0)f(t)$. Since we are assuming that $h'(0) - g'(0) = 1$, including these in the sum yields the final result of $f(t)$. Thus $p(D)(\phi * f)(t) = f(t)$ as needed.

The ideas in this section apply also to discrete signals. We can again define a linear time-invariant system via convolution. We have a stability criterion: $\sum_{n=-\infty}^{\infty} |h(n)| < \infty$. The material on constant coefficient linear difference equations parallels what we have done for constant coefficient ODE. In fact, most aspects of both these theories also extend to multi-variate signals.

EXERCISE 8.1. Verify Example 8.4.

EXERCISE 8.2. Suppose $f(x,t)$ is continuously differentiable and we put

$$v(t) = \int_{-\infty}^{t} f(x,t)\,dx$$

$$w(t) = \int_{t}^{\infty} f(x,t)\,dx.$$

Show that v, w are differentiable and find their derivatives.

EXERCISE 8.3. Fill in the details of the computations in Example 8.6.

EXERCISE 8.4. Assume $p(D) = (D - \lambda_1)(D - \lambda_2)$. Find the functions g and h from Example 8.6.

EXERCISE 8.5. Find the fundamental solution of $\left(\frac{d}{dt}\right)^m$ for m a positive integer.

EXERCISE 8.6. Give an example of a function h whose Fourier transform does not exist but whose Laplace transform does exist.

EXERCISE 8.7. Put $h(t) = \cos(t)$. Show that the system $f \mapsto f * h$ is not BIBO stable.

EXERCISE 8.8. Put $h(t) = \text{sinc}(t)$. Does convolution by h define a BIBO stable system?

EXERCISE 8.9. Let h be a Gaussian $e^{\frac{-t^2}{2}}$. Does convolution by h define a BIBO stable system?

EXERCISE 8.10. Develop the theory of linear time-invariant systems for discrete signals. Suggestion: read Section 2 of Chapter 5.

EXERCISE 8.11. In Example 8.3, consider the input of $e^{i\omega t}$. The output is then a multiple of $e^{i\omega t}$. Give a simple physical interpretation.

EXERCISE 8.12. Put $f(t) = e^{st}$. Find y such that $y = (I + G)^{-1}G(f)$. Here $G(f) = h * f$. Express the answer in terms of the transfer function (the Laplace transform of h).

EXERCISE 8.13. Suppose G is a linear map from a space to itself. Consider the map

$$T = G - G^2 + G^3 - G^4 + \cdots$$

assuming the series converges. Show that $T = (I + G)^{-1}G$. Show that $y = T(f)$ satisfies $G(f - y) = y$.

EXERCISE 8.14. A coin lands on heads with probability p. One keeps flipping the coin until the number of either heads or tails leads by two. Find the probability $H(p)$ that heads wins. Given an equal number of heads and tails, heads wins either by winning two consecutive flips or by reaching equality again in two flips and then winning. This approach leads to an equation for $H(p)$ analogous to the feedback equation. Alternatively, write out an infinite series to determine $H(p)$, analogous to Exercise 8.12.

EXERCISE 8.15. One sometimes sees the notations $Z = R$ for a resistor, $Z = sL$ for an inductor, and $Z = \frac{1}{sC}$ for a capacitor. What is the precise meaning of these formulas? (Here s is the Laplace transform variable.)

CHAPTER 8

Appendix: The language of mathematics

1. Sets and functions

Readers of this book have different sorts of training. Engineers, physicists, and mathematicians consider many of the same ideas and techniques, but often express them in quite different languages. The purpose of this appendix is to establish some of the standard language, notations, and conventions of mathematics. The author hopes that readers will find it useful in understanding the text itself and that it will help scientists appreciate why mathematicians must be so fastidious.

The basic objects in mathematics are sets and functions. A **set** is a collection of objects. These objects are called the **elements** or **members** of the set. The following are examples of sets: the real number system $\mathbb{R}$, the collection of people who drove a car sometime in 2015, the collection of two-by-two matrices with complex entries, the twelve months of the year, and so on. The concept is useful precisely because it is so general.

The empty set $\varnothing$, consisting of no objects, is a set. When S is a finite set, it is common to list its elements using braces as in the following examples:

$$S = \{x_1, \ldots, x_N\}$$

$$M = \{\text{Jan., Feb., Mar., Apr., May, June, July, Aug., Sept., Oct., Nov., Dec.}\}.$$

When specifying a set, the order in which the elements are listed is irrelevant. When S is a set, and x is an element of S, we write $x \in S$. Two sets are equal if and only if they have the same elements, even if the descriptions differ. For example, put $S = \{1, 2, 4, 9\}$; let T be the set of real solutions to the equation $(x^2 - 3x + 2)(x^2 - 13x + 36) = 0$; let U be the set of decimal digits used to write the number fourteen hundred and ninety two. Then $S = T = U$. Nearly all scientific problems amount to providing as precise a description as possible of some important but hard-to-understand set.

We emphasize that a set is determined by its elements, not by the order in which they are listed. Ordered lists do arise throughout pure and applied mathematics. It is possible, but somewhat cumbersome, to express ordered lists in terms of sets. Consider for example an ordered pair (x, y) of real numbers. We can regard (x, y) as a set as follows:

$$(x, y) = S = \{\{x, y\}, \{x\}\}.$$

Here the set S consists of two elements. Its first element is the set $\{x, y\}$; its second element is the object which comes first in the ordered pair. This example indicates that basing everything on sets can be pedantic and counter-intuitive. It is therefore important to use precise language when expressing mathematics.

Let A and B be sets. We can form the set $A \cup B$, called the **union** of A and B; it is the collection of objects in either A or B or both. For example,

$$\{1, 2, 3\} \cup \{2, 3, 4, 5, 6\} = \{1, 2, 3, 4, 5, 6\}.$$

We can also form the set $A \cap B$, called the **intersection** of A and B; it is the collection of objects contained in both A and B. For example,

$$\{1, 2, 3\} \cap \{2, 3, 4, 5, 6\} = \{2, 3\}.$$

REMARK. Exercise 5.9 of Chapter 1 offers an interesting perspective on unions and intersections. Suppose P and Q represent logical statements; we assume that a statement must be either true or false and not both. We assign the value 1 to a true statement and the value 0 to a false statement. In this setting, we write P' for the logical statement *not P*. We compute the truth value of both P and Q (their intersection) by multiplying modulo 2. Adding modulo 2 gives the truth value of P or Q **but not both**. This kind of union is called the **exclusive or** and sometimes written $P \oplus Q$. In the language of sets,

$$P \oplus Q = (P \cap Q') \cup (P' \cap Q).$$

This concept arises in the analysis of circuits and in computer science.

We mention standard notation for several sets. In each of these cases the set has considerable additional structure, such as the ability to add or multiply, and the standard notations presume the additional structure. Thus the notation $\mathbb{R}$ means the real numbers as a set, but suggests that the operations of arithmetic, inequalities, and limits apply.

- The natural numbers: $\mathbb{N} = \{1, 2, 3, \dots\}$.
- The integers: $\mathbb{Z} = \{0, 1, -1, 2, -2, 3, -3, \dots\}$.
- The rational numbers: $\mathbb{Q}$ = the set of all real numbers that can be written $\frac{a}{b}$ for $a, b \in \mathbb{Z}$ and $b \neq 0$.
- The real numbers: $\mathbb{R}$.
- The complex numbers: $\mathbb{C}$.
- $\mathbb{R}^n$ is the set of ordered n-tuples of real numbers.

Suppose A is a set, and B is another set with the property that $b \in B$ implies $b \in A$. We call B a **subset** of A and write $B \subseteq A$. The empty set is a subset of any set. The collection of odd integers is a subset of $\mathbb{Z}$. Often one specifies a set T as a subset of a known set S, writing

$$T = \{x \in S : \ x \text{ satisfies some property}\}.$$

Above we regarded $\mathbb{Q}$ as a subset of $\mathbb{R}$. One might also regard $\mathbb{Q}$ as known and construct $\mathbb{R}$ from it.

The **Cartesian product** $A \times B$ of sets A, B is the set of ordered pairs (a, b) where $a \in A$ and $b \in B$. Cartesian products and their subsets are particularly useful for visualizing functions and relations, which we now define.

Let A and B be sets. A **function** $f : A \to B$ is a machine or rule which assigns to each $a \in A$ an element $f(a) \in B$. Often in elementary mathematics a function is given by a formula, but this viewpoint is inadequate. The set A is called the **domain** of f; the set B is called the **target** of f. The **graph** of a function $f : A \to B$ is the subset of $A \times B$ consisting of the pairs $(a, f(a))$ for $a \in A$. One could define a function $f : A \to B$ to be a subset T of $A \times B$ with the following property. For each $a \in A$, there is a unique $b \in B$ such that $(a, b) \in T$. One can picture this idea using the Cartesian product; above each a there is just one b.

REMARK. An arbitrary subset R of $A \times B$ is sometimes called a **relation**; when $A = B$ such a subset is called a *relation on A*. One says that a is related to b if $(a, b) \in R$. For example, $<$ defines a relation R on $\mathbb{R}$: the pair (x, y) is an element of R means that $x < y$. This example shows that relations need not define functions. Above each a we can have more than one b. The set $\{(x, y) : x \leq y\}$ for x, y real numbers is obviously an important subset of $\mathbb{R}^2 = \mathbb{R} \times \mathbb{R}$, but it does not define a function; above each x there are infinitely many y. An arbitrary subset R of $A \times B$ might have points a for which there is no b with $(a, b) \in R$. This situation also prevents R from defining a function. The set $\{(x, x) : x \neq 0\}$ does not define a function on $\mathbb{R}$ because there is no point $(0, y)$ in this set. The set $\{(x, x)\}$ does of course define a function, namely the identity function.

EXAMPLE 1.1. We give a wide-ranging collection of examples of functions. Example 1.4 discusses additional properties of these functions.

(1) Let A be a set. Put $f(a) = a$ for all $a \in A$. Then $f : A \to A$ is a function, called the *identity*.

(2) Let $A = \mathbb{R}$ and let B be the subset of $\mathbb{R}$ consisting of non-negative numbers. Put $f(x) = |x|$, where $|x|$ denotes the absolute value of x. Then $f : A \to B$ is a function.

(3) Let V be the set of polynomials in one real variable. Let $D : V \to V$ denote differentiation. In other words, $D(p) = p'$. Then D is a function whose domain and target are each V.

(4) Let A be the set of rational numbers x with $0 \le x \le 1$. Let $B = \{0, 1, 2, 3, 4, 5, 6, 7, 8, 9\}$. Define $f(x) = n$ if n is the smallest decimal digit occurring in the base ten expansion of x. We regard 1 as the infinite decimal .999.... Otherwise we use the expansion ending in all zeroes rather than the expansion ending in all nines. Then $f : A \to B$ is a function; here are some ot its values:

$$f(0) = 0, \quad f(1) = 9, \quad f\left(\frac{1}{4}\right) = 0, \quad f\left(\frac{1}{3}\right) = 3, \quad f\left(\frac{1}{7}\right) = 1, \quad f\left(\frac{5}{11}\right) = 4.$$

(5) Put $A = \{z \in \mathbb{C} : z \neq 1\}$. Let $B = \mathbb{C}$. Put $f(z) = \frac{1}{1-z}$. Then $f : A \to B$ is a function.

(6) Let $[0, 1] = \{x \in \mathbb{R} : 0 \le x \le 1\}$. Put $f(x) = \frac{2x}{x+1}$. Then $f : [0, 1] \to [0, 1]$ is a function.

(7) Define $g : [0, 1] \to \mathbb{R}$ by $g(x) = \frac{2x}{x+1}$. This function g is **NOT** the same function as the previous f, because its target differs. This distinction matters, as subsequent examples show.

(8) Let A be the set of functions from $\mathbb{R}$ to $\mathbb{C}$. Let $B = \mathbb{C}$. For $x \in \mathbb{R}$, define δ_x by $\delta_x(f) = f(x)$. Then $\delta_x : A \to \mathbb{C}$ is a function, called the *Dirac delta function*. Its domain is neither $\mathbb{R}$ nor $\mathbb{C}$.

(9) Let A be the set of functions $g : \mathbb{R} \to \mathbb{C}$ for which $\int_{-\infty}^{\infty} |g(x)|\, dx < \infty$. Let B denote the set of continuous functions $h : \mathbb{R} \to \mathbb{C}$. Define $\mathcal{F}(g) = \hat{g}$ by

$$\hat{g}(\xi) = \frac{1}{\sqrt{2\pi}} \int_{-\infty}^{\infty} e^{-ix\xi} g(x)\, dx.$$

Then $\mathcal{F} : A \to B$ is a function, called the *Fourier transform*. One can define $\mathcal{F}$ on a larger set of functions, but doing so requires considerable technical care. Also, one must verify that $\hat{g}$ is continuous in order to claim that B is a valid target for $\mathcal{F}$.

(10) Let $A = \{1, 2, \ldots, n\}$. A **permutation** is a bijective function (See Definition 1.1) from A to itself.

EXAMPLE 1.2. Let M denote the months in the year. For $m \in M$, let $f(m)$ be the number of days in m. This f is **NOT** a function; its value on February is not well-defined, because of leap year.

EXAMPLE 1.3. Suppose $z^2 = w$ for a complex number z. One often writes $z = \pm\sqrt{w}$ but this rule assigning two outputs to w is not a function from $\mathbb{C}$ to itself. It can be interpreted as a function if one specifies the target appropriately. This issue arises throughout complex variable theory.

REMARK. Often a mathematical object is described in more than one way. Mathematicians use the word *well-defined* to mean that it makes logical sense to define the object. For example, suppose a function is defined on the union of two sets by possibly different rules. To check that it is well-defined, one must verify that the two rules agree on the intersection. See Example 1.7.

We can consider the functional relationship $f(a) = b$ as an equation for the unknown a. Especially in differential equations one is interested in the existence and uniqueness of solutions. These ideas lead to the mathematical notions of *injective*, *surjective*, and *bijective*.

Definition 1.1. Let $f : A \to B$ be a function. Then

- f is called **injective** or *one-to-one* if $f(x) = f(y)$ implies $x = y$.
- f is called **surjective** or *onto* if, for each $b \in B$ there is an $a \in A$ with $f(a) = b$.
- f is called **bijective** or *a one-to-one correspondence* if f is both injective and surjective.

REMARK. Suppose $f : A \to B$ is a function. Many authors use the terms *image* of f or *range* of f for $\{f(a) : a \in A\}$, and write this set as $f(A)$. Thus f is surjective if and only if $f(A) = B$. Other authors use range synonymously with target. In this book, range and image have the same meaning.

EXAMPLE 1.4. We revisit many of the functions in Example 1.1.

- The identity function is both injective and surjective.
- The absolute value function on $\mathbb{R}$ is not injective because, for example, $|2| = |-2|$. When the target is the non-negative reals, it is surjective. Had the target been $\mathbb{R}$, then it would not be surjective.
- Differentiation is not injective; $D(p) = p' = q' = D(q)$ if p and q differ by a non-zero constant, but $p \neq q$. Differentiation on polynomials is surjective; given a polynomial q, we define p by $p(x) = \int_0^x q(t)\, dt$. Then $Dp = q$.
- The third example is surjective but not injective; for example $f\left(\frac{1}{7}\right) = 1 = f\left(\frac{1}{9}\right)$. It is worth noting that this function would not be surjective if its domain did not include 1. The only decimal expansion whose smallest digit is 9 is .999.... Hence 9 would not be a value of f.
- The $\frac{1}{1-z}$ example is injective but not surjective; the value 0 is not in the image. In Chapter 2 we consider the Riemann sphere; we add the point ∞ to $\mathbb{C}$. This function extends to be a bijection of the Riemann sphere. Again we see that one must carefully specify the domain and target of a function.
- The function f defined by $f(x) = \frac{2x}{x+1}$ from $[0, 1]$ to itself is bijective. Given $y \in [0, 1]$ we can check that $\frac{y}{2-y} \in [0, 1]$ and $f(\frac{y}{2-y}) = y$.
- If we change the target to $\mathbb{R}$, then $\frac{2x}{x+1}$ does not define a surjective function. If, for example, $1 < y < 2$, then we cannot solve $g(x) = y$ for $x \in [0, 1]$.
- The Dirac delta function is ubiquitous in physics and engineering, and it appears in several places in this book. It is incorrect to regard it as a function whose domain is $\mathbb{R}$; logical errors can result. By regarding its domain as a set of functions, one avoids these errors but keeps the intuition. See [H] for interesting comments about the intuition of the engineer Oliver Heaviside.

  Given the emphasis on linear systems in this book, we note that the delta function is linear:

$$\delta_x(f + g) = \delta_x(f) + \delta_x(g)$$

$$(c\, \delta_x)(f) = \delta_x(cf).$$

- This book has an entire section devoted to the Fourier transform, which is also linear. Considerable technical issues arise. The Fourier transform is such an important concept that one must extend its definition well beyond what we do in this book. See for example [F1] and [T].
- Permutations arise everywhere. One can prove the following statement. Every injective function from a set to itself is also surjective if and only if the set is finite (or empty). One could define a permutation of a finite set to be an injective function, and it would automatically be bijective. Similarly, one could define a permutation of a finite set to be a surjective function and it would automatically be bijective.

  There are several notations for permutations. Consider for example an ordered list of five objects, and the permutation that reverses the order. One might write this function as 54321, an abbreviation for the image. One might write it using cycle structure notation as $(15)(24)$. This notation means that 1 gets mapped to 5 which gets mapped to 1, and 2 gets mapped to 4 which gets mapped to 2, and 3 is fixed. Both notations mean the function $\sigma : \{1,2,3,4,5\} \to \{1,2,3,4,5\}$ for which

$$\sigma(1) = 5, \; \sigma(2) = 4, \sigma(3) = 3, \sigma(4) = 2, \sigma(5) = 1.$$

This example illustrates that the same idea can be expressed in many different notations. In solving a problem one should use notation in which the idea is most transparent; Einstein in particular emphasized choosing the frame of reference in which computations were easiest. This theme recurs throughout this book.

When $f : A \to B$ is surjective, for each $b \in B$ there is a solution to $f(a) = b$. When $f : A \to B$ is injective, and there is a solution to $f(a) = b$, then that solution is unique. When f is bijective, for each b there is a unique solution. The example $f(x) = \frac{2x}{x+1}$ from $[0,1]$ to itself illustrates the situation nicely.

Next we recall the notion of **composition** of functions. Suppose $f : A \to B$ and $g : B \to C$ are functions. The composition $g \circ f : A \to C$ is defined by

$$(g \circ f)(a) = g(f(a)).$$

Starting with $a \in A$, the function f takes us to $f(a) \in B$. Then g takes us to $g(f(a)) \in C$. The function $g \circ f$ takes us from A to C directly.

It is easy to show that composition of functions is **associative**. In other words, $(h \circ g) \circ f = h \circ (g \circ f)$ when these expressions are defined. Even when $A = B = C$, however, the order of composition matters: $g \circ f$ in general does not equal $f \circ g$. This statement should surprise no one; for example, putting on your shoes first and then putting on your socks is quite different from putting on your socks first and then putting on your shoes. We also give two simple mathematical examples.

EXAMPLE 1.5. Suppose $f : \mathbb{R} \to \mathbb{R}$ is given by $f(x) = x^2$ and g is given by $g(x) = x + 1$. Then $(g \circ f)(x) = x^2 + 1$ whereas $(f \circ g)(x) = (x+1)^2 = x^2 + 2x + 1$.

EXAMPLE 1.6. Suppose A, B are n-by-n matrices. We identify A with the function from $\mathbb{R}^n$ to itself given by $\mathbf{x} \mapsto A\mathbf{x}$, and similarly with B. Composition is then given by the matrix product. But $AB \neq BA$ in general. See Chapter 1 and the section on Quantum Mechanics in Chapter 7 for more information.

Our next example concerns the notion of a function being well-defined.

EXAMPLE 1.7. For $|z| < 1$, put $g(z) = \sum_{n=0}^{\infty} z^n$. For $|z + 1| < 1$ put $h(z) = \sum_{n=0}^{\infty} \frac{(z+1)^n}{2^{n+1}}$. Since g and h have different domains, they are not the same function. On the intersection of the two domains, they agree; see Exercise 1.7. Hence there is a function defined on a larger set which agrees with g where it is defined and agrees with h where it is defined. In fact, both agree with the function f from (5) in Example 1.1. Here $f(z) = \frac{1}{1-z}$ on $\{z \in \mathbb{C} : z \neq 1\}$. The function f is called an *analytic continuation* of g or of h.

We recall the definition of equivalence relation from Chapter 1.

Definition 1.2. An **equivalence relation** on a set S is a relation $\sim$ such that

- $x \sim x$ for all $x \in S$. (reflexive property)
- $x \sim y$ implies $y \sim x$. (symmetric property)
- $x \sim y$ and $y \sim z$ implies $x \sim z$. (transitive property)

An equivalence relation partitions a set into **equivalence classes**. The equivalence class containing x is the collection of all objects equivalent to x. In Chapter 1 we discussed two crucial equivalence relations in linear algebra, **row-equivalence** and **similarity**. Similar matrices describe the same science; see Section 12 of Chapter 1. In Chapter 6 we say that two functions are equivalent if they agree except on a small set (a set of measure zero). Thus an element of the Hilbert space L^2 is an equivalence class of functions, rather than a function itself.

One subtle example of an equivalence class concerns the definition of the real number system. What is a real number? Assume that we have somehow defined rational numbers. One precise way to proceed is to define a real number to be an equivalence class of Cauchy sequences of rational numbers. This book presumes that the real numbers are known. It is important to realize, however, that one can start with the natural numbers $\mathbb{N}$ as given, construct the integers $\mathbb{Z}$ from $\mathbb{N}$, then construct the rationals $\mathbb{Q}$ from $\mathbb{Z}$, and finally construct $\mathbb{R}$ from $\mathbb{Q}$. Each of these constructions involves equivalence classes.

Next we briefly discuss inequalities. The real numbers system $\mathbb{R}$ is equipped with a relation $\leq$. Intuitively, $a \leq b$ if and only if a is to the left of b on the number line (or $a = b$). Alternatively, $a \leq b$ if and only if $b - a$ is non-negative. Note that the sum and product of *positive* numbers is positive.

In working with inequalities, one must be careful to avoid certain errors. Suppose we have expressions $a(x)$ and $b(x)$ and we know that $a(x) \leq b(x)$ for some set of x. Given a function f, we cannot necessarily say that $f(a(x)) \leq f(b(x))$. For example, if $a \leq b$, then $e^a \leq e^b$ holds for all real numbers a, b. On the other hand, $a^2 \leq b^2$ need not be true: $-2 < 1$ but $(-2)^2 > 1^2$. We do preserve inequalities when we compose with an increasing function. Since $x \mapsto x^2$ is increasing for $x \geq 0$, we can square both sides of the inequality $a \leq b$ if we have the stronger information $0 \leq a \leq b$. Since $x \mapsto e^x$ is increasing for all real x, we can exponentiate the inequality without worry.

The situation differs for complex numbers. It does not make sense to say that $z \leq w$ if z and w are complex. Readers should not confuse complex numbers with their magnitudes. It does make sense to say that $|z| \leq |w|$, because the magnitude is a real number. It is complete nonsense to say that $5i > i$ (here $i^2 = -1$) yet many scientists carelessly say so. Although $5 > 1$ makes sense and is true, we cannot multiply both sides by i. Doing so can introduce spectacular errors; doing it again gives $-5 > -1$.

Things become even more subtle when we work with matrices. Let A be a symmetric matrix. It is quite common to say A is positive-definite, and even to write $A > 0$. See Chapter 6. In finite dimensions, this statement is equivalent to saying that the *eigenvalues* of A are positive; of course the *entries* in the matrix need not be positive. With real numbers, the so-called *trichotomy property* holds. If $x \in \mathbb{R}$, then exactly one of the three assertions $x < 0$, $x = 0$, $x > 0$ is true. For matrices, one of the analogues might hold (all the eigenvalues are negative, all are 0, all are positive) but for most matrices, none of these statements is true!

One final remark about inequalities helps distinguish advanced math from elementary math. To prove the equality $x = y$ for real numbers x, y it is often true that one proves two inequalities: $x \leq y$ and $y \leq x$. In many examples, no amount of algebraic manipulations using $=$ alone can work. In a similar fashion, to show that sets A and B are equal, one often shows both containments: $A \subseteq B$ and $B \subseteq A$.

We close this section by discussing the limit of a sequence of complex numbers. This notion is defined in Definition 2.1 from Chapter 2. The definition might seem pedantic to some readers, but its precision is required to do any meaningful analysis. We are given a sequence $\{z_n\}$ of complex numbers: such a thing is in fact a function from $\mathbb{N}$ to $\mathbb{C}$. Its value at n is the complex number z_n. We are given a complex number L. What does it mean to say that $\{z_n\}$ converges to L? The answer is subtle. At first we imagine that a demanding taskmaster insists that our measurements are eventually all in some prescribed small region containing L. We can satisfy the taskmaster by picking n large enough. That statement is not good enough! We must allow the thought experiment of an arbitrarily demanding taskmaster. No matter how small the value ϵ is given, we must always be able to succeed. In other words, as the definition says, **for all** $\epsilon > 0$ **there is an** N **such that....** The key quantifier is **for all**. This kind of precision is required to do calculus.

The definition of the limit of a sequence makes us naturally ask the following question about a function f defined near a point $p \in \mathbb{C}$. Suppose $\{z_n\}$ converges to p. Does the sequence $\{f(z_n)\}$ converge to $f(p)$? The function f is **continuous** at p if the answer is yes for every sequence converging to p.

EXERCISE 1.1. Verify that $A \cap (B \cup C) = (A \cap B) \cup (A \cap C)$.

EXERCISE 1.2 (Inclusion-Exclusion). For a finite set S, let $\#(S)$ denote the number of elements in S. For finite sets A, B, C establish the following:

$$\#(A \cup B) = \#(A) + \#(B) - \#(A \cap B).$$

$$\#(A \cup B \cup C) = \#(A) + \#(B) + \#(C) - \#(A \cap B) - \#(A \cap B) - \#(B \cap C) + \#(A \cap B \cap C).$$

Generalize to n finite sets $A_1, \ldots, A_n$.

EXERCISE 1.3. Show that a set with n elements has 2^n subsets.

EXERCISE 1.4. Show that the composition of injective functions is injective. Show that the composition of surjective functions is surjective.

EXERCISE 1.5. Define $f : \mathbb{N} \times \mathbb{N} \to \mathbb{N}$ by $f(m, n) = 2^{m-1}(2n - 1)$. Prove that f is bijective.

EXERCISE 1.6. Verify Example 1.7. (Use the geometric series.)

EXERCISE 1.7. Verify the linearity of the delta function from Example 1.4.

2. Variables, constants, and parameters

The words variable, constant, and parameter arise throughout the sciences. Their usage changes with the context, and hence seeking precise definitions of the terms is hopeless. Nonetheless, the following remarks and examples might be useful.

The word **variable** often means an element of the domain of a function. The word **parameter** often refers to a constant term in the definition of a function, especially when the constant has a physical or geometric meaning, and it is possible to study how things vary when the parameter changes. Whether a scientific number is a constant or a parameter depends. For example, the acceleration g due to gravity is regarded as a constant when we solve elementary projectile problems, but g might be regarded as a parameter when studying rocket science.

EXAMPLE 2.1. Consider the function $x \mapsto e^{\lambda x}$. Here x is a real variable and λ is a (possibly complex) constant. As we vary λ we get different functions of x, and we then refer to λ as a parameter.

We next make a simple observation about what constitutes a variable.

EXAMPLE 2.2. Consider a function $f : \mathbb{R}^3 \to \mathbb{R}$. In this book the variable is a vector $\mathbf{x}$ in $\mathbb{R}^3$. Many calculus books instead regard f as a function of the three variables x, y, z. The resulting formulas, especially those using partial derivatives, get muddied and destroy geometric intuition! Things become even more complicated when $F : \mathbb{R}^3 \to \mathbb{R}^3$ is a vector field. Calculus books think of F as three functions (P, Q, R) of three variables. It is easier to regard F as a single (vector) function of one (vector) variable.

The next example indicates how a good choice of the variable can clarify what is happening.

EXAMPLE 2.3. For $\mathbf{w} \in \mathbb{R}^3$ with $\mathbf{w} \neq 0$, put $f(\mathbf{x}) = \mathbf{x} \cdot \mathbf{w}$. This notation seems to the author to be easier to understand than writing $f(x, y, z) = Ax + By + Cz$, where $\mathbf{w} = (A, B, C)$. Consider maximizing f on the unit sphere. The Cauchy–Schwarz inequality instantly yields

$$\mathbf{x} \cdot \mathbf{w} \leq ||\mathbf{x}|| \, ||\mathbf{w}|| = ||\mathbf{w}||.$$

Thus the maximum is the length of $\mathbf{w}$. We can write the right-hand side as $\sqrt{A^2 + B^2 + C^2}$, but we lose geometric intuition and perhaps miss the very simple proof. The author has seen dozens of students solve this problem in a complicated fashion using Lagrange multipliers. The main point here is to regard both the variable $\mathbf{x}$ and the parameter $\mathbf{w}$ as vectors, rather than as triples of numbers.

The next two examples consider contexts where one varies parameters.

EXAMPLE 2.4. Consider the normal probability distribution. Its density function f, often called a *Gaussian*, is given by the formula

$$f(x) = \frac{1}{\sqrt{2\pi}\sigma} e^{\frac{-(x-\mu)^2}{2\sigma^2}}.$$

Here $f : \mathbb{R} \to [0, \infty)$ is a function, with the following meaning: the probability that a normal random variable (yet another use of the term variable!) with mean μ and variance σ^2 lies in the interval $[a, b]$ is given by $\int_a^b f(x)\, dx$. The mean and variance are **parameters**. The mean μ can be any real number; the variance σ^2 can be any positive real number.

In many problems these parameters are known, or estimated from measured data. One can, however, vary parameters. In other words, we study how things change as these parameters change. For example, in Chapter 5, Section 4, we prove the Fourier inversion formula (a crucial result in mathematics, physics, and signal processing) by introducing a Gaussian with variance $\frac{1}{\epsilon^2}$ in an integral, in order to force rapid convergence. For $\epsilon > 0$, we can justify an interchange of order in a multiple integral. After doing so, we let ϵ tend to 0. The variance is therefore tending to infinity; we are spreading things out. We can also interpret the Dirac delta function as a similar sort of limit. See Chapter 5.

EXAMPLE 2.5. Consider a linear ordinary differential equation of the form $Lu = f$. As discussed in Chapter 1, the general solution is given by a particular solution plus an arbitrary solution of the homogeneous equation $Lu = 0$. One can find the particular solution by **variation of parameters**. For example, consider the simple equation $y'(t) - \lambda y(t) = f(t)$. The solution to the homogeneous equation is given by $y(t) = ce^{\lambda t}$ for an arbitrary real constant c. In order to solve the inhomogeneous equation, we treat the constant c as a function $t \mapsto c(t)$. Thus we try $y(t) = c(t)e^{\lambda t}$ and see what happens. We obtain

$$y'(t) = (\lambda c(t) + c'(t))\, e^{\lambda t}$$

and using the equation we get (an easier) differential equation for c:

$$c'(t) = e^{-\lambda t} f(t),$$

which we solve by direct integration. The idea of course is to regard the constant c as a parameter, then to vary this parameter, treating it as an unknown function. In this simple setting the terms variable, constant, and parameter are blurred. Is the variable t? Is the variable the unknown function y? Is the variable the unknown function c? One cannot answer these questions.

In this book we often regard functions as variables, and operations on these functions as functions. Thus for example the Fourier transform is a function; its domain and target are both spaces of functions. Many of the operations, including the Laplace and Fourier transforms, are linear, and many involve complex variable methods. Hence the book title is *Linear and Complex Analysis for Applications*.

References

[A] Ahlfors, Lars V., Complex Analysis: An introduction to the theory of analytic functions of one complex variable, 3rd edition, International Series in Pure and Applied Mathematics, McGraw-Hill Book Co., New York, 1978.

[B] Boas, Ralph P., Invitation to Complex Analysis, 2nd edition, revised by Harold P. Boas, Mathematical Association of America, 2010.

[CGHJK] Corless R. M., Gonnet, G. H., Hare, D. E. G., Jeffrey, D. J., Knuth, D. E., On the Lambert W-function, Advances in Computational Mathematics (1996), Volume 5, Issue 1, pp. 329-359.

[CV] Cinlar, E. and Vanderbei, R., Real and Convex Analysis, Undergraduate Texts in Mathematics, Springer, New York, 2013.

[D1] D'Angelo, J., Inequalities from Complex Analysis, Carus Mathematical Monograph No. 28, Mathematics Association of America, 2002.

[D2] D'Angelo, J. An Introduction to Complex Analysis and Geometry, Pure and Applied Mathematics Texts, American Math Society, Providence, 2011.

[D3] D'Angelo, J., Hermitian Analysis: From Fourier series to Cauchy–Riemann geometry, Birkhäuser-Springer, New York, 2013.

[Dar] Darling, R. W. R., Differential Forms and Connections, Cambridge University Press, Cambridge, 1994.

[Dau] Daubechies, Ingrid, Ten lectures on wavelets, CBMS-NSF Regional Conference Series in Applied Mathematics, 61, Society for Industrial and Applied Mathematics (SIAM), Philadelphia, PA, 1992.

[E] Epstein, C., Introduction to the Mathematics of Medical Imaging, Prentice Hall, Upper Saddle River, New Jersey, 2003.

[Fa] Falconer, K., Fractal Geometry, 2nd edition, Wiley, West Suffolk, England, 2003.

[F1] Folland, G. , Real Analysis: Modern techniques and their applications, 2nd edition, John Wiley & Sons, New York, 1984.

[F2] Folland, G., Fourier Analysis and its Applications, Pure and Applied Mathematics Texts, American Math Society, Providence, 2008.

[Fos] Fossum, R., The Hilbert Matrix and its Determinant, http://s3.amazonaws.com/cramster-resource/10732_n_23997.pdf.

[Fr] Frankel, T., The Geometry of Physics, Cambridge University Press, Cambridge, 1997.

[G] Greenberg, M., Advanced Engineering Mathematics, Prentice Hall, Upper Saddle River, New Jersey, 1998.

[Gr] Griffith, D., Introduction to Quantum Mechanics, 2nd edition, Pearson, Upper Saddle River, New Jersey, 2014.

[GS] Goldbart, P. and Stone, M., Mathematics for Physics, Cambridge University Press, Cambridge, 2009.

[He] Herman, R., An Introduction to Fourier Analysis, CRC Press, Boca Raton, Florida, 2017.

[H] Hunt, Bruce J., Oliver Heaviside: A first-rate oddity, Phys. Today 65(11), 48(2012), 48-54.

[HK] Hoffman, K. and Kunze, R., Linear Algebra, 2nd Edition, Prentice Hall, Upper Saddle River, New Jersey, 1971.

[HH] Hubbard, J. H. and Hubbard, B., Vector Calculus, Linear Algebra, and Differential Forms, Prentice Hall, Upper Saddle River, New Jersey, 2002.

[HPS] Hoel P., Port S., and Stone C., Introduction to Probability Theory, Houghton Mifflin, Boston, 1971.

[KM] Kudeki, E. and Munson Jr., D., Analog Signals and Systems, Illinois ECE Series, Pearson, Upper Saddle River, New Jersey, 2009.

[M] Mallat, Stephane, A Wavelet Tour of Signal Processing, 3rd edition, Academic Press, New York, 2008.

[MH] Mathews, J. and Howell, R., Complex Analysis for Mathematics and Engineering, Jones and Bartlett Publishers, Sudbury, 1996.

[N] Nievergelt, Yves, Wavelets Made Easy, Birkhäuser Boston, Inc., Boston, MA, 1999.

[R] Ross, K., Elementary Analysis: The Theory of Calculus, Springer-Verlag, New York, 1980.

[RS] Reed, M. and Simon, B., Methods of Modern Mathematical Physics. I. Functional analysis, 2nd edition, Academic Press, Inc. [Harcourt Brace Jovanovich, Publishers], New York, 1980.

[Sp] Spivak, M., Calculus on Manifolds: A modern approach to classical theorems of advanced calculus, W. A. Benjamin, Inc., New York, 1965.

[S] Steele, J. Michael, The Cauchy-Schwarz Master Class, MAA Problem Book Series, Cambridge University Press, Cambridge, 2004.

[SS] Stein, E. and Shakarchi, R., Fourier Analysis: an Introduction, Princeton University Press, Princeton, New Jersey, 2003.

[St] Strang, G., Linear Algebra and its Applications, 3rd edition, Harcourt Brace Jovanovich, San Diego, 1988.

[T] Taylor, M., Partial Differential Equations, Springer, New York, 1996.

[V] Velleman, D., If the IRS had discovered the quadratic formula, Math Horizons, April 2007, p. 21.

Index